After the Fires

After the Fires

The Ecology of Change in Yellowstone National Park

Edited by Linda L. Wallace

Yale University Press

New Haven and London

Set in Garamond and Stone Sans types by The Composing Room of Michigan, Inc.
Printed in the United States of America by Sheridan Books, Ann Arbor, Michigan

Library of Congress Cataloging-in-Publication Data
After the fires : the ecology of change in Yellowstone National Park /
edited by Linda L. Wallace.
p. cm.
Includes bibliographical references and index (p.).
ISBN 0-300-10048-5 (cloth : alk. paper)
1. Forest fires—Yellowstone National Park. 2. Forest fires—Environmental aspects—Yellowstone National Park. 3. Fire ecology—Yellowstone National Park. 4. Yellowstone National Park. I. Wallace, Linda L., 1951–
SD421.32.Y45 A46 2004
634.9′618′0978752—dc22

2003024155

A catalogue record for this book is available from the British Library.

The paper in this book meets the guidelines for permanence and durability of the Committee on Production Guidelines for Book Longevity of the Council on Library Resources.

10 9 8 7 6 5 4 3 2

This book is dedicated to the memory of Jay E. Anderson (1937–2002). Jay's love of the Yellowstone Ecosystem and its surroundings was surpassed only by his love of life, family, and friends. His knowledge and understanding of Rocky Mountain ecosystems guided students, colleagues, and land managers throughout the many years that he served as a professor at Idaho State University. A vigorous outdoorsman, he counted cycling and kayaking among his favorite activities. Jay was the recipient of many professional awards, a reflection of his curiosity and enthusiasm for studying the fascinating landscapes of the region. His chapter in this volume is a clear example of the high quality of his work. All of us benefited from his collegiality and are grateful to have known him as a friend.

Contents

Preface

In 1988 the world experienced its first "prime-time" forest fire, as television, radio, and print media covered the fires in Yellowstone National Park and the surrounding areas. Needless to say, the coverage seldom focused on the scientific issues of fire in the Northern Rocky Mountains. Rather, most of the information given to the public was chosen for its sensational nature.

In the years following the fires, there has been a great deal of continued interest from the general public, the media, and the scientific community in how the park has fared. Again, most of the information available to the general public, while technically correct, does a superficial job of discussing the scientific work behind it. This volume is written to remedy that situation.

Most people now agree that the park was not destroyed by the 1988 fires. However, concern about fire "destruction" is still high given the large fires experienced in the United States (primarily in the western part of the nation) every summer. To ease those concerns, it is critical that we know how the experiences and knowledge gained in Yellowstone can be applied to other fire-affected regions. This volume shows us how that can be done for both terrestrial and aquatic ecosystems.

Immediately after the 1988 fires, a large number of scientists convened at the University of Wyoming research facility located between Yellowstone and Grand Teton National Parks. We planned a concerted, integrated research effort to help us fully comprehend how large-scale fires influenced the ecosystems and landscapes of the Greater Yellowstone Ecosystem. Unfortunately, that effort did not result in research funding, but the scientists involved in the planning received individual grants and proceeded to do the work, maintaining contact among ourselves. Ultimately, we collaborated to develop this volume to report on what research has uncovered about the effects of the fires. Our research continues to the present day.

The results of these studies will be of interest to a number of readers, including researchers in ecology, geography, limnology, paleoecology, and geology, as well as land managers and anyone who loves and cares for Yellowstone National Park. As some of the chapters are somewhat technical, researchers will find them of greatest interest. However, the take-home messages presented in all chapters and the epilogue will be of great interest to anyone who reads this volume. Further information on the scope of the text and its organization is available in Chapter 1.

Any effort of this scale requires the work of a large number of individuals. My sincerest thanks go to them. Included in this list are Jean Thomson Black of Yale University Press, whose belief in this effort never faltered, and the members of the Samuel Roberts Noble Foundation, whose endowment to me helped fund a portion of the expenses of preparing this volume. Numerous reviewers commented on the individual chapters and their help is deeply appreciated. Finally, all of the authors in this volume learned more about fire and about Yellowstone from Dr. Don Despain, park ecologist, than one could possibly imagine. He warrants our deepest and sincerest thanks.

Part I **Historical and Geological Perspective**

In the beginning there was no fire, and the world was cold, until the Thunders (Ani'-Hyun'tikwala'ski), who lived up in Galun'lati, sent their lightning and put fire into the bottom of a hollow sycamore tree which grew on an island.

—*Cherokee tale in Alianor True,* Wildfire: A Reader

Chapter 1 The Fires of 1988: A Chronology and Invitation to Research

Linda L. Wallace, Francis J. Singer, and Paul Schullery

The goal of this volume is to take a comprehensive look at the large-scale fires of 1988 that swept through the Greater Yellowstone Ecosystem (GYE), most notably in Yellowstone National Park (YNP). We will not examine the political ramifications of the fires, but rather will look at the fires as they affected the geology, ecology, and structure of this system. Research work is continuing as this volume is being written and printed. However, the work presented represents a little more than one decade of research by scientists from across the United States in a number of different disciplines. Although many of these individuals have published results of studies in separate papers, this volume gives us an opportunity to synthesize these results and compare findings in a cohesive and comprehensive manner.

The goals of this introduction are twofold. First we will briefly describe the chronology of the 1988 fires and discuss what areas were burned and the extent of fire in those regions. Second, we will set the stage for the research work and describe the different ecological scales that the different research projects represent.

GENERAL DESCRIPTION OF THE GREATER YELLOWSTONE ECOSYSTEM

The GYE is centered on YNP, a truly unique volcanic landscape. The original goals of Yellowstone, founded in 1872, were to preserve unique thermal features, including geysers, hot springs, and thermal pools. Later, it became obvious that one of the other key factors attracting visitors to YNP was the wildlife. In 1883, wildlife was protected from hunting in the park and became a major management focus as well. As our knowledge of ecosystem science expanded, management's concerns expanded to include the habitat of important wildlife species and the health of forage species, trees, and soils. Fishing and aquatic systems management gradually joined traditional wildlife management as important focuses of administrative action. By 1910, U.S. Army administrators were beginning to think of protecting native fish species from further nonnative fish introductions. Increased attention to aquatic systems followed the creation of the National Park Service in 1916, though intensive manipulation of many park fisheries continued until midcentury (Varley and Schullery 1998).

Yellowstone National Park can be divided into two major areas, the northern range and the subalpine plateau (Fig. 1.1). The northern range is deeply dissected low to midelevation (2400–2800 m) sagebrush grasslands (55 percent grasslands) and forests and acts as the winter range to populations of elk (17,000–23,000) and bison (500–2000) (Houston 1982). Soils are andesetic in origin and are hence fairly nutrient rich. The subalpine plateau is mid- to high elevation and is largely forested with smaller grassland openings (less than 5 percent grasslands). However, some important grassland areas such as Hayden Valley and Pelican Valley also occur here. The largest thermal areas are found here as well. Soils are rhyolitic in origin and are fairly nutrient poor. This region is typically the summer home for very large populations of elk and bison. In excess of 32,000 elk from eight different herds migrate each summer to the park's central plateau (Singer and Mack 1999). Many fewer individuals overwinter on the plateau because of deep snow and frigid temperatures. Some notable exceptions are animals overwintering in the thermal basins as well as a small bison herd that persists in the deep snows of Pelican Valley (Meagher 1973, Singer 1991).

AN ABBREVIATED HISTORY OF THE 1988 FIRES

The fires of 1988 resulted from a combination of drought, above-average temperatures, and numerous dry thunderstorms with lightning strikes and high

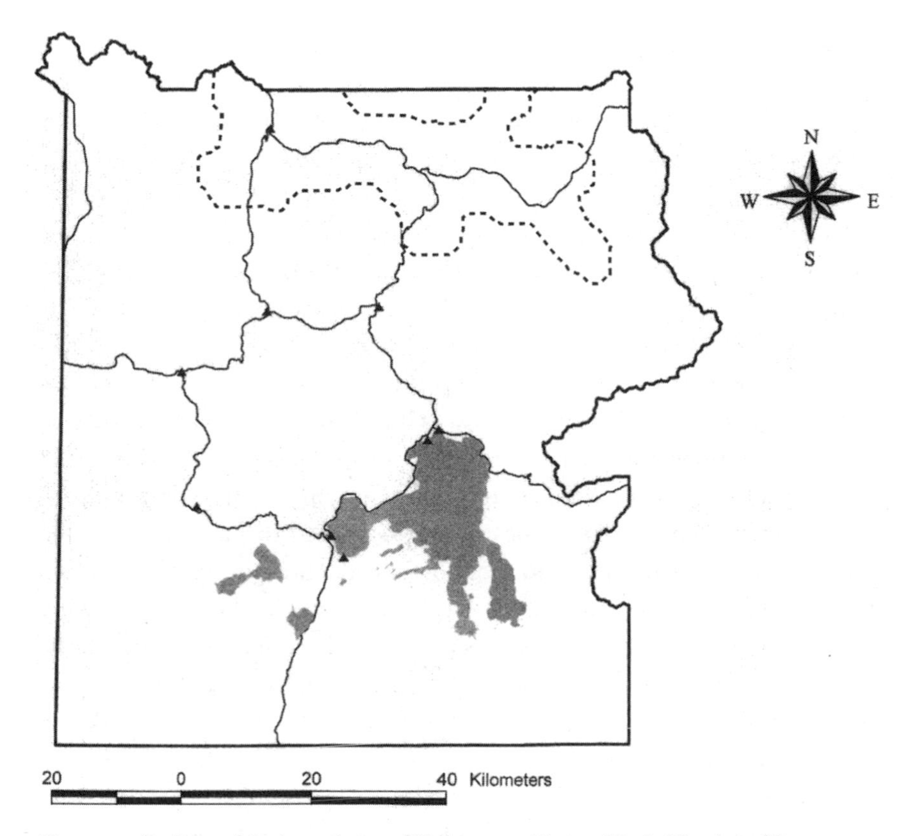

Figure 1.1. Outlines of the boundaries of Yellowstone National Park. The dotted lines within the park boundaries denote the approximate boundary of the northern winter range. Areas outside of this area to the south are part of the subalpine plateau summer range of elk and bison. Solid lines indicate the highway system inside the park, and triangles mark the location of developed areas (Old Faithful, Fishing Bridge, and the like). Gray areas are lakes.

winds. This meteorological combination, considered to be a 300–400-year event, resulted in the burning of 1,405,775 ha in the GYE (Despain and others 1989, Schullery 1989). About one-third of the 590,000 ha summer range for elk and bison burned, and 22 percent of the 134,000 ha winter range for the northern Yellowstone elk and bison herd burned in this series of fires. The meteorology of 1988 should not be viewed in isolation, however. Prior to the drought of the summer of 1988, the GYE had experienced dry winter and spring conditions in both 1986 and 1987. This helped set the stage for the extremely low fuel moistures that subsequently occurred by July 1988. Fuel moisture levels were only 2–3 percent (Schullery 1989), lower than values found in household furnishings!

High temperatures further exacerbated the low precipitation in 1988. June and July averages were 8.6 and 2.7° C above normal, respectively. Fire starts were frequent during July–September because a series of abnormally dry cold fronts passed through the park, igniting fires but bringing almost no rain. Winds during the passage of these fronts were as high as 64–96 km hr^{-1} (Schullery 1989). This resulted in fires spotting 2 km ahead of fire fronts with fires moving as fast as 3 km hr^{-1}. Fire fronts advanced as much as 8–16 km per day (Schullery 1989).

In 1971, the National Park Service (NPS) instituted a policy in which backcountry and high-elevation fires would not be suppressed if they were of lightning origin and did not threaten critical structures, habitat, or human life. This had resulted in 235 fires burning only 13,823 ha during this entire time, with the largest fire burning only 2,995 ha (Schullery 1989). During 1988, a number of fires were allowed to burn within the park until July 21, when park staff was convinced that fire conditions were too extreme to allow further burning. All existing and new fires were suppressed after this date, but firefighting efforts had little influence. An enormous firefighting presence was in the park, involving firefighters from the NPS, the Forest Service, and the military. On the evening of September 14, 1988, approximately 1 cm of snow fell, effectively extinguishing the fires of 1988. Fire crews remained in the park doing mop-up work and later worked on revegetation of fire lines.

Given the scale of the fires, scientists felt that this represented an unprecedented opportunity for fire research. In October of 1988, the authors of the chapters in this volume gathered at the University of Wyoming Research Ranch in Grand Teton National Park and began planning coherent research projects. This volume represents our synthesis of these widely varying research efforts.

RESEARCH QUESTIONS AND SCALES

The volume is divided into four sections: Historical and Geological Perspective, Effects on Individuals and Species, Effects on Aquatic Systems, and Terrestrial Ecosystem and Landscape Perspective, plus an Epilogue. Using this rubric, we can first examine the historical context of fire in this ecosystem using two different geological tools, charcoal found in lake sediment cores (Chapter 2) and dated debris-flow layers (Chapter 3). Further, both of these techniques can tell us about how this ecosystem responds to fire via difference vegetation structure and debris-flow movement of nutrients from terrestrial to aquatic habitats. Further, both chapters point out the importance of differences in regional climate in the park relative to fire events.

One of the biggest concerns of the public during and after the fires of 1988 was how individual plants and animals fared. Thinking hierarchically (Ahl and Allen 1996), we know that the patterns seen at the community and ecosystem levels are the result of mechanistic responses at the individual and population levels. Therefore, it is important to know how forest trees (Chapter 4) and grassland species (Chapter 5) responded. Some of the greatest public concern was for large animals, particularly elk. Elk mortality and population responses after the fires took some surprising turns (Chapter 6).

Although responses in aquatic systems can readily fit into the hierarchies above, we chose to separate aquatic systems into a special section for several reasons. First, the effects of fire on aquatic species are primarily indirect, with only a few areas subject to direct heating from flames (Chapter 7). There also is a great deal of public interest, particularly by sportsmen, in how the fisheries of the GYE were affected by the fires. Because the effects of the fires on aquatic systems are indirect, it is important to understand streams and rivers from an ecosystem perspective and to see how changes in streamside cover, stream flow characteristics, and streambed morphology can influence species composition and survival after fires (Chapter 8). Further, basic changes in community structure and food webs in streams present a unique challenge to life not seen in terrestrial environments (Chapter 9). Finally, public interest in how fires affect water availability and quality downstream from burned watersheds is quite high in the arid West (Chapter 10).

In order to make adequate predictions of what life will be like in the GYE after the fires of 1988, it is imperative that we look at past fires and see how community and ecosystem parameters responded to those (Chapter 11). Also, it is important that we look at all components of the ecosystem, including those which are less "glamorous" but highly essential in terms of nutrient retention and release. Again, in the arid West, coarse woody debris is important as a nutrient sponge as well as being potential fuel for future fires (Chapter 12). In the brief time between the 1988 fires and the present, plant communities have already established and have grown. The GYE is an extremely heterogeneous environment. Are there rules governing how these communities respond in the different growth environments of the GYE (Chapter 14)? Plant communities provide essential habitat for the megaherbivores (elk and bison) of the GYE as well. Although we know numbers and how the populations of these animals have changed since the fires, it is difficult to determine the mechanisms behind these changes. Using simulation models and comparing their results with reality can yield important insights as to the mechanisms governing ungulate response to fire (Chapter 13).

Figure 1.2. Map of Yellowstone National Park with the fire-affected areas colored dark gray. Numbers on the map show where research was done for the corresponding numbered chapters. Note that several chapters had multiple research sites, that most of the sites studied in Chapter 11 were south of the boundaries of YNP, and that the majority of the sites in Chapter 10 were north of the YNP boundaries.

Given the vast amount of research effort and thought that these fifteen chapters represent, we now have the opportunity to synthesize these results into a more cohesive picture of how cold, arid montane systems respond to large-scale fires. Research efforts covered nearly the entirety of the GYE (Fig. 1.2), which enables us to be more confident in our understanding of whole-system responses. Several fire seasons since 1988 have captured both public and political attention because of the scale and intensity of the fires that burned large areas of the West. What can we tell them from the Yellowstone experience? How are the ecosystems of these areas going to respond? How are the members of those ecosystems going to respond? It is important that we understand both the mechanisms and scales of system response to fire (Chapter 15) in order to effectively analyze what our policy response to fire has been in the past and might be in the future.

References

Ahl, V., and T. F. H. Allen. 1996. Hierarchy theory: A vision, vocabulary, and epistemology. Columbia University Press, New York.

Despain, D., A. Rodman, P. Schullery, and H. Shovic. 1989. Burned area survey of Yellowstone National Park: The fires of 1988. Yellowstone National Park, Wyoming.

Houston, D. B. 1982. The northern Yellowstone elk: Ecology and management. MacMillan, New York.

Meagher, M. 1973. The Bison of Yellowstone National Park. National Park Service Science Monograph Series no. 1. Washington, D.C.

Schullery, P. 1989. Yellowstone fires: A preliminary report. Northwest Science 63: 44–55.

Singer, F. J. 1991. The ungulate prey base for wolves in Yellowstone National Park. Pages 323–348 in R. B. Keiter and M. S. Boyce, eds., The greater Yellowstone ecosystem: Man and nature in America's wildlands. Yale University Press, New Haven.

Singer, F. J., and J. Mack. 1999. Ungulate prey for carnivores with predicted effects from restoration of wolves and the fires of 1988. Pages 189–237 in T. Clar, A. P. Curlee, S. C. Minta, and P. M. Kareiva, eds., Carnivores in ecosystems. Yale University Press, New Haven.

True, A. 2001. Wildfire, A Reader. Island Press, Washington, D.C.

Varley, J., and P. Schullery. 1998. Yellowstone Fishes. Stackpole Books, Harrisburg, Pennsylvania.

Chapter 2 Postglacial Fire, Vegetation, and Climate History of the Yellowstone-Lamar and Central Plateau Provinces, Yellowstone National Park

Sarah H. Millspaugh, Cathy Whitlock, and Patrick J. Bartlein

The fires of 1988 were unique in the history of Yellowstone National Park (YNP), because during that summer a relatively small number of fires occurred over an enormous region (Schullery 1989). Previous fires in YNP were confined to particular regions and comparatively small sizes (Chapters 4, 14). Although the 1988 event has been considered unprecedented on short time scales, knowledge of fire history is required to evaluate its uniqueness over longer periods. In particular, the 1988 fires raise important questions about the natural range of variability of large fires: What is the long-term frequency of fires in different parts of Yellowstone National Park, and how often and under what conditions is fire occurrence synchronized across regions? How have fire regimes responded to different climate and vegetation conditions in the past, and what do past fire-climate interactions suggest about future fire regimes given the nature of projected changes in regional climate? These questions can be answered only by looking at the prehistoric fire record.

The Yellowstone climate at present is characterized by two geographically delineated precipitation regimes (Despain 1987, Whitlock and Bartlein 1993). These regimes are a local manifestation of large-scale climate features that determine the seasonal pattern of precipitation in the western United States (Tang and Reiter 1984, Mock 1996). One regime receives a significant amount of precipitation in summer as a result of summer storms generated by the onshore flow of moisture from the Gulf of California and Gulf of Mexico into the southwestern United States, the Great Plains, and the southern Rocky Mountains. This type of monsoonal circulation extends to the northern part of YNP and the eastern Snake River Plain, where convectional precipitation occurs along orographic barriers. The other regime, with relatively dry summers, occurs in areas under the influence of the northeastern Pacific subtropical high-pressure system and accounts for summer aridity in the Pacific Northwest, the western Snake River Plain, and central and southern YNP. In both regions precipitation in winter is generated by individual mountain ranges intercepting storm systems moving inland from the Pacific Ocean.

Paleoenvironmental data from the northern and central Rocky Mountains suggest that the contrast between summer-wet and summer-dry precipitation regimes was greater during periods of higher-than-present insolation (Whitlock and Bartlein 1993, Whitlock and others 1995, Fall and others 1995). In this chapter, we examine the influence of such climate changes on the Holocene fire and vegetation history of two geovegetation provinces of YNP that are influenced by different precipitation regimes at present (see Despain 1990 for a description of YNP geovegetation provinces). The environmental history, which is obtained from an analysis of charcoal and pollen in radiocarbon-dated lake-sediment cores at two sites, reveals the long-term interactions among fire, climate, and vegetation within YNP over the past ca. 15,000 cal years.

Slough Creek Lake (Lat. 44°57′N, Long. 110°21′W, elevation 1884 m) is located in the Yellowstone-Lamar Province and lies in a kettle-hole depression in a broad valley that is underlain by andesitic volcaniclastic rocks and carbonate rocks and shale. Most of the soils derive from glacial till deposited during the Pinedale Glaciation (Pierce 1979). The Yellowstone-Lamar Province receives precipitation in spring and summer, as well as during winter (Despain 1987, Whitlock and Bartlein 1993). The late spring and summer precipitation maximum (May–June) is produced by a combination of isolated upper-level low-pressure systems and convection from increased land surface temperatures (Mock 1996). In July, surface warming causes moist air from the Gulf of California to flow northward into the interior of the continent along the western edge of an upper-level subtropical ridge (Higgins and others 1997). Climate records from the Lamar Ranger Station near Slough Creek Lake indicate a mean

January temperature of −25°C and an average July temperature of 15.3°C from 1948 to 1972. Average annual precipitation from 1948 to 1972 was 36 cm (Dirks and Martner 1982). The *Pseudotsuga* parkland surrounding Slough Creek Lake is maintained by low-intensity surface fires that are generally lethal to understory vegetation but do not kill the mature trees. Dendrochronological records suggest that fires prior to about 1890 occurred every twenty to fifty years in the *Pseudotsuga* parkland (Houston 1973, Barrett 1994). Between about 1850 and present day, there were only a few small fires (Houston 1973), with the exception of 1988, when 34 percent of the Slough Creek Lake catchment burned.

Cygnet Lake (Lat. 44°39′N, Long. 110°36′W, elevation 2530 m) lies in the Central Plateau Province (Fig. 2.1). The climate of the Central Plateau Province is characterized by relatively dry summers and wet winters. Summer conditions are produced by large-scale subsidence associated with the northeastern subtropical high-pressure system. In winter, storms following the path of the jet stream bring moisture from the Pacific Ocean via the Columbia River Basin and the Snake River Plain (Bryson and Hare 1974). Data from Lake Station at the north end of Yellowstone Lake indicate a mean January temperature of −11.8°C and a mean July temperature of 12.8°C from 1948 to 1970. Snowfall averages 508 cm for the same period (Dirks and Martner 1982). At present, the plateau region supports uniform forest of *Pinus contorta,* despite climate conditions that are suitable for *Picea engelmannii, Abies lasiocarpa,* and *Pinus albicaulis.* In the past few centuries, large (>2500 ha), infrequent (200–400 yr mean fire interval), stand-replacing fires have resulted in a forest mosaic composed of different stand ages of *P. contorta* and little species diversity (Romme and Despain 1989).

FIELD AND LABORATORY METHODS

The sediments of Yellowstone's lakes provide an opportunity to reconstruct the vegetation and fire history of the region back to the time of late-Pleistocene deglaciation. The temporal framework for the paleoenvironmental reconstructions is established from a series of AMS radiocarbon dates on plant remains and the ages of volcanic ashes in the core that come from known eruptions. Stratigraphic pollen records provide information on changing forest composition (see Bennett and Willis 2002). Analysis of particulate charcoal in the cores provides information on past fire occurrence (see Whitlock and Larsen 2002). Studies in YNP indicate that large particles of charcoal (>225 microns in diameter) are not transported far from their sources and offer information on local fire history (for example, Whitlock and Millspaugh 1996). Variations in charcoal influx

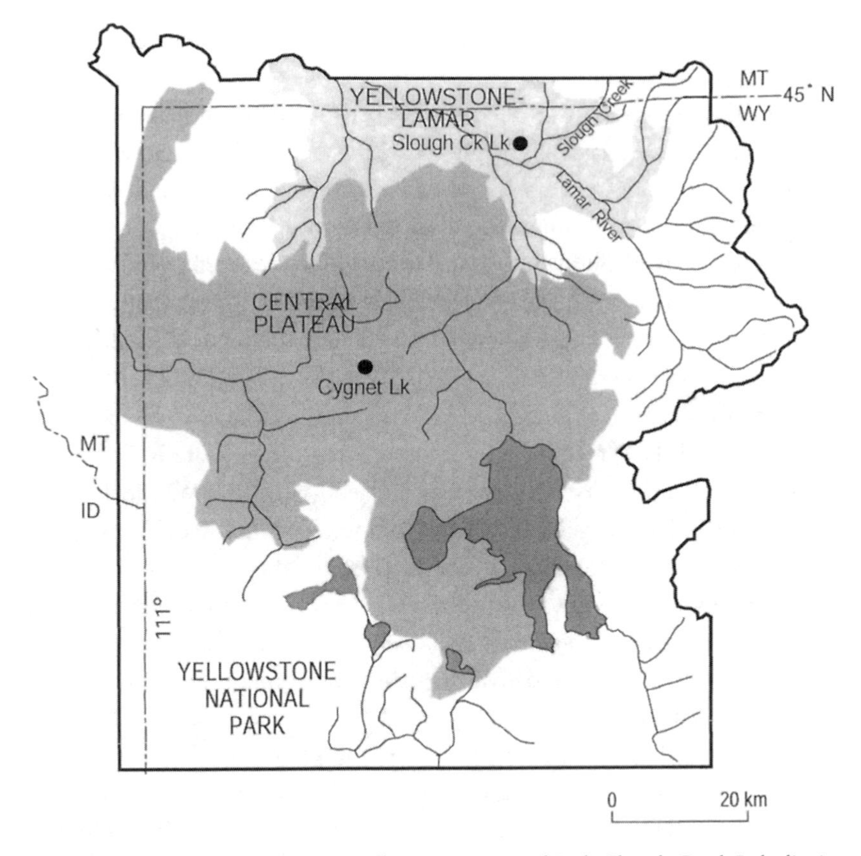

Figure 2.1. Location of study sites in Yellowstone National Park. Slough Creek Lake lies in the Yellowstone-Lamar Province, and Cygnet Lake is located in the Central Plateau Province.

(CHAR, particles cm^{-2} yr^{-1}) in contiguous core sections are a record of past fire events.

Charcoal records lack the temporal resolution of dendrochronological fire reconstructions, and core intervals with high levels of charcoal may represent one or more fires occurring over the time that an interval was deposited. In Yellowstone, each 1-cm sample spans about ten years (Millspaugh 1997). Charcoal records also lack the spatial specificity of tree-ring data, and past fires are, at best, located within or near a particular watershed (whereas tree-ring records can identify the specific trees that were burned). The strength of charcoal data, however, is that they provide a fire reconstruction that spans thousands of years beyond the age of living trees and offers information on the role of fire during major reorganizations of vegetation and climate in the past. Charcoal data also register past stand-replacing fires, which are poorly described by most tree-ring records (Agee 1993).

The methods used in this study are described elsewhere and will not be discussed in detail here (See Millspaugh 1997, Long and others 1998, Millspaugh and others 2000). Briefly, sediment cores were taken from the centers of Slough Creek and Cygnet lakes using a modified Livingstone sampler. A series of radiocarbon ages were obtained on plant macrofossils collected from each core, and the dates were converted to calendar years (cal yr B.P., Stuiver and others 1998). Weighted third-order polynomial regressions (constrained to pass through age 0) were used to interpolate between the radiocarbon dates at each site and establish age-versus-depth relations (Table 2.1). Subsamples of 0.5 cm^3 volume were collected at 10- to 20-cm intervals for pollen analysis. Pollen percentage data, described by Whitlock and Bartlein (1993) and Whitlock (1993), provided a record of postglacial vegetation change. Macroscopic charcoal particles were tallied from contiguous 1-cm intervals of the cores after they were washed through 0.125- and 0.250-mm-mesh sieves. Charcoal concentration (particles cm^{-3}) was interpolated to pseudoannual values; these were integrated over ten-year time intervals and divided by the deposition time to determine the average CHAR per decade.

Background CHAR is the slowly varying component of the charcoal data (Long and others 1998, Millspaugh 1997). These levels were determined by use of a locally weighted running mean with a moving window. The likely source of background CHAR is (1) charcoal sequestered in the watershed and littoral zone of the lake and later introduced to the lake, (2) local charcoal that varies depending on standing biomass and fuel load (in other words, charcoal production), and (3) charcoal that is transported to the lake from distant fires (Long and others 1998). Charcoal peaks were identified as levels of high charcoal accumulation rates above the background level. Peaks consisted primarily of charcoal from local (within the watershed) or extralocal (close to the watershed) fire events as well as natural, random variations in accumulation. The spacing of charcoal peaks provides an estimate of the frequency of fire events (one or more fires occurring within a ten-year period).

CLIMATE, VEGETATION, AND FIRE VARIATIONS ON ORBITAL TIME SCALES

On time scales of 10^3–10^4 years (hereafter referred to as orbital time scales), an important large-scale control of climate has been the variation in the seasonal

Table 2.1. Uncalibrated and calibrated age determinations for Slough Creek Lake

Depth (SCL#88A) (m)	Material	Lab No.	Calibrated Age ± 1 σ (^{14}C yr B.P.)	Calibrated Calendar Age & 1 σ range[a] (cal yr B.P.)	Weighting[b]
0					1.00
0.85–0.95	charcoal	AA-4519	1840 ± 70	1635 (1735) 1863	0.10
2.28–2.29	charcoal	Beta-87009	3130 ± 60	3266 (3354) 3386	0.10
3.05–3.15	charcoal	AA-4520	4550 ± 80	5047 (5288) 5317	0.10
3.46–3.47	Mazama O		6845 ± 50[c]	7578 (7631) 7664	1.00
4.86–4.88	charcoal	Beta-87010	9710 ± 60	10889 (10954) 10980	0.10
5.64–5.65	Glacier Peak B	Pitt-0554	11800 ± 190[d]	13522 (13755) 14008	0.10
5.95–6.05	charcoal	AA-4523	12060 ± 130	13873 (14065) 14272	0.10

[a]Minimum of 1 σ calibrated age range (calibrated calendar age) maximum of 1 σ calibrated age range (1 σ = square root of sample SD^2 + curve SD^2, Stuiver and Reimer 1993; Stuiver and others 1998).
[b]Value assigned to a calibrated age depending on our confidence in accuracy of date (in other words, ages of 0 and Mazama O tephra assigned highest weighting). Weighting used in determination of weighted 3rd-order polynomial regression.
[c]From Sarna-Wojcicki (1983).
[d]From Whitlock (1993).

cycle of insolation (for example, incoming solar radiation) caused by the tilt of the earth's axis and the timing of perihelion (Kutzbach and Guetter 1986). In the past 20,000 years, the contrast in insolation between winter and summer was greatest in the early Holocene (between approximately 11,000 and 8000 cal yr B.P.) because the tilt of the earth's axis was greater then and perihelion occurred in the northern hemisphere summer rather than winter, as at present. At the maximum amplification, ca. 10,000 cal yr B.P., insolation at the latitude of YNP was 8.5 percent greater in summer and 10 percent less in winter than at present (Berger 1978).

In what is now the western United States, greater summer insolation directly increased temperatures and decreased effective moisture in the continental interior. Indirectly, it affected atmospheric circulation patterns that controlled levels of summer precipitation (Thompson and others 1993, Bartlein and others 1998). Increased temperature contrasts in the early Holocene between the land and ocean strengthened the northeastern Pacific subtropical high-pressure system, and this intensified summer-drought conditions in what is now the northwestern United States, including southern and central YNP (Whitlock and Bartlein 1993). The amplification of the land-to-ocean temperature contrast in summer also intensified monsoonal circulation during the early Holocene (Bartlein and others 1998). Moist air flowed northward into the interior of the continent, reaching the basins of Wyoming. The Yellowstone-Lamar Province probably experienced wetter summers between 11,000 to 8000 cal yr B.P. than at present because of this increased onshore flow. The influence of the subtropical high-pressure system and monsoonal circulation weakened progressively in the middle and late Holocene, when the seasonal contrasts in insolation decreased to present levels. As a result, conditions probably became progressively cooler and moister in the current southern and central YNP and drier in the northern YNP (Whitlock and Bartlein 1993).

The postglacial vegetation history of YNP is understood from an examination of several pollen records in the region (Waddington and Wright 1974, Baker 1976, Gennet and Baker 1986, Whitlock 1993, Whitlock and Bartlein 1993, Whitlock and others 1995). At Slough Creek Lake, pollen data from the late-glacial period features high amounts of *Artemisia* and herbaceous taxa, which suggest an early period of tundra vegetation. *Betula* percentages increase between approximately 12,600 and 12,400 cal yr B.P., followed by high values of *Picea* at about 12,000 cal yr B.P. (Fig. 2.2). These taxa suggest the presence of *Picea* parkland with riparian areas of *Betula*. The climate was cooler and effectively wetter than today, but warmer than before. After ca. 11,000 cal yr B.P.,

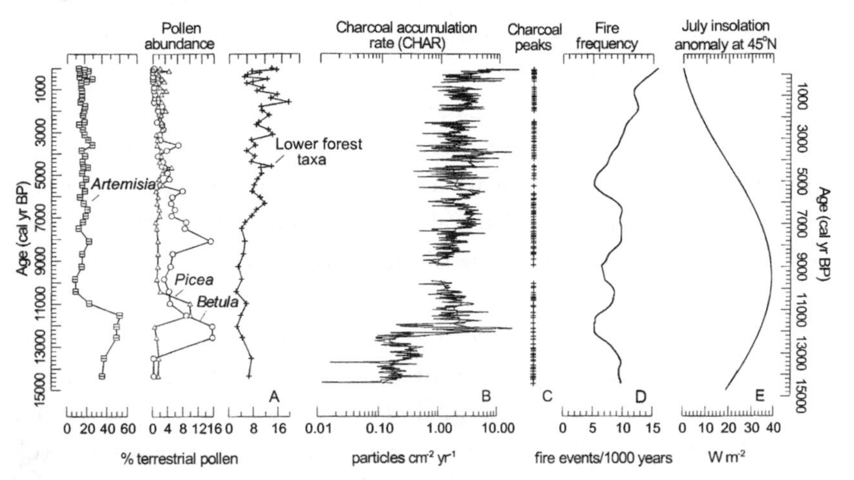

Figure 2.2. Comparison of pollen and charcoal data from Slough Creek Lake with the July insolation anomaly at Lat. 45°N. (A) Pollen percentages, including those of lower forest taxa *(Poaceae, Pseudotsuga,* and *Populus)* (data from Whitlock and Bartlein 1993). (B) Log-transformed CHARs decomposed into background (slowly varying line) and peaks using a window width of 500 years and a threshold ratio of 1.00. (C) CHAR peaks. (D) Fire frequency, represented as fire events/1000 years. (E) July insolation, calculated as the difference from present of average daily insolation at Lat. 45°N in mid-July (Berger 1978).

Picea parkland was replaced by a mixed forest with *Pinus contorta, Pinus flexilis* and/or *Pinus albicaulis,* and *Juniperus.* Most of these conifers grow at higher elevations today, and the pollen assemblage suggests wetter-than-present conditions in the early Holocene. A gradual increase in Poaceae, *Pseudotsuga,* and *Populus tremuloides* after about 7000 cal yr B.P. indicates the development of *Pseudotsuga* parkland and implies greater aridity than before.

The fire history generally matches the changes in vegetation over the past 11,000 years. Between about 11,000 and 8000 cal yr B.P., fire frequency in the *Pinus-Juniperus* forest was relatively low, ranging from six to ten events/1000 years (Fig. 2.2). Summer convective storms would have been frequent at this time as a result of increased surface heating associated with high summer insolation. Summer precipitation apparently depressed the fire activity as fuels were too moist to burn. Between approximately 8000 and 6000 cal yr B.P., fire frequency increased to ten events/1000 years. After a brief period (between approximately 6000 and 5000 cal yr B.P.) when fire frequency declined to six events/1000 years, the number of fires increased steadily through the late Holocene to approximately seventeen events/1000 years at present. The general trend of increasing fire frequency in the late Holocene suggests progressively drier summers. Thus the development of low-elevation parkland and high fire activity in the late Holocene is consistent with a model of decreased onshore flow associated with declining summer insolation. In the past 2000 years, fire frequency reached its highest level of twelve to seventeen events/1000 years.

Pollen data from Cygnet Lake suggest that tundra or grassland grew in what is now southern and central YNP following deglaciation (Whitlock 1993). In areas of southern/central YNP underlain by nonrhyolitic soils, *Picea* parkland developed between approximately 13,400 and 12,400 cal yr B.P. This vegetation was replaced by *Picea-Abies-Pinus albicaulis* forest between approximately 12,400 and 11,000 cal yr B.P. In contrast, the pollen record from Cygnet Lake indicates that *Picea, Abies,* and *P. albicaulis* were uncommon in the Central Plateau Province, probably because they were unable to establish on infertile rhyolite substrates. Instead, tundra or meadow communities persisted until about 11,300 cal yr B.P. when *Pinus contorta* forest was established (Fig. 2.3). *P. contorta* forest became widespread in southern/central YNP (on both rhyolitic and nonrhyolitic substrates) after about 11,000 cal yr B.P.

At Cygnet Lake, the frequency of local fires was initially low (approximately four events/1000 years) at about 17,000 cal yr B.P. when tundra prevailed (Fig. 2.3; Millspaugh and others 2000). July insolation at that time was similar to or slightly higher than present; however, summer conditions were cooler than today as a result of continental-scale circulation patterns and locally retreating glaciers. A gradual rise in fire frequency from four to six events/1000 years between 17,000 and 11,700 cal yr B.P. was most likely a response to increasing summer insolation. During this period, increasing background CHAR probably corresponded with a change from open vegetation to parkland as *P. contorta* colonized the Central Plateau Province.

The incidence of fires increased rapidly around Cygnet Lake after approximately 11,700 cal yr B.P. when summers became warmer and drier in the Central Plateau Province with the expansion of the northeastern Pacific subtropical high-pressure system (and the direct effects of greater insolation). *Pinus contorta* forest was present in the watershed by about 11,300 cal yr B.P. At the time of the early-Holocene insolation maximum, effectively drier summers than before increased fire frequency near Cygnet Lake to fifteen events/1000 years. After 9900 cal yr B.P., fire frequency gradually decreased to the present levels (< three events/1000 years). The trend of declining fire frequency in the middle and late Holocene paralleled decreasing summer insolation and a shift to cooler and effectively wetter conditions than before. Reduced drought in the

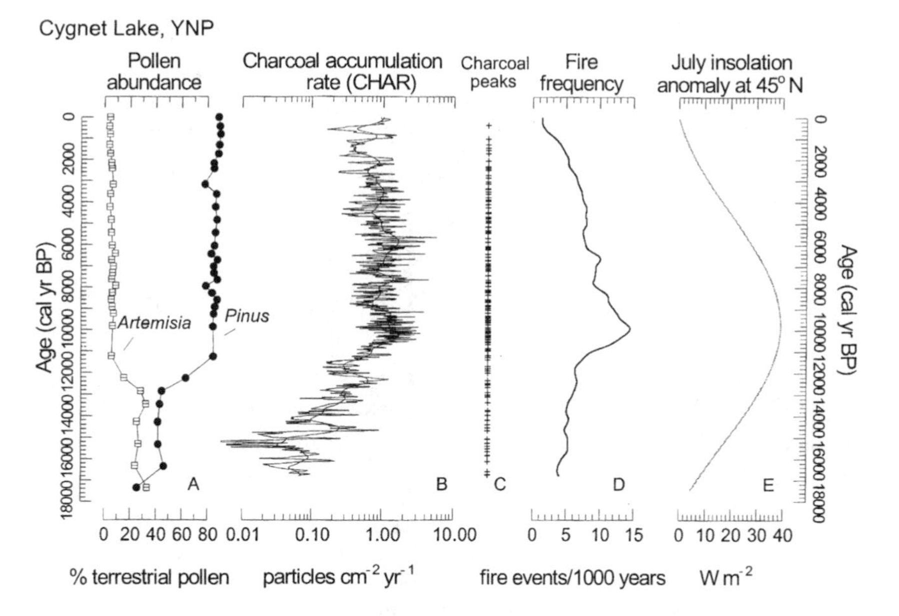

Figure 2.3. Comparison of pollen and charcoal data from Cygnet Lake with the July insolation anomaly at Lat. 45°N (Millspaugh and others 2000). (A) Pollen percentages of *Artemisia* and *Pinus* (data from Whitlock 1993). (B) Log-transformed CHARs decomposed into background (slowly varying line) and peaks using a window width of 750 years and a threshold ratio of 1.00. (C) CHAR peaks. (D) Fire frequency, represented as fire events/1000 years. (E) July insolation, calculated as the difference from present of average daily insolation at Lat. 45°N in mid-July (Berger 1978).

late Holocene shortened the fire season (for example, by influencing the probability of ignition, fuel moisture, and fire weather) to its present length (July to mid-October) in any given year. Background CHAR remained relatively stable through the Holocene despite variations in the stand-age distribution (and thus above-ground biomass) of the *P. contorta* forest. In the past two millennia, fire frequency has been lower than at any time since the establishment of *P. contorta* forests. Protracted periods without fire probably have allowed *P. contorta* stands to mature and the forest to become more homogeneous. This type of landscape pattern has helped maintain the current fire regime, which features infrequent severe fires.

The period of maximum fire frequency in the Holocene has therefore differed between the Yellowstone-Lamar Province and the Central Plateau Province (Fig. 2.4). The fire record from Slough Creek Lake indicates high fire occurrence in the Yellowstone-Lamar Province in recent millennia as a result of increasing

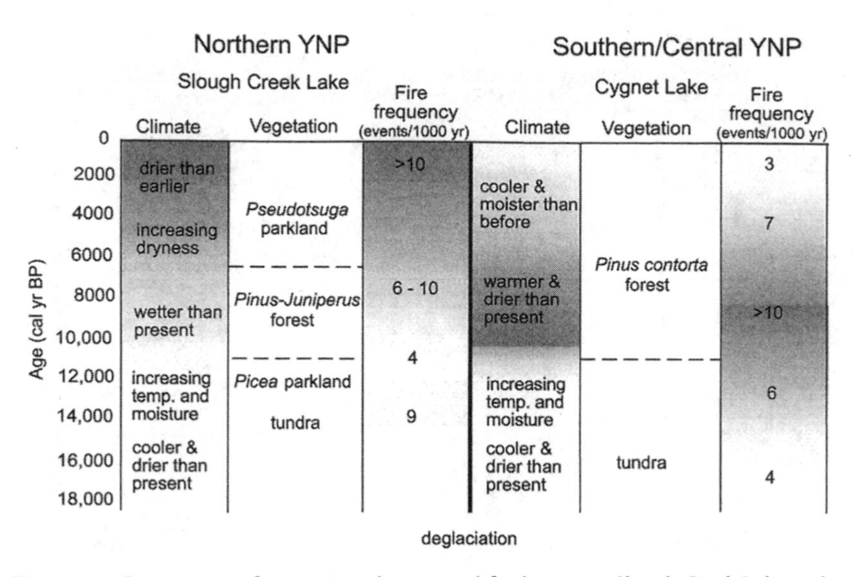

Figure 2.4. Comparison of vegetation, climate, and fire history at Slough Creek Lake and Cygnet Lake. Shading indicates the intensity of aridity (darker shading = greater aridity).

drought conditions. In contrast, the fire record from Cygnet Lake suggests that fire frequency was highest on the Central Plateau between approximately 11,000 and 8000 cal yr B.P. when drought conditions prevailed there.

CLIMATE, VEGETATION, AND FIRE LINKAGES ON ANNUAL-TO-CENTENNIAL TIME SCALES

Embedded within the long-term climate changes that differentiate the history of the northern and southern/central YNP are shorter (i.e., annual-to-centennial) climate variations. Climate intervals, such as the Medieval Warm Period (1000 to 650 cal yr B.P., Hughes and Diaz 1994) and the Little Ice Age (about 500 to 100 cal yr B.P., Jones and Bradley 1992), have been identified in several regions (Esper and others 2002). For example, evidence from moraine sequences in the Rocky Mountains suggests that the Little Ice Age was cold enough to produce glacial advances in many mountain ranges (for example, Carrara 1987). Climate variations on this time scale are attributed to variations in volcanic aerosol production (Bryson and Goodman 1980), solar activity (Lean and Rind 2001, Stuiver and Braziunas 1992), oceanic thermohaline circulation (Rind and Overpeck 1993), and atmospheric composition (Crowley 2000).

Comparison of the fire histories at Slough Creek and Cygnet lakes on this

Table 2.2. Periods of high fire occurrence based on charcoal records from the Central Plateau and Yellowstone-Lamar provinces and records of debris-flow activity from the Yellowstone-Lamar Province

Charcoal records from Cygnet Lake and Slough Creek Lake (this chapter) (cal yr B.P.)	Fire-related debris-flow activity from Yellowstone-Lamar Province (Meyer and others 1995) (cal yr B.P.)
7000–6500	
6250–5750	6450– 5950
5400–5000[a]	5450–4350
4250–3500	3900–3150
2500–2000	2750–1600, major peak: 2300–2050
1000–500	1300–750, major peak: 900–750

[a]High fire occurrence in Yellowstone-Lamar Province only.

time scale reveals 500-year periods when fire occurrence was high in both the Yellowstone-Lamar and the Central Plateau provinces (Table 2.2), including the Medieval Warm Period. These intervals were centuries of severe drought and widespread fires that overrode the climate trends expressed on orbital time scales. The fact that widespread fires correspond with periods of fire-related debris-flow activity (Meyer and others 1992, 1995, Chapter 3) suggests that convective summer storms destabilized steep burned slopes during large-fire years.

Climate variations on annual-to-decadal time scales are governed by ocean-atmospheric interactions, such as ENSO and the Pacific Decadal Oscillation (McCabe and Dettinger 1999, Mantua and others 1997). The response of fire occurrence to these climate variations is recorded by instrumental, dendrochronological, and sedimentological data from YNP. Instrumental data suggest that summer temperatures have increased and winter precipitation has decreased since 1895 in YNP. This trend toward aridity has been paralleled by an increase in burned area during recent fires (Balling and others 1992). In the Yellowstone-Lamar Province, dendrochronological evidence suggests that large fires burned in *Pseudotsuga* parkland in A.D. 1740, 1758, 1776, and 1855 (Barrett 1994, Houston 1973), whereas few small fires occurred between 1855 and 1988. Dendrochronological data from the Central Plateau Province indicate that, before 1988, large fires took place from about A.D. 1690 to 1710 and 1730 to 1750 (Romme and Despain 1989). A composite 750-year record from several small lakes on the Central Plateau shows widespread fires in about A.D. 1450, 1560, 1700, and 1988, and very few fires from A.D. 1220 to 1440 and 1700 to 1988

(Millspaugh and Whitlock 1995). A 600-year-long record from Cygnet Lake corroborates the evidence that fires burned on the Central Plateau in A.D. 1550 to 1560, 1730, and 1988 (Millspaugh 1997). The near-absence of fires between A.D. 1730 and 1988 on the Central Plateau and between A.D. 1855 and 1988 in the Yellowstone-Lamar Province may be related to cooler, wetter conditions at the end of the Little Ice Age.

Comparison of fire history on multiple time scales sheds light on the linkages in the two provinces among climate, fire, and vegetation. On orbital time scales, trends in fire frequency between the two provinces differed in response to variations in summer drought intensity. These droughts result from changes in atmospheric circulation governed by large-scale controls, such as changes in summer insolation. Similarly, shifts in the composition of vegetation in the Yellowstone-Lamar Province and other nonrhyolite provinces on this time scale have been a response to large-scale changes in temperature and effective moisture. In the Central Plateau Province, where edaphic controls limited the vegetation response to large-scale climate changes, the response is registered solely as changes in fire frequency. Variations in fire may have caused shifts in the distribution of stand ages, but the composition of *Pinus contorta* forest remained unchanged.

On centennial time scales, synchronous fire periods reflect times of widespread drought that overrode the opposing controls of fire regimes on orbital time scales. On shorter time scales, vegetational controls are increasingly more important. For example, fire records from Slough Creek Lake, Cygnet Lake, and other small lakes (Millspaugh and Whitlock 1995) show that fire patterns were highly variable at the decadal scale, and only rarely did more than one site record a particular fire. This asynchrony probably reflects the interactions between climate and ecological controls on short time scales. In the case of individual fire years, the distribution of old and young stands may have constrained or enhanced fire spread in the past (Knight 1987, Romme and Despain 1989, Chapters 4, 14). Between 1972 and 1987, fires were confined to 300-year-old forests, but "died out" when they reached 100-year-old patches of forest (Despain and Sellers 1977, Turner and Romme 1994). In contrast, the 1988 fires burned through forest stands of all ages because of extreme weather conditions.

POTENTIAL RESPONSE OF FIRE REGIMES AND VEGETATION TO GLOBAL WARMING

Early Holocene fire regimes, while not offering an exact analogue for future conditions, provide insights into how fire incidence might respond to climate changes in the future. Climate model simulations that incorporate increased levels of greenhouse gases show that many regions of the western United States will experience warmer drier summer conditions and warmer wetter winters in the next century (Rind and others 1990, Giorgi and others 2001). The model experiments further indicate that higher temperatures and greater atmospheric loading will result in increased thunderstorm activity (Overpeck and others 1990) and a consequent increase in lightning fires (Price and Rind 1994).

High-resolution climate and vegetation models portray potential environmental changes in the YNP region as a result of a doubling of atmospheric CO_2 (Bartlein and others 1997, Thompson and others 1998, Shafer and others 2001). In general, most model simulations suggest that mean July temperatures will increase and mean July precipitation will decrease throughout the region. Studies of soil moisture anomalies indicate drier-than-present conditions during the summer fire season, comparable to those portrayed in simulations for 6000 cal yr B.P. (Whitlock and others 2003). These anomalies suggest drier fuels and increased likelihood of fire spread in montane forests. The record of high fire occurrence at 6000 cal yr B.P. in what is now central Yellowstone may provide an analogue for future fire conditions in that region.

Model simulations also suggest an expansion of low-elevation and disturbance adapted taxa into montane forests in the Yellowstone region and a loss of taxa that presently grow at high elevations (Bartlein and others 1997). With warmer summers and more fires, the mosaic of the *P. contorta* forests on the Central Plateau might shift from middle-to-old-aged stands to younger and more patchy stands (Romme and Turner 1991, Turner and Romme 1994). Projected vegetation changes would maintain *P. contorta* where it now grows and also allow it to dominate on more fertile nonrhyolite substrates and at higher elevations (Bartlein and others 1997).

In the future, warmer, drier summers in the Yellowstone-Lamar Province may result in increased fire frequency and an expansion of *Pseudotsuga* parkland to middle elevations in YNP (Bartlein and others 1997). However, if higher summer temperatures result in greater onshore flow, summer conditions in the Yellowstone-Lamar Province may become wetter and fire frequency may decrease as it did in the early Holocene. A decrease in fire frequency in this province would enable less fire-tolerant species, such as *Abies* and *Picea,* to colonize the mesic habitats within this province. See Chapter 13 for a discussion of how these possible changes would influence ungulate survival.

In conclusion, records from Cygnet and Slough Creek lakes suggest that past fire regimes in YNP varied in response to climate changes on multiple time scales. On time scales of 1000 years and longer, fire regimes in the Central Plateau and the Yellowstone-Lamar provinces have tracked shifts in atmospheric circulation that resulted from changes in summer insolation. Maximum fire frequency coincided with the timing of long-term drought in a particular province. The fire record from Cygnet Lake in the Central Plateau Province suggests that fire frequency was highest in the *P. contorta* forest between approximately 11,000 and 8000 cal yr B.P. and the changes in fire frequency occurred without significantly altering forest composition. In contrast, the record from Slough Creek Lake indicates that the period of maximum fire frequency in the Yellowstone-Lamar Province occurred in the past 2000 years. Fire regimes in the Central Plateau and Yellowstone-Lamar provinces have been nonstationary on long time scales because fire frequency has varied continuously with Holocene climate changes. Superimposed on these long-term trends are variations in fire occurrence on submillennial time scales. Fire activity was high in both provinces during 500-year periods when summer drought was severe and widespread. One such period from about 500 to 1000 cal yr B.P. corresponds with the Medieval Warm Period, when fire-related debris flows were more frequent in YNP. The fire-drought linkage is also evident on decadal and annual time scales.

An examination of fire history of YNP does not tell us when or where to expect the next large fire, but it does tell us what unique combinations of vegetation and climate have given rise to heightened fire activity in the past. It also shows the influence of large-scale climate variations on the environmental history of the YNP region. It is interesting to note that an area as small as YNP exhibits a spatially variable climate history that in turn has shaped within-park differences in vegetation and fire regimes. The contrasts in fire regimes during the Holocene reflect the influence of two precipitation regimes and their expression within a topographically complex landscape. Occasionally, long-term differences have been overridden by climate variations on shorter time scales and resulted in widespread fire events in particular years, decades, and centuries. As the large-scale controls of climate continue to change, the complex interactions that have governed Yellowstone's fire, climate, and vegetation in the past will surely result in equally complex and intriguing patterns of vegetation change and fire activity in the future.

Acknowledgments

The research described in this chapter was supported by grants from the National Park Service, the National Science Foundation (ATM-0117160,

SES-9305436), the University of Oregon Graduate Research Fellowship, and Sigma Xi.

References

Agee, J. K. 1993. Fire ecology of the Pacific Northwest forests. Island Press, Washington, D.C.

Baker, R. G. 1976. Late Quaternary vegetation history of the Yellowstone Lake Basin, Wyoming. U.S. Geol. Surv. Prof. Pap. 729-E.

Balling, R. C., Jr., G. A. Meyer, and S. G. Wells. 1992. Climate change in Yellowstone National Park: Is the drought-related risk of wildfires increasing? Climate Change 22:34–35.

Barrett, S. W. 1994. Fire regimes on andesitic mountain terrain in northeastern Yellowstone National Park, Wyoming. Int. J. Wildl. Fire 4:65–76.

Bartlein, P. J., K. H. Anderson, P. M. Anderson, M. E. Edwards, C. J. Mock, R. S. Thompson, R. S. Webb III, C. Whitlock. 1998. Paleoclimatic simulations for North America over the past 21,000 years: Features of the simulation climate and comparisons with paleoenvironmental data. Quat. Sci. Rev. 17:549–585.

Bartlein, P. J., C. Whitlock, and S. L. Shafer. 1997. Potential future environmental change in the Yellowstone National Park Region. Conserv. Biol. 11:782–792.

Bennett, K. D., and K. J. Willis. 2002. Pollen. Pages 5–31 in J. P. Smol, H. J. B. Birks, and W. M. Last, eds., Tracking environmental change using lake sediments, vol. 3, Terrestrial, algal and siliceous indicators. Kluwer, Dordrecht.

Berger, A. L. 1978. Long-term variations of caloric insolation resulting from earth's orbital elements. Quat. Res. 9:139–167.

Bryson, R. A., and B. M. Goodman. 1980. Volcanic activity and climatic changes. Science 207:1041–1044.

Bryson, R. A., and F. K. Hare. 1974. The climate of North America. Pages 1–47 in R. A. Bryson and F. K. Hare, eds., World survey of climatology. Elsevier, Amsterdam.

Carrara, P. E. 1987. Holocene and latest Pleistocene glacial chronology, Glacier National Park, Montana. Can. J. Earth Sci. 24:387–395.

Crowley, T. J. 2000. Causes of climate change over the last 1000 years. Science 289:270–277.

Despain, D. G. 1987. The two climates of Yellowstone National Park. Proc. Montana Acad. Sci. 47:11–19.

———. 1990. Yellowstone vegetation: Consequences of environment and history in a natural setting. Roberts Rinehart, Boulder.

Despain, D. G., and R. E. Sellers. 1977. Natural fire in Yellowstone National Park. West. Wildlands 4:20–24.

Dirks, R. A., and B. E. Martner. 1982. The climate of Yellowstone and Grand Teton National Parks. Occ. Pap. no. 6, U.S. Nat. Park Serv., Washington, D.C.

Esper, J., E. R. Cook, F. J. Schweingruber. 2002. Low-frequency signals in long tree-ring chronologies for reconstruction past temperature variability. Science 274:2250–2253.

Fall, P. L., P. T. Davis, and G. A. Zielinski. 1995. Late Quaternary vegetation and climate of the Wind River Range, Wyoming. Quat. Res. 43:393–404.

Gennet, J. A., and R. G. Baker. 1986. A late-Quaternary pollen sequence from Blacktail Pond, Yellowstone National Park, Wyoming, USA. Palynology 10:61–71.

Giorgi, F., B. Hewitson, J. Christensen, M. Hulme, H. Von Storch, P. Whetton, R. Jones, L.

Mearns, C. Fu. 2001. Regional climate information: Evaluation and projections. Pages 583–638 in J. T. Houghton, Y. Ding, D. J. Griggs, M. Noguer, P. J. van der Linden, X. Dai, K. Maskell, and C. A. Johnson, eds., Climate change 2001: The scientific basis. Contribution of Working Group I to the Third Assessment Report of the Intergovernment Panel on Climate Change. Cambridge Univ. Press, Cambridge.

Higgins, R. W., Y. Yau, and X. L. Wang. 1997. Influence of the North American monsoon system on the United States precipitation regime. J. Climate 10:2600–2622.

Houston, D. B. 1973. Wildfires in northern Yellowstone National Park. Ecology 54:1111–1116.

Hughes, M. K., and H. F. Diaz. 1994. Was there a Medieval Warm Period, and if so, where and when? Climate Change 26:109–142.

Jones, P. D., and R. S. Bradley. 1992. Climatic variations over the last 500 years. Pages 649–665 in R. S. Bradley and P. D. Jones, eds., Climate since A.D. 1500. Routledge, London.

Knight, D. H. 1987. Parasites, lightning, and the vegetation mosaic in wilderness landscapes. Pages 59–83 in M. G. Turner, ed., Landscape heterogeneity and disturbance. Springer-Verlag, New York.

Kutzbach, J. E., and P. J. Guetter. 1986. The influence of changing orbital patterns and surface boundary conditions on climate simulations for the past 18,000 years. J. Atmosph. Sci. 43:1726–1759.

Lean, J., and D. Rind. 2001. Earth's response to a variable sun. Science 292: 234–236.

Long, C. J., C. Whitlock, P. J. Bartlein, and S. H. Millspaugh. 1998. A 9000-year fire history from the Oregon Coast Range, based on a high-resolution charcoal study. Can. J. For. Res. 28:774–787.

Mantua, N. J., S. R. Hare, Y. Shang, J. M. Wallace, and R. D. Francis. 1997. A Pacific interdecadal climate oscillation with impacts of salmon production. Bull. Amer. Meteor. Soc. 78:1069–1079.

McCabe, G. J., and M. D. Dettinger. 1999. Decadal variation in the strength of ENSO teleconnections with precipitation in the western United States. Int. J. Climatology 19:1399–1410.

Meyer, G. A., S. G. Wells, R. C. Balling, and A. J. T. Jull. 1992. Response of alluvial systems to fire and climate change in Yellowstone National Park. Nature 357:147–150.

Meyer, G. A., S. G. Wells, and A. J. T. Jull. 1995. Fire and alluvial chronology in Yellowstone National Park: Climatic and intrinsic controls on Holocene geomorphic processes. Geol. Soc. Amer. Bull. 107:1211–1230.

Millspaugh, S. H. 1997. Late-glacial and Holocene variations in fire frequency in the Central Plateau and Yellowstone-Lamar provinces of Yellowstone National Park. Ph.D. diss., University of Oregon, Eugene.

Millspaugh, S. H., and C. Whitlock. 1995. A 750-year fire history based on lake sediment records in central Yellowstone National Park, USA. The Holocene 5:283–292.

Millspaugh, S. H., C. Whitlock, and P. J. Bartlein. 2000. A 17,000-year history of fire for the Central Plateau of Yellowstone National Park. Geology 28:211–214.

Mock, C. J. 1996. Climatic controls and spatial variations of precipitation in the western United States. J. Climate 9:1111–1125.

Overpeck, J. T., D. Rind, and R. Goldberg. 1990. Climate-induced changes in forest disturbance and vegetation. Nature 343:51–53.

Pierce, K. L. 1979. History and dynamics of glaciation in the northern Yellowstone Park area. U.S. Geol. Surv. Prof. Pap. 729-F.

Price, C., and D. Rind. 1994. The impact of a 2x CO_2 climate on lightning caused fires. J. Climate 7:1484–1494.

Rind, D., R. Goldberg, J. Hansen, C. Rosenzweig, and R. Ruedy. 1990. Potential evapotranspiration and the likelihood of future drought. J. Geophys. Res. 95:9983–10004.

Rind, D., and J. Overpeck. 1993. Hypothesized causes of decade- to century-scale climatic variability: Climate model results. Quat. Sci. Rev. 12:357–374.

Romme, W. H., and D. G. Despain. 1989. Historical perspective on the Yellowstone fires of 1988. Bioscience 39:695–698.

Romme, W. H., and M. G. Turner. 1991. Implications of global climate change for biogeographic patterns in the Greater Yellowstone Ecosystem. Conserv. Bio. 5:375–386.

Sarna-Wojcicki, A. M., D. E. Champion, and J. O. Davis. 1983. Holocene volcanism in the conterminous United States and the role of silicic volcanic ash layers in correlation of latest-Pleistocene and Holocene deposits. Pages 52–77 in H. E. Wright, Jr., ed., Late-Quaternary environments of the United States. Vol. 2, The Holocene. University of Minnesota Press, Minneapolis.

Schullery, P. 1989. The fires and fire policy. Bioscience 39:686–695.

Shafer, S. L., P. J. Bartlein, and R. S. Thompson. 2001. Potential changes in the distributions of western North America tree and shrub taxa under future climate scenarios. Ecosystems 4:200–215.

Stuiver, M., and T. F. Braziunas. 1992. Evidence of solar variations. Pages 593–605 in R. S. Bradley and P. D. Jones, eds., Climate since A.D. 1500. Routledge, London.

Stuiver, M., and P. J. Reimer. 1993. Extended ^{14}C data base revised CALIB 3.0 ^{14}C age calibration program. Radiocarbon 35:215–230.

Stuiver, M., P. J. Reimer, E. Bard, J. W. Beck, G. S. Burr, K. A. Hughen, B. Kromer, G. McCormac, J. Van der Plicht, and M. Spurk. 1998. INT-CAL89 radiocarbon age calibration, 24,000–0 cal. B.P. Radiocarbon 40:1041–1083.

Stuiver, M., P. J. Reimer, and R. Reimer. 1998. CALIB Radiocarbon calibration ver. 4.0 [online]. Available from http://depts.washington.edu/qil/ [updated ver.4.4 2003].

Tang, M., and E. R. Reiter. 1984. Plateau monsoons of the Northern Hemisphere: A comparison between North America and Tibet. Mon. Weather Rev. 112:617–637.

Thompson, R. S., S. W. Hostetler, P. J. Bartlein, and K. H. Anderson. 1998. A strategy for assessing potential future changes in climate, hydrology, and vegetation in the western United States. U.S. Geol. Surv. Cir. 1153.

Thompson, R. S., C. Whitlock, P. J. Bartlein, S. P. Harrison, and W. G. Spaulding. 1993. Climatic changes in western United States since 18,000 yr B.P. Pages 468–513 in H. E. Wright, Jr., J. E. Kutzbach, T. Webb III, W. F. Ruddiman, and F. A. Street-Perrot, eds., Global climates since the last glacial maximum. University of Minnesota Press, Minneapolis.

Turner, M. G., and W. H. Romme. 1994. Landscape dynamics in crown fire ecosystems. Landscape Ecol. 9:59–77.

Waddington, J. C. B., and H. E. Wright. 1974. Late Quaternary vegetational changes on the east side of Yellowstone Park, Wyoming. Quat. Res. 4:175–184.

Whitlock, C. 1993. Postglacial vegetation and climate of Grand Teton and southern Yellowstone National Parks. Ecol. Monogr. 63:173–198.
Whitlock, C., and P. J. Bartlein. 1993. Spatial variations of Holocene climatic change in the Yellowstone region. Quat. Res. 39:231–238.
Whitlock, C., P. J. Bartlein, and K. J. Van Norman. 1995. Stability of Holocene climate regimes in the Yellowstone region. Quat. Res. 43:433–436.
Whitlock, C., and C. Larsen. 2002. Charcoal as a fire proxy. Pages 75–97 in J. P. Smol, H. J. B. Birks, and W. M. Last, eds., Tracking environmental change using lake sediments, vol. 3, Terrestrial, algal and siliceous indicators. Kluwer, Dordrecht.
Whitlock, C., and S. H. Millspaugh. 1996. Testing the assumptions of fire-history studies: An examination of modern charcoal accumulation in Yellowstone National Park, USA. The Holocene 6:7–15.
Whitlock, C., S. H. Shafer, and J. Marlon. 2003. The role of climate and vegetation change in shaping past and future fire regimes in the northwestern U.S. and the implications for ecosystem management. For. Ecol. Manag. 178:5–21.
Wright, H. E., D. E. Mann, and P. H. Glaser. 1984. Piston cores for peat and lake sediments. Ecology 65:657–659.
Zdanowicz, C. M., G. A. Zielinski, and M. S. Germani. 1999. Mount Mazama eruption: Calendric age verified and atmospheric impact assessment. Geology 27:621–624.

Chapter 3 Yellowstone Fires and the Physical Landscape

Grant A. Meyer

Extensive severe fires like those of 1988 in the Greater Yellowstone Ecosystem clearly have major ecological significance via changes in age structure and composition of vegetation, but their impacts on the physical landscape can be equally profound. Both transient and persistent alterations of terrestrial and aquatic ecosystems may result from postfire geomorphic processes. Fire is particularly important as a catalyst of landscape change in mountain regions, where high-severity burns markedly increase the potential for surface runoff, soil erosion, and landslides on steep slopes, resulting in debris flows and floods during intense storms and rapid snowmelt. Such events account for a large proportion of long-term sediment export in many mountain drainage basins (Scott and Williams 1978, Meyer and others 1995, Kirchner and others 2001). Increased sediment loading of streams often occurs after high-severity fires, and fine sediment in particular has the potential for strongly negative ecological impacts (Newcombe and MacDonald 1991). Therefore, it is important to understand the specific geomorphic effects of the 1988 Yellowstone fires, and place these in a broader and longer-term context.

In this chapter, erosion and sedimentation following the 1988 fires are compared with postfire activity in other mountain environments of the western United States, and the Holocene history of fire-related sedimentation in Yellowstone is examined. This information is used to address the following questions: Were the 1988 fires unusual in their physical landscape effects, for example, in producing extreme sediment yields and exceeding the natural range of disturbance over the last several thousand years? What are the roles of climate and management in the occurrence of high-severity fires and associated geomorphic change, especially in light of potential future warming? And what is the overall importance of fire in postglacial landscape and ecosystem change in Yellowstone?

GEOMORPHIC SETTING AND POSTFIRE RESPONSE IN YELLOWSTONE

Because of the diverse topography of Yellowstone National Park (Fig. 1.2), hydrologic and geomorphic response to the 1988 fires was highly variable. Much of the volcanic plateau in Yellowstone was constructed by sheets of welded Lava Creek tuff from the ca. 640,000-year-old Yellowstone caldera, or by voluminous rhyolite lava flows that largely filled the collapsed caldera (Christiansen 2001). Relief is generally small on the plateau, so postfire erosion potential is limited mainly to river canyons and steep flow margins (for example, Obsidian Cliff). In contrast, Yellowstone's mountain ranges have a high propensity for fire-induced erosion and sediment transport. The Absaroka Range along the eastern park boundary features summit elevations over 3000 m, local relief up to 1000 m, and steep, densely forested slopes along glacial trough valleys. Muddy, poorly cemented Eocene volcaniclastic rocks form most upper valley walls (Prostka and others 1975) and supply abundant weathered debris. Dense forests dominated by lodgepole pine *(Pinus contorta)* with Engelmann spruce *(Picea engelmannii)* and Douglas fir *(Pseudotsuga menziesii)* mantle most valley sides. Subalpine fir *(Abies lasiocarpa)* and whitebark pine *(Pinus albicaulis)* are present near upper treeline. Many small tributary basins in the Absaroka Range were almost entirely burned by severe fires in 1988, and numerous debris flows and flash floods occurred there in ensuing years (Meyer 1993, Bozek and Young 1994, Meyer and Wells 1997). Postfire erosional events were observed in several other mountainous areas of the park, including the Snake River drainage (Shroder and Bishop 1995) and Red Mountains in southern Yellowstone.

Steep sideslopes of canyons draining the volcanic plateau also produced postfire debris flows, as in the Grand Canyon of the Yellowstone River and Lava Creek Canyon (Fig. 1.2). Gibbon Canyon, where the Gibbon River is flanked by the caldera rim and steep-fronted rhyolite lava flows, was the site of repeated debris flows that covered the Madison-Norris road (Johansen 1991, Meyer and Wells 1997), which in part prompted rerouting of that road out of the canyon bottom.

GEOMORPHIC PROCESSES AFTER THE 1988 FIRES

Much postfire sediment transport and geomorphic change occurs through large storm-generated flows. These events range in sediment concentration from dense, slurry-like debris flows to sediment-charged flash floods (Pierson and Costa 1987). Often, this entire continuum is observed in a single event, but one end of the range may dominate (Meyer and Wells 1997). These events are initiated by two distinct sets of processes: (1) enhanced surface runoff, with sediment entrainment in extreme discharges and (2) saturation and *en masse* failure of colluvium (that is, slope-mantling debris and soil) (Swanson 1981, Meyer and Wells 1997, Cannon and others 1998). The first mechanism is most effective on steep slopes in high-severity burns, where the litter layer, understory, and fine fuels in the canopy have been largely consumed, leaving a bare soil surface with a mantle of fine ash. These slopes produce pervasive surface runoff during intense rainfall, which typically occurs in summer convective storms in the Yellowstone region. Surface runoff far exceeds that from unburned slopes because of reduced interception, smooth flow paths, and most importantly, decreased infiltration. Infiltration rates may be reduced by physical changes to the soil surface such as compaction and surface sealing by fine soil particles and ash (Megahan and Molitor 1975, Wells 1987). Water-repellent layers are sometimes created during fire when volatiles from combusted organic material condense as hydrophobic compounds within the upper few centimeters of the soil (DeBano and others 1979, Wells 1987, Robichaud 2000). Water-repellent soils were uncommon in Yellowstone after the 1988 fires, however (Shovic 1988), including in basins producing runoff-related debris flows and flash floods in 1989 (Johansen 1991, Meyer 1993, Meyer and Wells 1997). Cannon and others (1998) and Cannon (2001) also note that water-repellent soils are not a prerequisite for greatly enhanced postfire runoff and debris flows in western U.S. mountains.

Surface runoff on steep burned slopes causes sheetwash and especially rill erosion, with entrainment of fine sediment, ash, and charcoal from soil surfaces (Fig. 3.1). Muddy slope runoff converges to produce extreme discharges in small

Figure 3.1a. Severely burned slope in Gibbon Canyon showing pervasive rilling from an August 1989 thunderstorm and associated surface runoff.
b. Main channel in the middle section of the Gibbon Canyon basin in (a), showing erosion of stored alluvium down to underlying bedrock (Lava Creek tuff). Debris-flow levees and mud coatings first appear a short distance down the channel from this area.
c. Muddy, gravel-poor debris-flow deposit on the "12 km" alluvial fan (Meyer and Wells 1997) in the Slough Creek valley just north of Elk Tongue Creek. Abundant fine sediment and charcoal in this facies are derived primarily from thunderstorm-generated rill and sheetwash erosion in the severely burned basin above. A bouldery debris-flow lobe from the same event is visible in the forested area in the background. Photos by Grant A. Meyer.

steep channels that ordinarily carry limited or ephemeral flow. Sediment concentrations increase by scouring and entrainment of stored channel sediment (Meyer and others 1992, Meyer and Wells 1997). In many such events in Yellowstone, sediment concentration reached debris-flow conditions. Other events were dominated by more dilute high-energy flash flooding, but nearly all progressed through a wide range of flow conditions and declined in sediment concentration as flows waned. Debris flows were more likely in small, steep basins of less than about 2 km^2, especially where erosion of loose fine soil in the early postfire period generated a muddy matrix fluid. Similar runoff-related debris-flow generation has been well documented in severe burns in central Montana (Parrett 1987), California, and Colorado mountains (Cannon and others 1998, Cannon 2001).

Postfire debris-flow generation by runoff does not require exceptional magnitude or intensity of precipitation (for example, Wells 1987). No nearby precipitation data are available for Yellowstone postfire debris flows or flash floods, but witnesses typically reported that the associated thunderstorms lasted less than thirty minutes, with even shorter episodes of high-intensity rainfall (Johansen 1991, Meyer 1993). When rare, extreme storms coincide with recently burned mountainous areas, runoff and sediment yield may be enormous (Jarrett 1999), but such events may also involve saturation-induced slope failures (Scott 1971), as discussed below.

The second mechanism for initiation of postfire sediment transport events stems primarily from the decay of roots with time after burning, resulting in reduced strength in colluvium (Schmidt and others 2001). Shallow slides (often called "soil slips") may then occur when colluvium becomes saturated to a critical depth (Gray and Megahan 1981), typically during heavy rainfall in more prolonged storms and/or rapid snowmelt. Rapid infiltration is promoted under these conditions, especially if infiltration rates have recovered after the early postfire period. Reduced evapotranspiration after fire may also increase the probability of saturation and failure, which typically occurs in concave slope hollows that serve both to collect slope debris and concentrate shallow subsurface flow (Reneau and Dietrich 1987, Montgomery and others 2000). The period of increased susceptibility to failure after fire depends on rates of root decay and reestablishment, and is likely to vary with environment. In general, the highest probability of failure is observed four to ten years after burning or logging (Burroughs and Thomas 1977, Gray and Megahan 1981).

Although the saturation-failure mode of debris-flow generation is promoted by fire, colluvial failures often occur in major storms on unburned slopes as well, making a direct link to fire more difficult to establish. In some cases, the evidence for fire-induced instability is strong, as when a 1996–1997 winter storm and thaw produced numerous colluvial failures and debris flows in western Idaho (Meyer and others 2001). A large majority of failures on forested slopes occurred where ponderosa pine and Douglas fir stands were destroyed in the 1989 Lowman fire (Shaub and Donato 1999).

Loss of root strength appeared to be of relatively minor consequence to burned slopes in Yellowstone. Colluvium is generally thin and therefore more stable on the glacially scoured bedrock slopes, and erosion by postfire surface runoff may have limited colluvium buildup since deglaciation. Very few *en masse* slope failures were noted in the debris-flow basins investigated through 1991 by Johansen (1991) and Meyer and Wells (1997). In a 1991 aerial survey of the Lamar River basin in northeastern Yellowstone, I observed only a few shallow debris slides, and saw relatively few such failures in field reconnaissance in subsequent years. Some shallow slides initiated in burned areas after thawing of unusually heavy snowpacks as in 1996, but most occurred on slopes mantled by weathered shale debris.

Several larger, deep-seated rotational and planar landslides initiated in 1988 burned areas, but most were associated with the unusually large snowmelt runoff of 1996. One large slump in the severely burned upper Silver Creek basin near the Northeast Entrance was initiated during a heavy thunderstorm in July 1996, after snowmelt. Last-glacial till slid off an underlying older, clay-rich weathered diamicton and transformed into a massive debris flow that traversed the Silver Creek fan to Soda Butte Creek. Most other large landslides initiated in areas of unstable shales—for example, the Cambrian Park Shale in northeastern Yellowstone, or Cretaceous shales in the southern park. Shroder and Bishop (1995) investigated nineteen landslides in burned areas in shale bedrock along the Snake River near the South Entrance (Fig. 1.2). Most clearly predated the 1988 fires. They estimated that about half of these landslides were shallow enough for fire to have increased the potential for reactivation, but also found no direct evidence that postfire movement had occurred.

Rick Hutchinson (NPS-Yellowstone, written communication, 1991) documented a large landslide in a burned area on Barlow Peak in southern Yellowstone (Fig. 1.2). The slide initiated in mid-June 1991, and the saturated mass transformed into a rapid debris flow of more than 50,000 m^3 that temporarily dammed the Snake River. The slide surface was several meters below the rooting zone, and the detached mass involved incompetent Cretaceous mudrocks as well as overlying regolith. Loss of root strength was probably an insignificant

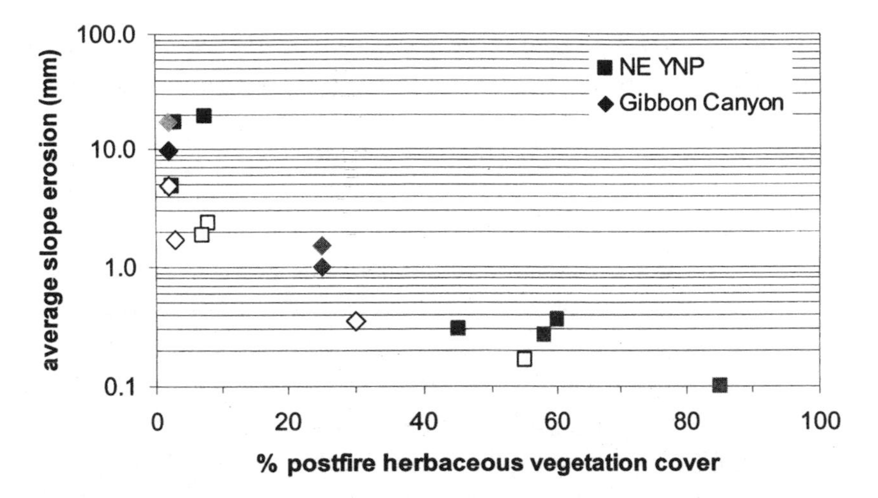

Figure 3.2. Average depth of erosion on burned and rilled slopes in Yellowstone debris-flow basins in relation to percentage ground cover by postfire herbaceous regrowth. Data are from 30-m cross-slope transects measured shortly after convective storms in the "12 km" basin in NE Yellowstone and the "La Familia" basin in Gibbon Canyon (Meyer and Wells 1997). Data are grouped according to an index of the potential erosional power of surface runoff, equal to the slope length in meters above the transect (a proxy for runoff discharge) multiplied by slope gradient at the transect (tan θ). Open symbols denote runoff power index < 15; gray symbols denote 15 < power index < 40; black symbols denote power index > 40 (maximum 98). Note that for a given vegetation cover, erosion is typically less at lower power index values. Erosion is small at 30 percent or greater vegetation cover.

factor, but loss of transpiration by trees may have been important in promoting saturation.

Enhanced postfire erosion was short-lived in most of Yellowstone. Regrowth of as little as 30 percent herbaceous plant cover dramatically reduced surface runoff and the potential for debris-flow generation (Fig. 3.2). In northeastern Yellowstone, revegetation had largely protected some severely burned basins from runoff by early summer 1989 (Meyer 1993). In most burned areas, runoff-related debris-flow and flood events became uncommon after 1991 (Meyer and Wells 1997). Factors such as the prefire plant community, soil texture, slope aspect, and microclimate created large differences in rates of revegetation. Dense and well-shaded north-facing slopes typically supported little understory vegetation before the 1988 fires; thus fewer seeds and rhizomes were present for revegetation. Some lower-elevation, south-facing slopes also revegetated more slowly because of drier conditions, as in Gibbon Canyon, where coarse-textured soils also have poor water-holding properties. Although declining in frequency, debris flows and floods were generated through 1998 by runoff from south-facing basins in Gibbon Canyon, where the primary postfire vegetation was sparse grass and lodgepole pine seedlings.

EROSION AND SEDIMENT YIELD FROM THE 1988 FIRES

Increased soil erosion is a direct consequence of severe fires, and erosion rates may be elevated by thirty times or more over prefire levels in mountainous terrain (Krammes 1960, Wohlgemuth 1999). An average of about 4–5 mm of soil was stripped from rilled upper basin slopes in a 1.6 km^2 debris-flow basin in northeastern Yellowstone, producing a mass of about 64 Mg/ha (Meyer and Wells 1997) (Fig. 3.2). This erosion is comparable to the 50 Mg/ha reported by Robichaud (1999) for the upper range of first-year postfire erosion in the Northern Rocky Mountains. It is modest, however, compared to first-year postfire soil erosion of 72–370 Mg/ha in central Arizona (Hendricks and Johnson 1944) and 250 Mg/ha in the Sierra Nevada (Rowe 1951). As much as 870 Mg/ha of erosion was estimated after a single postfire storm in central Montana (Schultz and others 1986). The large variability in erosion values is likely a function of differences in rainfall, soil, and slope conditions, as well as measurements averaged over different spatial scales.

Estimates of sediment yield from small drainage basins provide a different (but also scale-dependent) measure of postfire erosion, and include sediment derived from channel erosion as well as soil stripped from slopes. Few baseline data exist in Yellowstone, however, to compare sediment yields from small basins before and after the 1988 fires. Most reported postfire sediment yields in the western United States are for single debris-flow or flood events from small (0.1–3 km^2) mountain drainage basins. A preliminary comparison of published data shows that sediment yields following the 1988 Yellowstone fires were similar to postfire runoff-generated events elsewhere (Fig. 3.3). The first debris-flow event in the "La Familia" basin in Gibbon Canyon on August 10, 1989, produced the greatest mass of sediment per unit basin area of any runoff-generated event measured in Yellowstone (Johansen 1991, Meyer and Wells 1997). Subsequently, several smaller debris-flow events occurred in the same Gibbon Canyon basins, and repeated debris flows in the first few years after fire are not uncommon (Wells 1987). Most small basins that produced debris flows in northeastern Yellowstone experienced only one major event, however. Overall, the magnitude and frequency of sedimentation events after the 1988 Yellowstone fires appear to be typ-

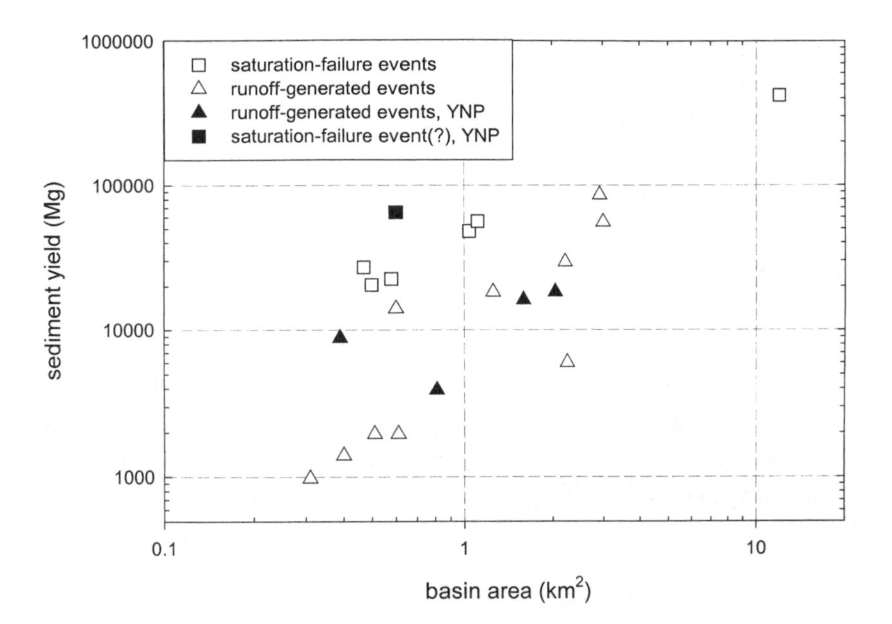

Figure 3.3. Sediment yield in single fire-related debris-flow and flood events from small mountain drainage basins in the western United States. Events where sediment is contributed primarily by saturation and failure of colluvium produced generally higher unit-area sediment yields than did runoff-erosion events. The single saturation-failure event measured in Yellowstone (YNP) is the Barlow Peak landslide, which was not clearly fire-related. Data from Scott 1971, Parrett 1987, Wells 1987, Johansen 1991, Wohl and Pearthree 1991, Meyer 1993, Meyer and Wells 1997, Cannon and others 1998, and Meyer and others 2001. The outlying point at 12 km^2 basin area is the Hidden Springs, California, debris flow described by Wells 1987; contribution of sediment by saturation-induced failures in this event is uncertain.

ical of runoff-generated events in severely burned basins in western U.S. mountains (Fig. 3.3).

Although the data are limited, postfire unit area sediment yields average about five times higher in events with saturation-induced failures than in those with runoff-generated erosion alone (Fig. 3.3). Saturation-induced events typically require unusual precipitation and/or snowmelt in addition to burning, but for most of the included events fire effects also were clearly involved. In Yellowstone, sediment yield from only one such event was measured at Barlow Peak, but fire was an uncertain causal factor (Hutchinson, written communication, 1991).

At the scale of larger river basins in Yellowstone, such factors as geomorphology, the mosaic character of the 1988 burns, and recycling of sediment from bank and channel storage complicate linkages between fire and sediment flux. Sediment yields from these larger basins are strongly correlated to stream discharge, and thus to variations in snowmelt runoff, which carries a large percentage of the suspended sediment flux. Sediment concentrations in stream waters provide a better measure of fire effects. About 38 percent of the 1709 km^2 Lamar River basin was burned (Burned Area Survey Team 1988), and suspended sediment concentrations near the Lamar-Yellowstone confluence showed clear increases after the 1988 fires (Ewing 1996). The primary increase in sediment concentrations from 1989 to 1992 occurred during summer after the primary snowmelt runoff period, when average concentrations were more than six times prefire levels. During snowmelt, however, no significant change in sediment concentration was observed (Ewing 1996). These observations highlight the importance of intense summer thunderstorm precipitation in postfire surface runoff and erosion in Yellowstone.

EFFECTS OF POSTFIRE SEDIMENT TRANSPORT ON STREAM SYSTEMS

High-energy flow, scour, and sediment deposition along debris-flow channels not only cause direct mortality of aquatic organisms but also produce persistent changes in channel form and lingering instability as incised channel walls continue to ravel and slump. Most debris-flow channels in Yellowstone, however, are small, steep channels with limited or ephemeral flow that harbor few or no fish. Effects in higher-order perennial stream channels that support major fish populations are less clear (see also Chapters 7, 8). The connection between debris-flow channels and larger streams is not necessarily direct. In glacial trough valleys of the Absaroka Range, tributary debris flows and flash floods commonly expand and decelerate over valley-side alluvial fans before reaching mainstem stream channels (Fig. 3.1c). Therefore much of the coarse sediment transported in post-1988 events was deposited on fans out of reach of mainstem streams (Meyer and others 1995). The majority of postfire sediment contributed to higher-order channels was probably sand and finer material transported in suspension or in "mudflows"—that is, highly mobile debris-flow phases with little gravel. Although short-term siltation of gravel-bed streams is likely (Minshall and others 1997), fire-related debris flows and flash floods are transient. Raveling of incised debris-flow channel walls probably contributes minor additional fine sediment after debris-flow activity has waned, but most Yellowstone stream channels have high stream power and can flush fine sediment rapidly. Fine sediment may persist longer in lower-gradient reaches such as in the Slough Creek meadows.

In narrow valleys like Gibbon Canyon, much debris-flow and flood sediment was deposited directly in the river channel. Gravel bars formed near major inputs, but general aggradation was not apparent in the steep Gibbon Canyon reach, suggesting that finer sediments were quickly moved downstream (see also Chapters 7 and 8). Some tributaries in the Lamar River system above Soda Butte Creek are incised within late Pleistocene glaciofluvial sediments along valley floors (Pierce 1974), thus have small alluvial fans and transport gravels more efficiently to mainstem channels. Minshall and others (1997) documented significant channel widening and lateral instability on a third-order segment of Cache Creek (a Lamar River tributary) that may have stemmed in part from increased postfire gravel loading. Channel widening and gravel deposition were apparent in the late 1990s along the Lamar River just above the Soda Butte confluence, but have not been quantified. In 2000 and 2001, however, Legleiter and others (2003) surveyed nearly 90 km of second- to fourth-order channels in northern Yellowstone and found that channels in basins with a greater proportion of burned area tended to be more incised than those in less burned drainages. They inferred that following early postfire sediment loading, decreased sediment inputs combined with continued high postfire discharges led to net incision. Interpretation of channel changes is further complicated by effects of the 1996 and 1997 floods, the largest on the Lamar River since 1918 (Meyer 2001). These floods were produced primarily by melting of unusually heavy snowpacks; Farnes (Chapter 10) estimated that fire effects increased 1997 peak daily discharge volumes on the Yellowstone River by less than 10 percent. In smaller drainages, however, postfire flood peaks are more likely to be elevated because of a greater proportion of steep, burned slopes and the potential for locally high snowmelt and precipitation rates.

THE ALLUVIAL-FAN STRATIGRAPHIC RECORD OF POSTFIRE RESPONSE

As described above, alluvial fans are common features along glacial trough valley sides in northeastern Yellowstone and other high-relief areas of the park where small tributary drainages join mainstem streams. These fans record the effects of large, severe fires as characteristic debris-flow and flash-flood deposits (Meyer and others 1992, 1995, Meyer and Wells 1997). In particular, muddy debris-flow facies containing abundant coarse, angular charcoal fragments are diagnostic fire-related deposits and are common both in post-1988 deposits (Fig. 3.1c) and in alluvial fans. Of five AMS ^{14}C ages on individual charcoal fragments from a single late-Holocene debris-flow deposit, four were statistically indistinguishable, supporting the inference that deposition occurred as a result of fire (Meyer and others 1992). Unlike debris flows, flood deposits often show no clear evidence of fire, as these sediments are deposited while most low-density charcoal is washed downfan into mainstem stream systems. Burned soil surfaces containing charred forest litter are also commonly found buried in alluvial fan deposits in Yellowstone. Where distinct and well preserved, these burned soil surfaces suggest deposition of overlying sediment shortly after fire. Meyer and others (1995) used this set of datable stratigraphic indicators to construct a record of fire-related sedimentation over the middle to late Holocene in northeastern Yellowstone (Fig. 3.4). Over the past 750 years, dated fire-related sedimentation events show good correspondence with major fires identified in Yellowstone tree-ring and lake-sediment charcoal records (Barrett 1994, Millspaugh and Whitlock 1995, Chapter 2).

The alluvial fan record shows large variations in fire-related sedimentation over the Holocene. The last 3500 years of record is most complete, and spectral analysis of this interval yields recurrence intervals of 300 to 450 years for episodes of fire-related sedimentation across the northeastern Yellowstone study area. These episodes are nested within major cycles of approximately 1300 years (Fig. 3.4) (Meyer and others 1995). The recurrence interval of 300–450 years is quite similar to the late-Holocene fire frequency of 2–5 fires per 1000 years inferred from the Cygnet Lake charcoal record in lodgepole pine forest at 2500 m elevation on the Central Plateau (Millspaugh and others 2000). Slough Creek Lake lies in northeastern Yellowstone, but in low-elevation (1884 m) sagebrush grasslands (Fig. 1.2; see also Chapter 2). In the Slough Creek Lake charcoal record, fire frequencies of approximately 9–17 events/1000 yr over the past 3500 years are considerably greater than the frequency of fire-related sedimentation in higher-elevation basins nearby. Tree-ring fire-scar records for the past 500 years also indicate a regime of frequent, low-severity fires in the sagebrush grassland and Douglas fir of Yellowstone's Northern Range prior to suppression (Houston 1973, Barrett 1994). In higher-elevation forests in Yellowstone, however, both alluvial-fan and lake-sediment charcoal records indicate a dynamic regime of infrequent large fires, where substantial variations in return intervals are likely influenced by climate.

Because the convective storms that typically generate fire-related debris flows are limited in spatial extent, postfire sedimentation at the scale of small tributary basins has a more stochastic character and longer recurrence intervals. In northeastern Yellowstone, intervals between dated fire-related units in individ-

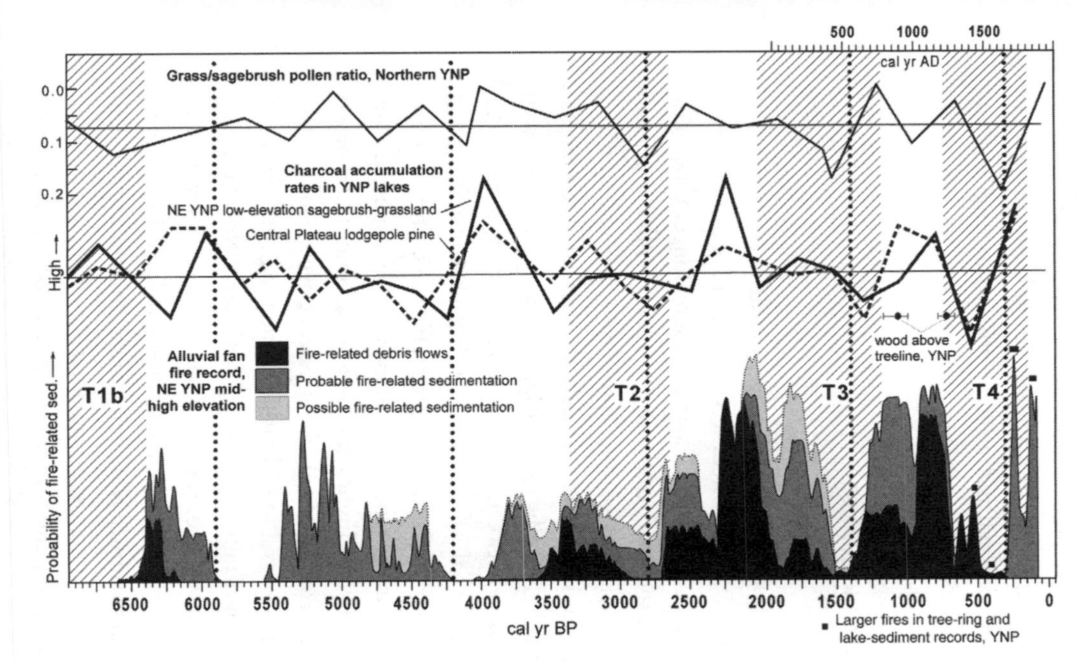

Figure 3.4. Comparison of records of middle- to late-Holocene fire-related alluvial-fan sedimentation (Meyer and others 1995) with other fire and climate proxy records. Scale in cal yr B.P. indicates ^{14}C ages calibrated to calendar year ages before present, where present is defined by convention as 1950 (Stuiver and Reimer 1993). The spectrum of probability of fire-related sedimentation is constructed by summation of calibrated probability distributions for radiocarbon dates on individual events. Note correspondence of fire-related sedimentation events with large fires identified in a high-resolution lake-sediment study over the past ~750 years (black squares; Millspaugh and Whitlock 1995). Charcoal accumulation rates in Yellowstone lakes averaged over 500 years (Millspaugh and others 2000, Chapter 2) are generally higher during major episodes of fire-related sedimentation in the past 3000 years. Wood above present treeline (K. L. Pierce, written communication, 1993) indicates warmer temperatures ca. A.D. 800–1200, also a time of increased fire sedimentation in the "Medieval Warm Period." Periods of overbank sedimentation on mainstem Soda Butte Creek terraces (diagonal hatch pattern, T1b–T4) correspond with generally wetter conditions indicated by high grass/sagebrush pollen ratios (note inverted scale; from Gennett and Baker 1986). Vertical dotted lines indicate times of maximum North Atlantic cooling identified by Bond and others (1997), each of which corresponds to a distinct minimum in fire-related sedimentation, strongly suggesting control of large fires by hemispheric- and millennial-scale climatic variations.

ual late-Holocene fan sections average about 900 years, but unrecognized fire-related deposits likely exist; therefore this represents a maximum recurrence interval (Meyer 1993). A section in Gibbon Canyon with five fire-related units yields a shorter late-Holocene recurrence interval of about 560 years, consistent with the drier environment and high geomorphic sensitivity of this area. Nonetheless, deposition of these Gibbon Canyon units was largely contemporaneous with times of fire-related sedimentation in northeastern Yellowstone overall.

FIRE, CLIMATE, AND GEOMORPHIC CHANGE IN THE HOLOCENE

Because the area burned in Yellowstone fires is strongly influenced by meteorological factors at annual scales (Balling and others 1992a, 1992b), links between fire-induced sedimentation and climate are likely to exist on longer timescales. Climate-related variations in the relative abundance of both pollen types and small mammal species in northern Yellowstone show that warmer and effectively drier conditions coincide with late-Holocene peaks in fire-related sedimentation (Gennett and Baker 1986, Hadly 1996). Major episodes of fire-related debris-flow deposition on fans between 2350–2050 and 900–750 cal yr B.P. match with peak rates of charcoal accumulation in Slough Creek Lake and Cygnet Lake (Millspaugh and others 2000, Chapter 2; Fig. 3.4). The latter episode (A.D. 1050–1200) corresponds to the height of the widely recognized "Medieval Warm Period" (Lamb 1977, Hughes and Diaz 1994, Stuiver and others 1995, Crowley 2000, Esper and others 2002). The incidence of extreme drought was much greater during this interval in California (Stine 1994), and Swetnam (1993) observed that giant sequoia groves of the western Sierra Nevada experienced highest fire frequencies from A.D. 1000 to 1300. During this and preceding warmer, drier intervals, northeastern Yellowstone alluvial fans built rapidly over floodplains of mainstem streams, aided by postfire sedimentation and generally reduced flows in those streams (Meyer and others 1995).

Relatively few fire-related sedimentation events in Yellowstone date within the Little Ice Age, variously considered to begin after A.D. 1200 or 1500 and to extend to about A.D. 1900 (Grove 1988, Davis 1988, Jones and Bradley 1995, Mann and others 1998, Luckman 2000, Esper and others 2002). Temperatures were not consistently cold in the Little Ice Age, and its expression is limited in some regions, but nonetheless much of the world was generally cooler during this time. This and other periods of minimal fire-related sedimentation in Yellowstone consistently match with times of increased ice rafting of debris and colder temperatures in the North Atlantic, where cold events at 300, 1400, 2800, 4200, and 5900 years ago are part of a persistent millennial-scale climatic cycle with pacing of 1374 ± 502 years over the full Holocene (Bond and others 1997). The same cycle is strongly expressed by the 1300-year variation in fire-related sedimentation revealed by spectral analysis. Greater snowmelt runoff and effectively wetter conditions during the late-Holocene cold periods are implied by the formation of broad floodplains along Soda Butte Creek, later preserved as stream terraces (Meyer and others 1995). Some cold events are also concurrent with cirque glacier advances or increased periglacial activity in the Rocky Mountains (Davis 1988, Armour and others 2002), but the timing of Neoglacial advances in the greater Yellowstone area remains poorly known.

Instrumental climate records since 1895 show that the 1988 fires occurred during the most severe summer drought of record, following a strong trend of increasing temperatures and summer drought in the Yellowstone area (Balling and others 1992a, 1992b). This trend is concurrent with overall global warming after the Little Ice Age minimum. Overall, fire-related sedimentation in Yellowstone is clearly linked to hemispheric- to global-scale temperature variations.

Although an overall increasing frequency of fire-related sedimentation over the Holocene is apparent in Figure 3.4, this is mostly a function of decreasing exposure and preservation of alluvial fan deposits with time before the present. Nonetheless, it is possible that some of this trend is real and represents an increase in extent and severity of burns, thus greater debris-flow activity toward the present. Such a trend is consistent with the decrease in frequency and increase in extent of fires over the Holocene inferred by Millspaugh and others (2000) for Yellowstone's Central Plateau. As noted previously, the Central Plateau is more similar in climatic and forest environment to the northeastern Yellowstone mountains than are the low-elevation sagebrush grasslands of Slough Creek Lake. The Central Plateau experienced 10–15 fires/1000 yr in the early Holocene, but burn area was probably restricted by short recurrence intervals and limited fuel. If similar fire regimes prevailed in northeastern Yellowstone, smaller and lower-severity burns may have limited the geomorphic response to fires. A gradual increase in recurrence intervals to between approximately 150 and 500 years in the past 3000 years probably resulted in much larger and more severe burns from greater accumulated fuels, as in 1988, with higher potential for debris-flow and flood generation.

Based on relative deposit thickness, approximately 30 percent of mid- to late-Holocene fan sediment in northeastern Yellowstone is fire-induced. This estimate is generally consistent with Swanson's (1981) conceptual view of how the relative importance of fire in sediment yield relates to climatic, ecological, and geomorphic controls (Fig. 3.5). In northeastern Yellowstone, the regime of severe but infrequent crown fires corresponds to a moderate fire index. With a moderate to high geomorphic sensitivity stemming from long, steep slopes that are commonly erodible even when unburned, a moderate fire-induced proportion of sediment yield is expectable. Steep chaparral landscapes of the rapidly uplifting California Transverse Ranges experience short intervals between severe fires, and highly erodible sediment-loaded slopes produce an estimated 70 percent of the total sediment yield in response to fire (Rice 1973, Scott and Williams 1978). In contrast, pine forests of the northern Midwest are characterized by more frequent fires than in Yellowstone (Clark 1988), but this low-relief region probably produces little postfire sediment compared to erosion and sediment transport in prolonged major floods, as in 1993 (Prestegaard and others 1994).

THE 1988 FIRES IN THE HOLOCENE CONTEXT: A CONSEQUENCE OF MANAGEMENT?

Malamud and others (1998) developed a mathematical model to support the idea that when fire suppression decreases fire frequency and allows fuel loading, individual fires increase in size. They termed this process the "Yellowstone effect," citing the 1988 fires as a prime example. The same concept was used by Pyne (1997) to explain the extensive 1988 Yellowstone burns, albeit with little analysis of historical or ecological data. Tree-ring studies clearly support the inference that suppression, grazing, and other land-use practices have increased the severity and extent of recent fires in some conifer ecosystems in the West, particularly in ponderosa pine forests (for example, Steele and others 1986, Barrett 1988, Grissino-Mayer and Swetnam 2000).

There is little evidence, however, that prior suppression was the primary factor in the magnitude and severity of the 1988 Yellowstone fires. Although suppression efforts began shortly after park establishment (Haines 1977), difficult access and limited technology hampered firefighting in much of Yellowstone, particularly during dry, windy conditions when fire growth can rapidly exceed the threshold of control, as in 1988. The use of aircraft after about 1945 greatly increased the speed and effectiveness of suppression, but the natural fire policy initiated in 1972 permitted fires that did not threaten developed areas (Schullery 1989). In most of Yellowstone's higher-elevation forest stands, the last significant fires occurred 130–400 years or more before 1988 (Romme and Despain

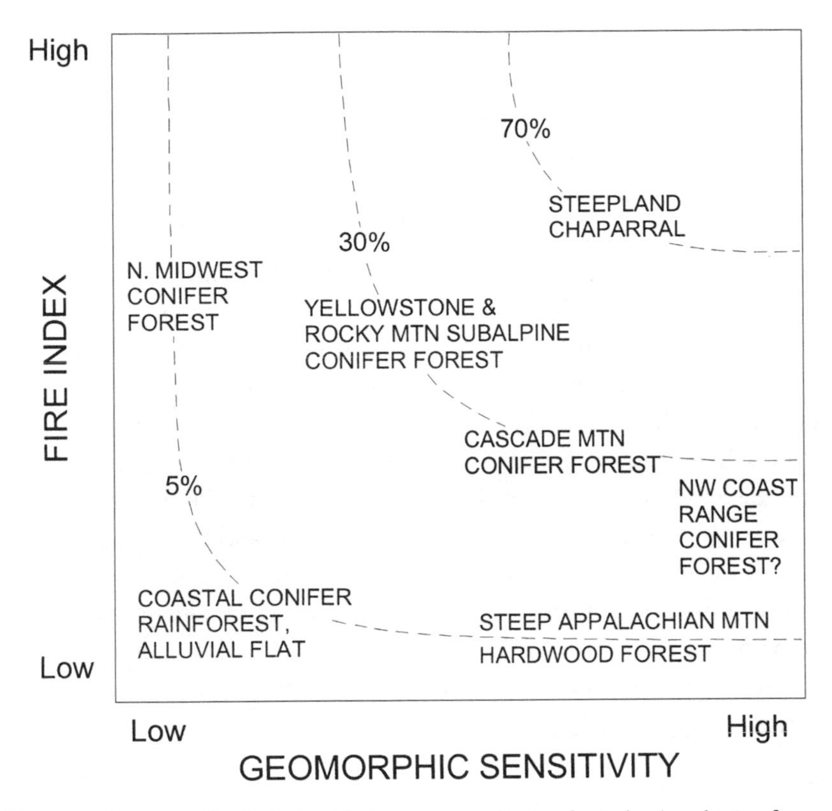

Figure 3.5. Conceptualized relationship between a qualitative fire index (combining fire frequency, severity, and magnitude) and geomorphic sensitivity of landscapes, showing the corresponding estimated percentage of long-term sediment yield induced by fire (dashed isolines). Modified from Swanson 1981.

1989, Barrett 1994). In contrast, suppression was effective for about 30 years, only 8–25 percent of the time since last burning. Europeans visiting Yellowstone in the 1800s frequently noted difficult travel through dense conifers and deadfall (Haines 1977), and photographs of higher-elevation forests in Yellowstone in the late 1800s typically show dense stands with abundant ladder fuels (Meagher and Houston 1998). Although total forest biomass and fuel certainly increased between park establishment in 1872 and 1988, these observations indicate relatively minor change in conditions for the initiation and spread of severe canopy fires over the preceding century. Tree-ring data suggest that fire suppression lengthened the period between fires primarily in the more accessible lower-elevation Douglas fir and sagebrush-grassland areas of northern Yellowstone, where fire return intervals of 15–50 years existed before park establishment (Houston

1973, Barrett 1994). After 1872, only fires in 1940 and 1988 burned significant areas of the Northern Range.

Both alluvial fan and lake-sediment records (Meyer and others 1995, Millspaugh and others 2000, Chapter 2) indicate that for most Yellowstone forests, late-Holocene fire recurrence intervals were similar to the interval preceding the 1988 fires (Romme and Despain 1989, Barrett 1994), clearly demonstrating that the period for fuel buildup prior to 1988 was not unusually lengthened by suppression as suggested by Malamud and others (1998). The extreme drought and high winds of summer 1988 followed a century of increasing temperatures and summer drought (Balling and others 1992a, 1992b). The above observations indicate that ecological, meteorological, and climatic controls far outweighed the effects of fire suppression in producing the great extent and severity of the 1988 burns. Thus the phrase "Yellowstone effect" in reference to the consequences of fire suppression does not accurately reflect the primary influences on the 1988 fires. Human activities may have influenced the extent and severity of the 1988 fires less directly, as a significant component of the rapid late-twentieth-century warming was probably induced by anthropogenic greenhouse gas emissions (Mann and others 1998, IPCC 2001, Meehl and others 2003, Chapters 2, 13, 14).

CONCLUSIONS AND IMPLICATIONS FOR YELLOWSTONE ECOSYSTEMS

Large stand–replacing fires and the debris flows and flash floods that follow have been major natural landscape-forming processes in the mountains of Yellowstone for at least 5000 years. Postfire debris flows and floods are likely to have short-term negative effects on aquatic and terrestrial biota along lower-order streams (Bozek and Young 1994), but their role in landscape and ecological change over longer timescales also warrants consideration (Bisson and others 2003). These flows often add large boulders and woody debris to channels that create habitat structures persistent over tens to hundreds of years or more (Benda and others 1998). The patchy and transient nature of postfire erosion in Yellowstone's natural landscapes limits streambed siltation impacts to stream ecosystems, as compared with chronic enhanced sediment inputs from debris flows in mountainous regions managed for industrial timber harvest (Montgomery and others 2000). Soil erosion may reduce productivity on rilled slopes, but the resulting organic-rich debris-flow deposits on alluvial fans were quickly vegetated and helped to generate a diverse postfire mosaic of vegetation.

Climate exerts a strong control on the probability of high-severity fires and the sediment transport events they induce in Yellowstone. Current models of long-term sediment flux in mountain drainage basins aptly incorporate fires and storms as stochastic processes (Benda and Dunne 1997, Miller and others 2003). Probability distributions of parameters in these models are based on past events, but climate change can substantially alter these distributions, presenting a problem for predicting future fire-induced sedimentation. If the rapid warming of the late twentieth century continues, fire frequency, severity, and extent may all increase before forest composition and structure can adapt to rising temperatures with more open, fire-resistant stands. In the transition period, a prolonged episode of fire-related sedimentation and geomorphic instability could result. Yellowstone ecosystems appear well adapted to disturbance pulses from infrequent catastrophic fires and their geomorphic effects, but rapid warming may cause more chronic sedimentation and disturbance through a series of burns with both high severity and increased frequency.

References

Armour, J., P. J. Fawcett, and J. W. Geissman. 2002. 15 k.y. paleoclimatic and glacial record from northern New Mexico. Geology 30:723–726.

Balling, R. C., Jr., G. A. Meyer, and S. G. Wells. 1992a. Climate change in Yellowstone National Park: Is the drought-related risk of wildfires increasing? Climatic Change 22: 34–35.

———. 1992b. Relation of surface climate and burned area in Yellowstone National Park. Agricultural and Forest Meteorology 60:285–293.

Barrett, S. W. 1988. Fire suppression's effects on forest succession within a central Idaho wilderness. Western Journal of Applied Forestry 3:76–80.

———. 1994. Fire regimes on andesitic mountain terrain in northeastern Yellowstone National Park, Wyoming. International Journal of Wildland Fire 4:65–76.

Benda, L., and T. Dunne. 1997. Stochastic forcing of sediment supply to channel networks from landsliding and debris flow. Water Resources Research 33:2836–2849.

Benda, L. E., D. J. Miller, T. Dunne, G. H. Reeves, and J. K. Agee. 1998. Dynamic landscape systems. Pages 261–288 in R. J. Naiman and R. E. Bilby, eds., River ecology and management: Lessons from the Pacific coastal Ecoregion. Springer, New York.

Bisson, P. A., B. E. Rieman, C. Luce, P. F. Hessburg, D. C. Lee, J. L. Kershner, G. H. Reeves, and R. E. Gresswell. 2003. Fire and aquatic ecosystems of the western USA: Current knowledge and key questions. Forest Ecology and Management 178:213–229.

Bond, G., W. Showers, M. Cheseby, R. Lotti, P. Almasi, P. deMenocal, P. Priore, H. Cullen, I. Hajdas, and G. Bonani. 1997. A pervasive millennial-scale cycle in North Atlantic Holocene and glacial climates. Science 278:1257–1266.

Bozek, M. A., and M. K. Young. 1994. Fish mortality resulting from delayed effects of fire in the greater Yellowstone ecosystem. Great Basin Naturalist 54:91–95.

Burned Area Survey Team. 1988. Preliminary survey of burned areas: Yellowstone National Park and adjoining national forests. Yellowstone National Park, Greater Yellowstone Post-fire Resource Assessment Committee, report on file.

Burroughs, E. R., Jr., and B. R. Thomas. 1977. Declining root strength in Douglas-fir after felling as a factor in slope stability. U.S. Department of Agriculture Forest Service Research Paper INT-190, 27 pp.

Cannon, S. H. 2001. Debris-flow generation from recently burned watersheds. Environmental and Engineering Geoscience 7:321–341.

Cannon, S. H., P. S. Powers, and W. Z. Savage. 1998. Fire-related hyperconcentrated and debris flows on Storm King Mountain, Glenwood Springs, Colorado, USA. Environmental Geology 35:210–218.

Christiansen, R. L. 2001. The Quaternary and Pliocene Yellowstone plateau volcanic field of Wyoming, Idaho, and Montana. U.S. Geological Survey Professional Paper 729-G, 145 pp.

Clark, J. S. 1988. Effect of climate change on fire regimes in northwestern Minnesota. Nature 334:233–235.

Crowley, T. J. 2000. How warm was the Medieval Warm Period? Ambio 29:51–54.

Davis, P. T. 1988. Holocene glacier fluctuations in the American Cordillera. Pages 129–157 in P. T. Davis and G. Osborn, eds., Holocene glacier fluctuations. Quaternary Science Reviews, vol. 7.

DeBano, L. F., R. M. Rice, and C. E. Conrad. 1979. Soil heating in chaparral fires: Effects on soil properties, plant nutrients, erosion, and runoff. USDA Forest Service Research Paper PSW-145, Pacific SW Forest and Range Experiment Station, Berkeley, Calif.

Esper, J., E. R. Cook, and F. H. Schweingruber. 2002. Low-frequency signals in long tree-ring chronologies for reconstructing past temperature variability. Science 295:2250–2253.

Ewing, R. 1996. Postfire suspended sediment from Yellowstone National Park, Wyoming. Water Resources Bulletin 32:605–627.

Gennett, J. A., and R. G. Baker. 1986. A late-Quaternary pollen sequence from Blacktail Pond, Yellowstone National Park, Wyoming, U.S.A. Palynology 10:61–71.

Gray, D. H., and W. F. Megahan. 1981. Forest vegetation removal and slope stability in the Idaho batholith. USDA-Forest Service, Intermountain Forest and Range Experiment Station, Research Paper INT-271, 23 pp.

Grissino-Mayer, H. D., and T. W. Swetnam. 2000. Century-scale climate forcing of fire regimes in the American Southwest. The Holocene 10:213–220.

Grove, J. M. 1988. The Little Ice Age. Methuen, New York.

Hadly, E. A. 1996. Influence of late-Holocene climate on northern Rocky Mountain mammals. Quaternary Research 46:298–310.

Haines, A. L. 1977. The Yellowstone story, vol. 1. Yellowstone National Park, Wyoming, Yellowstone Library and Museum Association, Yellowstone National Park.

Hendricks, B. A., and J. M. Johnson. 1944. Effects of fire on steep mountain slopes in central Arizona. Journal of Forestry 24:568–571.

Houston, D. B. 1973. Wildfires in northern Yellowstone National Park. Ecology 54:1111–1116.

Hughes, M. K., and H. F. Diaz. 1994. Was there a Medieval Warm Period, and if so, where and when? Climatic Change 26:109–142.

Intergovernmental Panel on Climate Change (IPCC). 2001. Climate change 2001: The scientific basis. J. T. Houghton, Y. Ding, D. J. Griggs, M. Noguer, P. J. van der Linden, and D. Xiaosu, eds. Cambridge University Press, Cambridge.

Jarrett, R. D. 1999. Geomorphic estimates of rainfall, floods, and sediment runoff: Applied to the 1996 wildfire, Buffalo Creek, Colorado. Geological Society of America Abstracts with Programs 31(7):A-313.

Johansen, E. A. 1991. Quantitative analysis of the variables causing debris flows. M.S. thesis, Montana College of Mineral Science and Technology, Butte.

Jones, P. D., and R. S. Bradley. 1995. Climatic variations over the last 500 years. Pages 649–665 in R. S. Bradley and P. D. Jones, eds., Climate since A.D. 1500. Routledge, New York.

Kirchner, J. W., R. C. Finkel, C. S. Riebe, D. E. Granger, J. L. Clayton, J. G. King, and W. F. Megahan. 2001. Mountain erosion over 10 yr, 10 k.y., and 10 m.y. time scales. Geology 29:591–594.

Krammes, J. A. 1960. Erosion from mountain side slopes after fire in southern California. U.S. Forest Service Research note 171. Pacific Southwest Forest and Range Experiment Station, Berkeley, Calif.

Lamb, H. H. 1977. Climate: Present, past, and future, vol. 2. Climatic history and the future. Methuen, London.

Legleiter, C. J., R. L. Lawrence, M. A. Fonstad, W. A. Marcus, and R. Aspinall. 2003. Fluvial response a decade after wildfire in the northern Yellowstone ecosystem: A spatially explicit analysis. Geomorphology 54:119–136.

Luckman, B. H. 2000. The Little Ice Age in the Canadian Rockies. Geomorphology 32:357–384.

Malamud, B. D., G. Morein, and D. L. Turcotte. 1998. Forest fires: An example of self-organized critical behavior. Science 281:1840–1842.

Mann, M., R. S. Bradley, and M. K. Hughes. 1998. Global scale temperature patterns and climate forcing over the past six centuries. Nature 392:779–788.

Meagher, M., and D. B. Houston. 1998. Yellowstone and the biology of time. University of Oklahoma Press, Norman.

Meehl, G. A., W. M. Washington, T. M. L. Wigley, J. M. Arblaster, and A. Dai. 2003. Solar and greenhouse gas forcing and climate response in the twentieth century. Journal of Climate 16:426–444.

Megahan, W. F., and D. C. Molitor. 1975. Erosional effects of wildfire and logging in Idaho. Pages 423–444 in Watershed management symposium. American Society of Civil Engineers, Irrigation and Drainage Division, Logan, Utah.

Meyer, G. A. 1993. Holocene and modern geomorphic response to forest fires and climate change in Yellowstone National Park. Ph.D. diss., University of New Mexico, Albuquerque.

———. 2001. Recent large-magnitude floods and their impact on valley-floor environments of northeastern Yellowstone. Geomorphology 40:271–290.

Meyer, G. A., J. L. Pierce, S. H. Wood, and A. J. T. Jull. 2001. Fires, storms, and sediment yield in the Idaho batholith. Hydrological Processes 15:3035–3038.

Meyer, G. A., and S. G. Wells. 1997. Fire-related sedimentation events on alluvial fans, Yellowstone National Park, U.S.A. Journal of Sedimentary Research A67:776–791.

Meyer, G. A., S. G. Wells, R. C. Balling, Jr., and A. J. T. Jull. 1992. Response of alluvial systems to fire and climate change in Yellowstone National Park. Nature 357:147–150.

Meyer, G. A., S. G. Wells, and A. J. T. Jull. 1995. Fire and alluvial chronology in Yellowstone National Park: Climatic and intrinsic controls on Holocene geomorphic processes. Geological Society of America Bulletin 107:1211–1230.

Miller, D., C. Luce, and L. Benda. 2003. Time, space, and episodicity of physical disturbance in streams. Forest Ecology and Management 178:121–140.

Millspaugh, S. H., and C. Whitlock. 1995. A 750-year fire history based on lake sediment records in central Yellowstone National Park, USA. The Holocene 5:283–292.

Millspaugh, S. H., C. Whitlock, and P. J. Bartlein. 2000. Variations in fire frequency and climate over the past 17,000 yr in central Yellowstone National Park. Geology 28:211–214.

Minshall, G. W., C. T. Robinson, and D. E. Lawrence. 1997. Postfire responses of lotic ecosystems in Yellowstone National Park, U.S.A. Canadian Journal of Fisheries and Aquatic Sciences 54:2509–2525.

Montgomery, D. R., K. M. Schmidt, H. Greenberg, and W. E. Dietrich. 2000. Forest clearing and regional landsliding. Geology 28:311–314.

Newcombe, C. P., and D. D. MacDonald. 1991. Effects of suspended sediments on aquatic ecosystems. North American Journal of Fisheries Management 11:72–82.

Parrett, C. 1987. Fire-related debris flows in the Beaver Creek drainage, Lewis and Clark County, Montana. U.S. Geological Survey Water-Supply Paper 2330:57–67.

Pierce, K. L. 1974. Surficial geologic map of the Abiathar Peak quadrangle and parts of adjacent quadrangles, Yellowstone National Park, Wyoming and Montana. U.S. Geological Survey Miscellaneous Geological Investigations Map I-646, 1:62,500.

Pierson, T. C., and J. E. Costa. 1987. A rheologic classification of subaerial sediment-water flows. Pages 1–12 in J. E. Costa and G. F. Wieczorek, eds., Debris flows/avalanches. Geological Society of America Reviews in Engineering Geology, vol. 7.

Prestegaard, K. L., A. M. Matherne, B. Shane, K. Houghton, M. O'Connell, and N. Katyl. 1994. Spatial variations in the magnitude of the 1993 floods, Raccoon River basin, Iowa. Geomorphology 10:169–182.

Prostka, H. J., E. T. Ruppel, and R. L. Christiansen. 1975. Geologic map of the Abiathar Peak Quadrangle, Yellowstone National Park, Wyoming. U.S. Geological Survey Geological Quadrangle Map GQ-1244, 1:62,500.

Pyne, S. J. 1997. World fire: The cultural history of fire on Earth. University of Washington Press, Seattle.

Reneau, S. L., and W. E. Dietrich. 1987. The importance of hollows in debris-flow studies: Examples from Marin County, California. Pages 165–180 in J. E. Costa and G. F. Wieczorek, eds., Debris flows/avalanches. Geological Society of America Reviews in Engineering Geology, vol. 7.

Rice, R. M. 1973. The hydrology of chaparral watersheds. Pages 27–33 in Living with the chaparral. Sierra Club Books, Riverside, Calif.

Robichaud, P. R. 1999. Fire and erosion: What happens after the smoke clears . . . Geological Society of America Abstracts with Programs 31(7):312.

———. 2000. Fire effects on infiltration rates after prescribed fire in Northern Rocky Mountain forests, USA. Journal of Hydrology 231:220–229.

Romme, W. H., and D. G. Despain. 1989. Historical perspective on the Yellowstone fires. Bioscience 39:695–699.

Rowe, P. B. 1951. Some factors of the hydrology of the Sierra Nevada foothills. Transactions American Geophysical Union 22:90–100.

Schmidt, K. M., J. J. Roering, J. D. Stock, W. E. Dietrich, D. R. Montgomery, and T. Schaub. 2001. The variability of root cohesion as an influence on shallow landslide susceptibility in the Oregon Coast Range. Canadian Geotechnical Journal 38:995–1024.

Schullery, P. 1989. The fires and fire policy. Bioscience 39:686–694.

Schultz, S., R. Lincoln, J. Cauhorn, and C. Montagne. 1986. North Hills fire erosion may explain formation of Montana landscape. Montana Agricultural Research 1986 Autumn: 9–13.

Scott, K. M. 1971. Origin and sedimentology of 1969 debris flows near Glendora, California. Pages 242–247 in U.S. Geological Survey Professional Paper 750-C.

Scott, K. M., and R. P. Williams. 1978. Erosion and sediment yield in the Transverse Ranges, southern California. U.S. Geological Survey Professional Paper 1030.

Shaub, S., and M. M. Donato. 1999. Landslides and wildfire: An example from the Boise National Forest. Geological Society of America Abstracts with Programs 31(7):A-441.

Shovic, H. F. 1988. Postfire soil research: Report to the Greater Yellowstone Postfire Ecological Assessment Workshop, November 17, 1988, Mammoth, Wyoming.

Shroder, J. F., Jr., and M. P. Bishop. 1995. Geobotanical assessment in the Great Plains, Rocky Mountains, and Himalaya. Geomorphology 13:101–119.

Steele, R., S. F. Arno, and K. Geier-Hayes. 1986. Wildfire patterns change in central Idaho's ponderosa pine–Douglas-fir forest. Western Journal of Applied Forestry 3:76–80.

Stine, S. 1994. Extreme and persistent drought in California and Patagonia during medieval time. Nature 369:546–549.

Stuiver, M., P. M. Grootes, and T. F. Braziunas. 1995. The GISP2 $\delta^{18}O$ climate record of the past 16,500 years and the role of the sun, ocean, and volcanoes. Quaternary Research 44:341–354.

Stuiver, M., and P. J. Reimer. 1993. Extended ^{14}C data base and revised CALIB 3.0 ^{14}C age calibration program. Radiocarbon 35:215–230.

Swanson, F. J. 1981. Fire and geomorphic processes. Pages 401–420 in H. A. Mooney, T. M. Bonnicksen, N. L. Christensen, J. E. Lotan, and W. A. Reiners, eds., Fire regimes and ecosystem properties. USDA Forest Service General Technical Report WO-26, Washington, D.C.

Swetnam, T. W. 1993. Fire history and climate change in giant sequoia groves. Science 262:885–889.

Wells, W. G. II. 1987. The effect of fire on the generation of debris flows. Pages 105–114 in J. E. Costa and G. F. Wieczorek, eds., Debris flows/avalanches. Geological Society of America Reviews in Engineering Geology, vol. 7.

Wohl, E. E., and P. A. Pearthree. 1991. Debris flows as geomorphic agents in the Huachuca Mountains of southeastern Arizona. Geomorphology 4:273–292.

Wohlgemuth, P. M. 1999. Rates and timing of postfire hillslope erosion in southern California steeplands. Geological Society of America Abstracts with Programs 31(7):441.

Part II **Effects on Individuals and Species**

If we see only a monotonous blanket of trees and grass, we miss so much of what is available to enjoy.
—*Don Despain,* Yellowstone Vegetation

Chapter 4 Establishment, Growth, and Survival of Lodgepole Pine in the First Decade

Jay E. Anderson, Marshall Ellis, Carol D. von Dohlen, and William H. Romme

Fire is among the most important factors affecting the evolution, development, structure, and overall ecology of forest ecosystems in the Northern and Central Rocky Mountains (Habeck and Mutch 1973, Houston 1973, Brown 1975, La Roi and Hnatiuk 1980, Romme 1982, Christensen and others 1989, Johnson 1992, Johnson and Wowchuk 1993). Mean fire return intervals range from a decade or two in low elevation ponderosa pine *(Pinus ponderosa)* and Douglas fir *(Pseudotsuga menziesii)* forests to hundreds of years in high-elevation subalpine forests of the region (Arno 1980, Houston 1973, Romme 1982, Romme and Despain 1989, Barrett 1994, Chapter 2). Lodgepole pine *(Pinus contorta* var. *latifolia),* the Rocky Mountain race of the most widely distributed conifer in western North America, dominates midmontane forests over much of this region, often forming monospecific stands (Wheeler and Critchfield 1985, Fahey and Knight 1986).

Lodgepole pine is among the most fire-resilient species of pines, a group characterized by precocious reproduction, small seed, and, in the case of *P. banksiana* and *P. contorta* var. *latifolia,* a high degree of cone serotiny (McCune 1988). Cones bearing viable seeds are produced at an early age, five to ten years in open stands and fifteen to twenty years in heavily stocked stands (Fowells 1965). Older trees bear serotinous cones or cones that open upon maturity (rarely both), and both types of trees typically are found in the same stand (Muir and Lotan 1985b). Trees possessing the serotinous phenotype produce only open cones until they are ca. twenty to sixty years old, after which a gradual transition to serotinous cones occurs (Perry and Lotan 1979, Lotan and Perry 1983). Serotinous cones are retained in the crown of the tree and remain closed until they are heated to 45–60°C (Perry and Lotan 1977, Johnson and Gutsell 1993), typically by fire. Lodgepole pine is a prolific seed producer, with estimates of annual yield ranging from 180,000 to 790,000 seeds/ha (Fowells 1965). Because serotinous cones are retained in the canopy for many years, mature trees may hold well over 1,000 closed cones, and the stored seed supply may number into the millions/ha (Clements 1910, Hellum and Wang 1985).

Differential patterns in lodgepole pine establishment after a fire may have long-lasting effects on stand structure, physiognomy, and dynamics. Because they are initiated by fire, lodgepole pine stands are often even aged (Moir 1969, Peet 1981). Stands of similar age, however, frequently differ in density, ranging from open stands of large trees to extremely dense, stunted "dog-hair" stands (for example, Fowells 1965, Lyon 1984). Such differences are common in the Greater Yellowstone Area. While conducting fire history studies in Yellowstone National Park, Romme and Despain (unpublished data) often observed adjacent lodgepole pine stands that differed in density and physiognomy, but the trees in both stands were the same age. The physiognomic differences were as great as one might see in stands that were one hundred years different in age. Such differences probably reflect the initial postfire establishment of lodgepole pine seedlings (Lyon 1976), but the site-specific processes that result in dramatic differences in seedling densities are not well understood (Horton 1953, Arnott 1973, Lyon 1976, 1984, Lyon and Stickney 1976).

In addition to their vast extent (Chapter 1), the 1988 Yellowstone fires were characterized by their heterogeneity, which produced a complex mosaic of burn severities at spatial scales from a few meters to kilometers. This mosaic included areas of (1) severe crown fires where most foliage and fine branches were consumed, (2) adjacent halos of moderate severity in which trees were killed from surface fires or heat from adjacent crown fires, but needles, fine branches, and cones remained in the canopy, and (3) stands that were unburned or experienced only light surface fires (Christensen and others 1989, Anderson and Romme 1991, Turner and others 1994). The scale and complexity of these fires provided a unique opportunity to study postfire regeneration of lodgepole pine forests (Christensen and others 1989, Knight and Wallace 1989).

OBJECTIVES OF OUR POSTFIRE STUDIES

Our studies of postfire responses of lodgepole pine forests in Yellowstone National Park were initiated in 1989 with the establishment of paired plots in seven stands at four sites burned the previous year. One plot was in a severe canopy burn; the paired plot was within an adjacent stand where the trees were killed but the fire did not ignite the tree crowns. Fire severity had a strong influence on seedling density in the first year after the fire; at all seven sites, lodgepole pine seedlings were far less numerous in the canopy-burn area (Anderson and Romme 1991). Lodgepole pine seedling density was not related to density of potential competitors (Anderson and Romme 1991). We postulated that within-site differences were primarily a consequence of loss of seeds related to fire severity, whereas differences among sites were a consequence of various factors, but primarily the availability of seeds at the time of the fire as determined by the proportion of serotinous trees in the stand.

To further test these hypotheses, we expanded the study in 1990 to include sites spanning the elevational range of lodgepole pine within Yellowstone National Park (Ellis and others 1994). The overall objective of our study was to assess the relative importance of key variables in determining initial postfire establishment and growth of lodgepole pine seedlings. We evaluated (1) burn severity and size of the burned patch, (2) site characteristics such as geologic substrate and climate, (3) prefire stand characteristics including tree density, stand age, and proportion of trees having serotinous cones, and (4) postfire abundance of potentially competing herbs and shrubs. Our observations in 1989 and various literature sources suggested that these were potentially the most important variables influencing early postfire establishment of lodgepole pine. We assessed the importance of these variables by a combination of extensive surveys of pine seedling densities under a range of environmental conditions and by experimental studies in which we manipulated densities of potential competitors.

STUDY AREA, SAMPLING DESIGN, AND METHODS

Yellowstone National Park occupies approximately 9,000 km^2 in the northwest corner of Wyoming and adjacent areas of Idaho and Montana. The park overlies a physiographically diverse landscape created by episodes of mountain building, volcanism, glaciation, and erosion. Our studies were conducted primarily within the Central Plateaus Geovegetation Province (Despain 1990, see map in Chapter 1), where elevations typically are between 2300 and 2500 m. These high plateaus are underlain by Quaternary rhyolite, which weathers to a coarse, sandy soil having poor water-holding capacity and low nutrient levels (Despain 1990). Lodgepole pine forests cover most of the area. Mean annual precipitation ranges from about 500 mm to more than 600 mm (Despain 1987, Marston and Anderson 1991). Mean annual temperatures are near the freezing point (Marston and Anderson 1991). For further details on the physical environment and vegetation, see Despain (1990).

Extensive observations of burned areas in 1988 and 1989 indicated that three burn severity classes and one unburned class could be readily distinguished in the field and in remote imagery (Anderson and Romme 1991):

Severe canopy burn: trees killed; needles and fine branches consumed by hot crown fire; most litter and duff on forest floor consumed.
Moderate burn: trees killed by crown scorch from surface fire or adjacent crown fire; needles in canopy not consumed; much litter and duff on forest floor consumed; needle fall produced litter layer on forest floor during 1989–1990.
Light burn: little tree mortality; light and scattered surface fire; most litter and duff not consumed.
Unburned: no evidence of recent fire.

Johnson (1992) defined fire severity primarily in terms of duff consumption rather than the canopy characteristics that we emphasized. Although postfire conditions of the forest floor clearly are important in plant regeneration, duff consumption cannot be measured readily over a very large landscape. The fire severity classes that we used can be precisely mapped with Landsat (Turner and others 1994). Moreover, these severity classes are strongly correlated with percent tree mortality as well as important forest floor characteristics, including percent mineral soil and total plant cover in the first postfire year (Turner and others 2000).

PERMANENT PLOTS AND SAMPLING REGIMES FOR SEEDLING ESTABLISHMENT

In August 1989 and June 1990, we identified potential study sites where areas subjected to severe canopy burns were immediately adjacent to areas of moder-

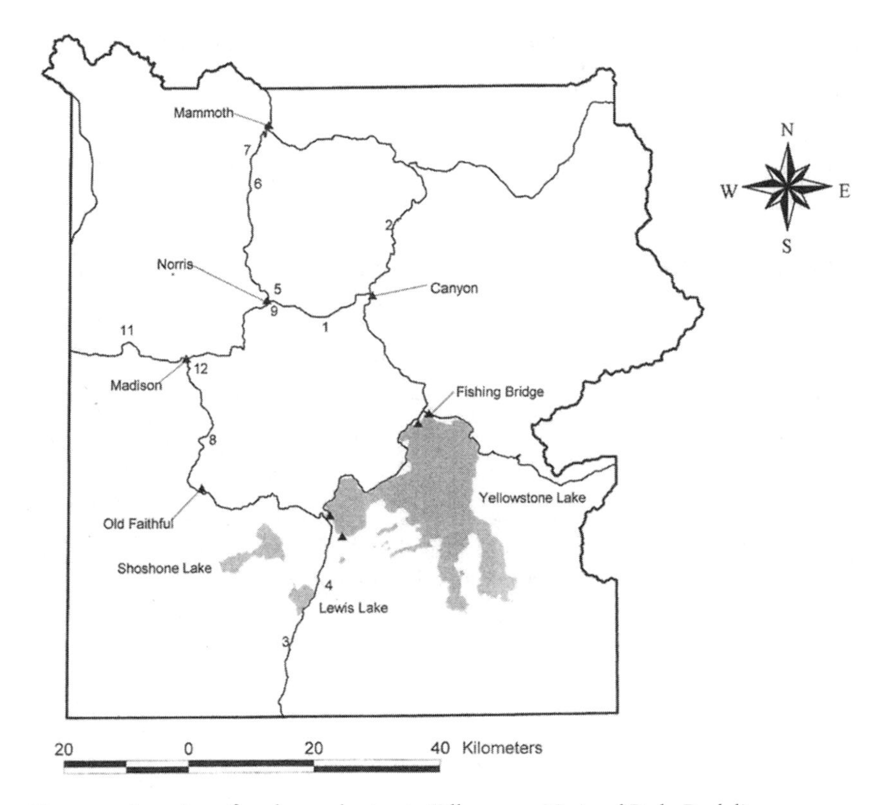

Figure 4.1. Location of twelve study sites in Yellowstone National Park. Dark lines are major highways; large lakes are shown in gray.

ate burn severity. Such areas were located by visually scanning burned sites from roadways and by examining burn severity maps on the park's GIS. Moderate burns were readily distinguished by the persistence of rust-colored needles on the dead trees. Permanent plots were established at four sites in 1989 (Anderson and Romme 1991) and at eight additional sites in 1990 (Fig. 4.1). Permanent plots were chosen subjectively to represent the elevational range and diversity of forests dominated by lodgepole pine on the central plateaus. At the time of the 1988 fires, each site supported a mature, nearly monospecific stand of lodgepole pine. Locations and characteristics of the study sites are summarized in Table 4.1.

At each of the twelve sites, paired plots were established. One member of each pair was in a severe canopy burn; the paired plot was in an adjacent area subjected to a moderate burn. The paired plots were within 60 m of each other and were subjectively chosen to be as similar as possible in relation to aspect, slope,

Table 4.1. Characteristics of study sites in Yellowstone National Park, Wyoming

Site	Elev. (m)	Substrate	Stage	Serotiny (%)
1 Solfatara Plateau	2490	rhyolite	LP 3/2	0
2 Mount Washburn	2560	andesite	LP 4	11
3 Lewis River	2500	rhyolite	LP 2/3	0
4 Lewis Lake	2350	rhyolite	LP 2/3	0
5 Norris Junction East	2270	alluvium	LP 2	0
6 Indian Creek	2260	basalt	LP 2	48
7 Swan Lake Flat	2270	basalt	LP 2	13
8 Whiskey Flat	2290	rhyolite	LP 1/2	48
9 Norris Junction South	2320	rhyolite	LP 2	4
10 West Yellowstone	2040	alluvium	LP 2	15
11 Madison River	2100	tuff	LP 1	45
12 Madison Junction	2120	rhyolite	LP 1	44

Note: Substrate indicates parent material as shown by USGS (1972). Stage refers to the cycle of lodgepole pine stand development as described by Romme and Despain (1989) and Despain (1990). Where two stages are shown, the stand was intermediate but tended toward the first stage listed. Serotiny is the proportion of trees in the stand having serotinous cones.

substrate, and prefire stand structure. Each plot consisted of a permanently marked, 50-m transect.

In addition to the paired transects, a third transect was established in the nearest unburned or lightly burned stand at each site. These stands were used to estimate the proportion of prefire serotinous trees in the area. Seedling density and stand structure data were also collected in these stands. To determine whether seed dispersal might limit seedling densities in severe burn areas, we established a "remote canopy burn" plot at ten of the study sites. This fourth transect was at least 100 m from the nearest unburned or moderately burned stand that could serve as a seed source.

Lodgepole pine seedlings and cones on the ground were counted in 50 contiguous 1-m^2 quadrats along each transect. This sampling protocol was chosen to minimize the number of permanent plot markers employed within the park (in other words, it was not necessary to mark each individual 1-m^2 quadrat) and because it was very time efficient. At transects where mean seedling densities fell below 0.5 m^{-2}, a strip transect of either 2 m or 5 m in width along the permanent transect line was inventoried. (This occurred only on the canopy- and remote canopy-burn transects at sites 1, 2, and 3.) At two sites (11 and 12) where seedling densities were extremely high, quadrats were sampled at every other meter along the transect for a total of 25 1-m^2 quadrats. Seedlings of other conifer species were noted when encountered but were generally rare and were not included in density calculations. Lodgepole pine seedlings were recounted each summer through 1994 to monitor changes in seedling density.

The effects of substrate characteristics and microtopography on seedling establishment were assessed on moderate- and canopy-burn plots at sites 6, 7, 11, and 12 with a fine-scale sampling frame that subdivided m^2-plots into 0.01-m^2 quadrats. Within these quadrats, the incidence and size of lodgepole seedlings, the type of substrate (unburned litter or duff, burned litter/duff, charcoal, moss, mineral soil), configuration of microtopography (flat, convex, deeply concave, or slight depression), and total cover of other green plants were recorded. Twenty-five 1-m^2 quadrats were sampled at each site.

Two sample, independent *t* tests were used to examine differences between moderate- and canopy-burn transects for lodgepole seedling densities at each study site. For these analyses, each 1-m^2 quadrat was treated as an independent sample. Although it might be argued that the contiguous quadrats were not independent, the assumption of independence seemed reasonable because the scale of a quadrat was large compared to that of an individual seedling, or even a clump of seedlings. This scale was also large compared to the scale at which characteristics of the soil surface in burned areas appeared to change (typically a decimeter or two). Thus, it is likely that the distribution of seedlings or other entities sampled was essentially random with respect to quadrat placement. For many comparisons, the differences were so large and so consistent across sites that statistical tests were unnecessary.

PREFIRE STAND CHARACTERISTICS AND STAND SEROTINY

Prefire stand structure at each study site was reconstructed by estimating age, density, and the frequency distribution of stem sizes for all transects. Stand age was estimated by coring the nearest canopy-dominant tree at 5-m intervals along each 50-m permanent transect (ten trees total). After sanding the cores, annual rings were counted using a dissecting microscope, and transect age was taken as the mean age of those ten trees. Stand age for the entire site was estimated as the mean of the trees cored on all of the site's transects. Binoculars were used to count the number of cones remaining in the crown of each stand dominant tree that was cored in the moderate- and canopy-burn transects. The proportion of trees bearing serotinous cones at each site was estimated using binoculars to note the presence or absence of serotinous cones on a minimum of one hundred live trees at the unburned transect in each study site. Correlation analyses were used to test for relationships between mean lodgepole pine seedling density in the moderate- and canopy-burn plots and the corresponding proportion of serotinous trees in the stand.

COVER ESTIMATES

Coverage of vascular plant species and of nonplant entities such as rock, litter, mineral soil, burned litter/duff, charcoal, and so on, were estimated in 1990 on all transects by point interception using a 0.5-m × 1.0-m point-sighting frame with thirty-six points (Floyd and Anderson 1982). Cover data were collected within the same 1-m^2 quadrats used for seedling density data, with the long axis of the frame adjacent to the transect line. Vascular plant intercepts were tallied by species, and each cover category was converted to a percentage by dividing the total number of intercepts in that category by the total number of points sampled. Data were summarized by quadrat and by transect. Correlation analyses were used to examine the relationship between seedling density and cover parameters for each burned transect. For these within-site analyses, the m^2 quadrat was the sample unit.

EFFECTS OF FIRE SEVERITY ON SEEDLING GROWTH

To compare growth of lodgepole pine seedlings between moderate- and severe-burn plots, we measured stem diameter and height of seedlings at six sites in September 1991, at site 11 in September 1996, and at site 7 in September 1997. Diameter and height were measured on fifty seedlings in the moderate- and severe-burn plots at sites 7, 8, and 11 and on twenty-five seedlings in each burn category at sites 9, 10, and 12. Seedlings were selected systematically along a 50-m transect laid parallel to the permanent transect by sampling the seedling closest to each meter mark ($n = 50$) or every other meter mark ($n = 25$). Stem diameter was measured at ground level to the nearest 0.01 mm. Seedling height was measured to the nearest mm from the ground to the tips of the longest needles, holding the needles upright and parallel to the stem. At sites 7, 8, and 11, aboveground biomass was determined on all seedlings measured. Seedlings were clipped at ground level, dried in an oven at 75°C, and weighed to the nearest 0.1 g.

To obtain additional data for examining seedling size class distributions, two hundred seedlings were measured in the moderate- and severe-burn plots at sites 7, 9, 10, 11, and 12 in June 1992. Diameter and height were measured as explained above on the closest seedling in each of four quadrats corresponding to each meter mark along the 50-m permanent transect.

Two sample independent *t*-tests were used to examine differences in height, diameter, and aboveground biomass between burn categories for each site. Simple linear regression was used to examine the relationships between aboveground biomass and either seedling height or diameter. Within sites, regression coefficients for biomass versus height or diameter for seedlings in moderate- and severe-burn plots were compared with a *t* statistic. Across sites, the slopes of these relationships were compared using GLM ANOVAs (General Linear Models Analysis of Variance).

QUANTIFICATION OF COMPETITIVE NEIGHBORHOODS

In August and September 1991, we assessed the abundance of potential competitors in the immediate neighborhood of each of twenty-five target seedlings in the moderate- and severe-burn plots at sites 7, 8, 9, 10, and 12. Target seedlings in each plot were those nearest the tape at odd-numbered intervals along the permanent 50-m transect. Excavations at several sites indicated that the root systems of three-year-old lodgepole pine seedlings rarely extended beyond about 0.3 m horizontally, so we limited our sample of potential competitors to those within a circle measuring 0.62 m in diameter. A bicycle wheel rim was used to define this circle, with the target seedling at its center. Target seedling height and stem diameter at ground level were measured as described above, and the following additional data were recorded:

1. distance to the nearest lodgepole pine neighbor,
2. identity of and distance to the nearest interspecific neighbor,
3. numbers and identities of all other plants within the circle,
4. total number of lodgepole pine seedlings within the circle, and
5. total number of all vascular plants within the circle.

In all instances, plants that were dead, dying, or much smaller than the target seedling were assumed to represent little competitive threat and were excluded from the analyses.

Simple linear regression was used to examine the relationships between target seedling height or diameter versus:

1. total number of vascular plants,
2. total number of lodgepole pine seedlings,
3. total number of potential interspecific competitors,
4. distance to the nearest lodgepole pine neighbor, and
5. distance to the nearest interspecific neighbor.

Two sample independent *t*-tests were used to examine differences between burn categories for each of these five categories.

MANIPULATIONS OF THE COMPETITIVE ENVIRONMENT

At the beginning of the growing season in 1991 (late May, early June) we set up field experiments at sites 7, 8, and 11 in which the competitive environment of target lodgepole pine seedlings was manipulated (Ellis 1993). At each site, ten blocks with similar lodgepole pine seedling densities were selected within an area of moderate burn severity. In each block, four 1-m^2 plots were subjectively selected and permanently marked. The total number of lodgepole pine seedlings was recorded for each plot. Four target seedlings per plot were chosen by dividing the plot into quarters and then selecting a seedling near the center of each quarter. We then randomly assigned one of the following treatments to each plot within a block:

1. removal of all intraspecific and interspecific competitors,
2. removal of all intraspecific competitors,
3. removal of all interspecific competitors, or
4. no removal of competitors (control).

Therefore each of the three sites contained ten replicates of the four treatments. Mean values for stem diameter and height of the four target seedlings in each plot were used for statistical analyses of treatment effects. (In other words, the plot was the sample unit.) Potential competitors were clipped at ground level and removed from the appropriate plots. Plots were maintained by clipping any resprouting or colonizing vegetation every two weeks in 1991 and periodically in subsequent years. Target seedling performance was tracked over the course of the 1991, 1992, 1993, and 1994 growing seasons by measuring seedling height and stem diameter, as described earlier.

The results were analyzed with two-way, repeated-measures ANOVAs, with treatment and date as factors. For each of the four treatments, the mean height or diameter of all four target seedlings was taken as the value for any given block, and the mean of all ten blocks was taken as the site value at each measurement date. Thus measures were repeated on blocks. Student-Newman-Keuls multiple range tests, based on the repeated-measures ANOVA, were used to test for significant differences among treatment means.

FACTORS AFFECTING INITIAL SEEDLING ESTABLISHMENT

Of the four variables identified earlier as potentially important in influencing initial lodgepole pine seedling establishment, the interacting effects of fire severity and level of prefire stand serotiny were by far the most important. Substrate characteristics and abundance of herbs and shrubs had little influence on pine seedling densities.

FIRE SEVERITY AND PREFIRE STAND SEROTINY

Our data show clearly that fire severity and the level of stand serotiny are of primary significance in determining postfire densities of lodgepole pine seedlings. Both factors affect postfire seed supply.

Seedling densities across the twelve study sites and burn severity classes were highly variable, differing by more than four orders of magnitude from a low of 80/ha at the remote canopy-burn plot on the Solfatara Plateau to 1.9×10^6/ha at the moderate-burn plot at site 11 near the Madison River Bridge (Table 4.2). In general, seedling densities were highest at low- and mid-elevation sites and lowest at the high-elevation sites. In all but four canopy-burn plots there were more seedlings than required to replace the stand that burned, and in most cases the ratios of postfire seedling densities to prefire stand densities were much higher than one (Table 4.3). Across moderate-burn plots, postfire seedling densities were not correlated with prefire stand densities ($P = 0.20$), but for the canopy-burn plots, postfire seedling densities were positively correlated with density of the stand that burned ($r = 0.58$, $P = 0.047$). This indicates a tendency for dense stands to be replaced by dense stands and for sparse stands to be replaced by sparse stands, at least in severely burned areas.

Densities of lodgepole pine seedlings in 1990 were from four to twenty-four times as high in moderate-burn plots as in paired canopy-burn plots, and the differences were highly significant ($P < 0.005$) at all twelve study sites (Fig. 4.2). With the exceptions of sites 8 and 11, seedling densities were two- to sevenfold

Table 4.2. Mean stand age at time of fire, prefire stand density, postfire density of *P. contorta* seedlings and cones on the ground, and mean postfire number of cones remaining in crown of burned *P. contorta* trees for each of the study plots at twelve study sites in Yellowstone National Park listed in Table 4.1

	Study Site											
	1	2	3	4	5	6	7	8	9	10	11	12
Plot	Mean Stand Age (years)											
Unburned	243	204	163	165	116	197	138	99	116	67	111	109
Moderate	165	195	171	187	119	205	131	96	111	118	108	96
Canopy	131	196	163	170	116	191	129	99	111	115	103	98
R. Canopy	127	176	153	137	NA	178	125	101	113	NA	110	103
	Prefire Stand Density (trees/ha)											
Unburned	1390	1180	728	880	1250	666	710	4920	1620	1546	3133	9400
Moderate	780	1400	1160	1090	990	1490	970	5160	2140	2360	4580	16900
Canopy	1440	1390	2400	1170	1140	1380	990	4880	2300	1230	3120	12480
R. Canopy	1390	1240	1590	980	NA	1740	1050	4660	2410	NA	2580	8880

	Postfire Seedling Density (seedling/ha)											
Moderate	5600[a]	13200[a]	38600[a]	76800[a]	45000[a]	83800[a]	216300[a]	164600[a]	98600[a]	83200[a]	1906000[a]	663600[a]
Canopy	870	3300	1640	7800	6800	12200	37800	26600	9400	21200	226400	145600
R. Canopy	80	1900	440	3800	NA	6000	9400	29000	5400	NA	363600	80400
	Postfire Cones on Ground Cones (cones/m2)											
Moderate	2.98[a]	3.4[a]	6.86[a]	9.88[a]	3.92[a]	10.84[a]	6.23[a]	4.24[a]	5.34[a]	8.72[a]	30.2[a]	10.74[a]
Canopy	0.11	0.95	0.028	0.38	0.52	7.28	1.28	0.98	0.4	0.5	6.12	2.28
R. Canopy	1.12	0.21	0.24	0.06	NA	1.68	0.68	1.34	0.86	NA	8.14	2.5
	Postfire Cones in Canopy (cones/tree)											
Moderate	83	152[a]	342[a]	553[a]	161	409	346	246	258	299	424	171[a]
Canopy	48	58	24	82	200	267	228	167	251	240	315	63

Notes: Data were taken in 1990 from *P. contorta* forests that burned in 1988 (see Table 4.1). Burn severity classes are described in Methods; R. Canopy —remote severe canopy burn. NA indicates that there was no remote canopy burn plot at that site.

[a]Indicates significant difference ($P < 0.05$) between moderate burn and canopy burn plots.

Table 4.3. Ratios of postfire seedling to prefire stand densities in moderate burn, canopy-burn, and remote canopy-burn plots at twelve study sites in Yellowstone National Park

Site	Location	Moderate	Canopy	Remote Canopy
1	Solfatara Plateau	7.7	0.4[a]	0.1[a]
2	Mount Washburn	9.3	2.4	1.1
3	Lewis River	33.6	0.7[a]	0.3[a]
4	Lewis Lake	65.8	7.2	3.3
5	Norris Junction East	45.4	5.9	NA
6	Indian Creek	56.7	8.7	3.4
7	Swan Lake Flat	214.0	37.4	9.5
8	Whiskey Flat	31.9	5.1	6.2
9	Norris Junction South	45.8	4.1	2.2
10	West Yellowstone	35.2	17.1	NA
11	Madison River	402.0	76.3	141
12	Madison Junction	38.8	13.4	9.3

Notes: NA = No remote canopy plot.

[a]Transects with insufficient seedlings to replace the prefire stand.

lower at the remote canopy-burn plot than at the canopy-burn plot (Table 4.2). These data indicate either that establishment of lodgepole pine seedlings in canopy-burn plots was enhanced by dispersal of seeds from the adjacent moderately burned stand or that heat release and seed mortality were greater near the center of a patch of crown fire than near the edge. The high seedling densities at the remote canopy-burn plots at sites 8 and 11 probably reflect high levels of serotiny in the stands that burned; in both cases the proportion of serotinous trees was near 50 percent (Table 4.1).

Our results confirm that seedling densities will be lower in areas subjected to severe crown fires than in areas that received only surface fire and are consistent with the scenario presented by Brown (1975) based on studies by Muraro (as cited by Brown 1975): A hot surface fire with some crowning kills all of the trees and opens serotinous cones, releasing abundant seed on a mineral seedbed, which results in a dense stand. Where fire intensity is sufficient for a severe crown fire, however, some seeds are either consumed or exposed to lethal temperatures, the supply of viable seed is limited, and a relatively low-density stand develops. Thus Brown (1975) acknowledged that crown fires may destroy large quantities of stored seed, but noted, "The extent to which high-intensity fires reduce quantities of viable seed is speculative."

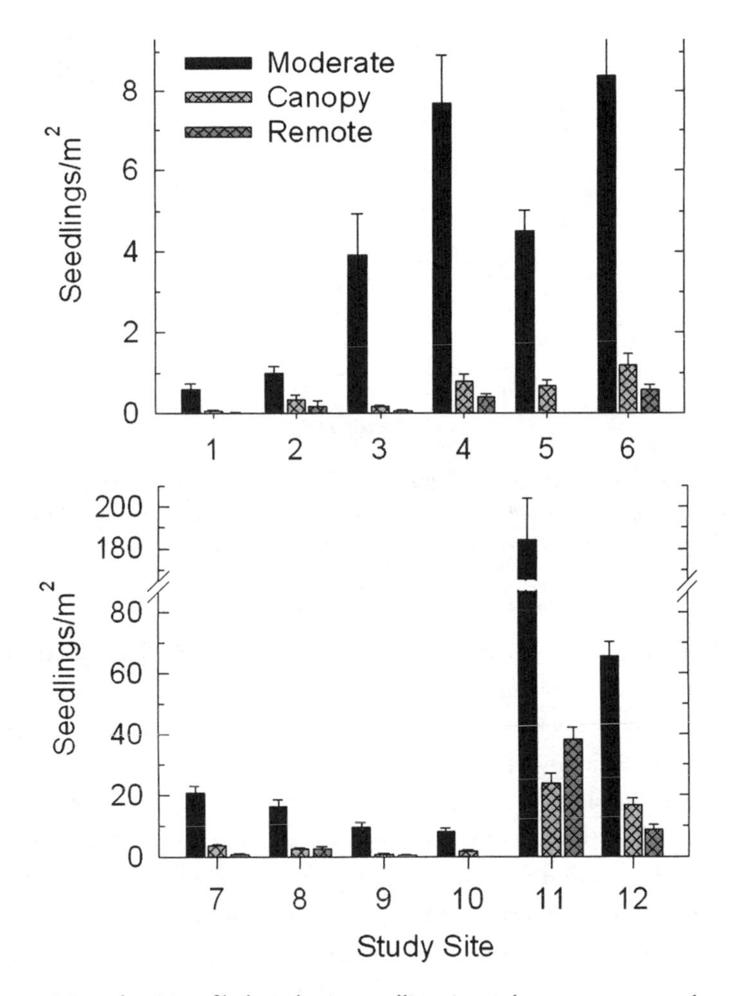

Figure 4.2. Mean densities of lodgepole pine seedlings in moderate-, canopy-, and remote canopy-burn plots at twelve study sites in Yellowstone National Park. Data were taken in 1990 from areas burned in 1988. Error bars represent one standard error.

Our data provide quantitative estimates of the extent to which viable seed may be lost in a crown fire. Across the twelve sites, the average reduction of seedling density on canopy-burn plots relative to that on the paired moderate burn plots (Table 4.2) was 84 percent (S.E. = 1.83; range: 75 to 96 percent). That the proportional reduction on canopy-burn plots relative to their moderate-burn counterparts was nearly constant across plots is shown by the parallel slopes of the relationship between $\log_{10}$ seedling density and percentage of serotinous trees in the stand (Fig. 4.3). Assuming that conditions for seedling establishment were comparable on the moderate- and canopy-burn areas, which seems reason-

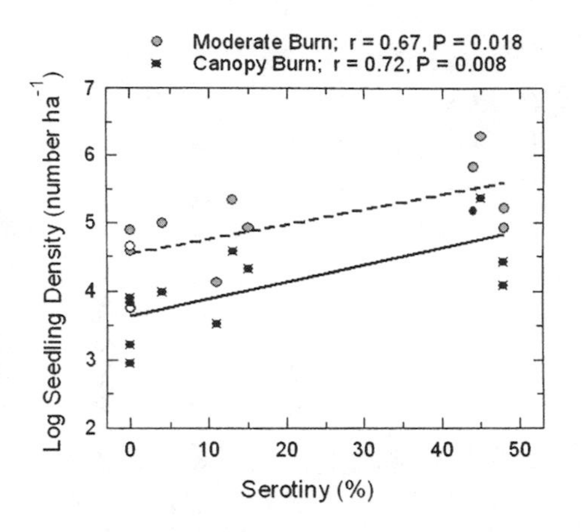

Figure 4.3. Correlation between log density of lodgepole pine seedlings in moderate- or canopy-burn plots and the proportion of serotinous trees in the adjacent unburned stand for twelve study sites in Yellowstone National Park. All data are from 1990.

able given the results of a companion study of seedbed characteristics (J. Anderson and C. von Dohlen, unpublished data), we suggest that a severe crown fire is likely to destroy 80 to 88 percent (95 percent C.I.) of the seed stored in a lodgepole forest canopy relative to the proportion that survives a hot surface fire.

Our data also show that if there is even a modest proportion of serotinous trees in the stand, the seed supply will be adequate to replace the stand under severe crown fire conditions. "Overstocking" can be thought of as a consequence of selection for sufficient storage and protection of seed to ensure regeneration of a stand under such conditions. Contrary to the suggestion of Despain and others (1996) that serotinous cones "would be of little advantage, however, if conditions common in burning tree crowns killed the seed," the advantage of serotiny is that some seeds will survive even under the most severe crown fires, as explained below.

The conclusion that severe crown fires may affect the simultaneous release and destruction of much of the seed rain in a lodgepole pine forest runs counter to several previous studies. Muir and Lotan (1985a) suggested that few seeds are burned by stand-replacing fires, and LaMont and others (1991) argued that evidence of heat-killed canopy-stored seeds in serotinous species is extremely rare and occurs only in what they cite as the unusual event of cone ignition. However, despite evidence showing high survival of *Pinus* seeds in thickly scaled serotinous cones exposed to 250°C for five seconds (LaMont and others 1991), there is ample evidence that exposure of seeds to temperatures in excess of 80°C will markedly decrease viability (Lotan 1975, Knapp and Anderson 1980, Hellum and Wang 1985). Johnson and Gutsell (1993) demonstrated that, in addition to temperature, other variables including time of exposure, the heat transfer properties of the cone, the height of cones in the canopy, and the rate of fire spread must all be considered to accurately predict the conditions under which serotinous cones will be opened or will ignite. Johnson and Gutsell's (1993) model indicates a correspondence between the minimum combinations of fire intensities and rates of spread sufficient to kill trees or to open serotinous cones. Their model also predicts the combinations of those variables that result in cone ignition. Our data suggest that a threshold between those extremes may be crossed when fire moves into the canopy. Our moderate-burn plots apparently were of sufficient intensity to kill all of the trees and open cones but cause little seed mortality. In contrast, fire in the canopy resulted in consistently high seed mortality.

We did find some evidence that cones were consumed in crown fires. With the exception of site 6 ($P = 0.30$), the mean number of cones on the ground in moderate-burn plots was significantly higher than that in canopy-burn plots ($P < 0.005$ in all eleven cases; Table 4.2). Across the moderate-burn plots, lodgepole pine seedling density was strongly correlated with the number of cones on the ground ($r = 0.92$, $P < 0.001$). The correlation across canopy-burn plots was not as strong ($r = 0.53$, $P = 0.07$), but indicated that the relationship also held for areas subjected to severe crown fires. Numbers of cones remaining in the crown of ten canopy-dominant trees at the moderate- and canopy-burn plots were highly variable, and it was difficult to get accurate counts on trees that had more than three hundred cones. Variability among trees was especially high in areas where trees with and without serotinous cones were sampled. Trees having the serotinous trait may have a thousand or more cones in the crown. As a consequence of the high variability among trees, significant within-site differences were found at only four sites (Table 4.2). Three of these were high-elevation sites where no trees had serotinous cones. Despite the within-site variability, there is a clear trend across sites: The mean number of cones per tree was higher in the moderate-burn than in the canopy-burn plot at eleven of the twelve sites; the probability of this occurring by chance is 0.006.

Of course, it is unlikely that seed losses result just from cone ignition; the severity and duration of thermal stress within the range of extremes between cone opening and ignition may be more important causes of seed mortality than consumption of seeds per se. Despain and others (1996) studied video footage of the 1988 Yellowstone fires and estimated that the most frequent time required to completely burn a tree crown was fifteen to twenty seconds. They then simulated exposure to crown fire by exposing serotinous and nonserotinous cones to flames from dry lodgepole pine branch wood. They reported that maximum germination corresponded exactly with the most frequent flaming duration for crown fires and that 70–80 percent of the canopy seed bank in a lodgepole pine stand would survive the most common crown fires. They acknowledge, however, that their fire did not provide the preheating and postheating produced by a natural crown fire, and they noted that seed survival would have been reduced to 25–30 percent with exposure times of the maximum duration observed in the videotapes (forty to sixty seconds). We suspect that many seeds released from serotinous cones as the flame front passes may be subsequently exposed to lethal temperatures or consumed by materials burning on the forest floor.

The proportion of serotinous trees in the unburned stands at the twelve sites ranged from zero to 48 percent (Table 4.1). Serotiny was highest among the low- and mid-elevation sites; no serotinous trees were found at four sites, three of which were at high elevations (Table 4.1). Prefire stand serotiny was strongly correlated with $\log_{10}$ seedling density of both moderate- and canopy-burn plots (Fig. 4.3). We conducted extensive searches of unburned stands in the vicinity of the three high-elevation sites where no serotinous trees were found in the original unburned plots. No serotinous trees were found in any of the three areas. That serotiny is very rare at such high-elevation sites was apparent from sampling on the Solfatara Plateau, where searching every unburned stand within 3 km of the study site failed to produce a single tree with serotinous cones (Ellis 1993).

Lotan (1975) and Muir and Lotan (1985a, b) suggested that differences in prefire serotiny levels could be a major factor determining postfire seedling densities and could also provide an index of seed availability. Our data provide strong support for those suggestions. We found that densities of lodgepole pine seedlings increased exponentially with increasing levels of serotiny, and the relationship held with a similar slope for both moderate- and canopy-burn plots (Fig. 4.3). The strength of the correlations indicate that 45–50 percent of the variation in seedling density could be accounted for by variation in stand serotiny. That the relationship is exponential means that relatively small differences in serotiny will be magnified in terms of seedling density. Several studies show that there can be substantial differences in the level of serotiny over relatively short distances (for example, Lotan 1975, Muir and Lotan 1985a). Tinker and others (1994) surveyed serotiny in nine patches that burned in the Yellowstone fires. They reported that serotiny was most variable at intermediate scales

(1–10 km) and relatively homogeneous at fine scales (< 1 km). For most of our sites, the unburned stand in which we inventoried serotiny was adjacent to the moderate-burn plot, so we are confident that the estimate would hold for the burned plots. At a few sites, however (for example, sites 9 and 10), the nearest unburned patch was 0.1 km or more from the moderate- and canopy-burn plots. In such cases, local variation in serotiny may have contributed to the variability in seedling densities not accounted for by our serotiny samples.

The proportion of serotinous trees in a stand appears to depend strongly on the nature of the most recent disturbance (Perry and Lotan 1979, Muir and Lotan 1985b). Muir and Lotan (1985b) found that stands originating after fire had a large percentage of trees with serotinous cones, whereas stands originating after other disturbances had a large proportion of open cones.

A second factor that appears to be important in determining the level of serotiny is related to elevation, probably through effects on fire frequency. Lotan (1967), Brown (1975), and Muir and Lotan (1985a, b) have suggested that the proportion of serotinous trees decreases with increasing elevation in the Northern Rocky Mountains. Muir and Lotan (1985b) found no correlation between elevation and level of serotiny, but all of their study sites were below 2300 m. Lotan (1968) found that serotiny was lower on the high-elevation Moose Creek Plateau (adjacent to the western boundary of Yellowstone National Park) than in nearby stands at lower elevations in the Island Park area. Tinker and others (1994) reported a strong negative correlation between incidence of serotiny and elevation for thirteen stands in Yellowstone, and, as we also found, serotiny levels were very low at elevations above 2300 m (Chapter 12). For our twelve sites, a negative correlation between percent serotinous trees and elevation was marginally significant ($t = 2.03$, $P = 0.07$); however, inclusion of four additional sites where we inventoried seedling densities and serotiny (data not shown) resulted in a significant relationship ($r^2 = 0.31$, $t = 2.53$, $P = 0.024$). There can be little doubt that the incidence of serotiny decreases with elevation in the Greater Yellowstone area.

We postulate that the low incidence of serotiny in higher-elevation forests is a consequence of long fire intervals. The high-elevation forests in Yellowstone may have mean fire intervals of more than three hundred years (Romme 1982, Romme and Despain 1989, Chapters 2, 3, 12). Stands may consist of mixed ages, as occurred at site 1, indicating that they have gradually become restocked since the last fire or that there has been recruitment of young trees as gaps have been created by the deaths of old individuals. Mature spruce-fir forests in Yellowstone typically have a sizable complement of uneven aged *P. contorta* that have colonized gaps created as old lodgepoles have died (D. Despain, personal communication, see also Perry and Lotan 1979, p. 966). We propose the following scenario to account for the low incidence of serotiny at high elevations: Long fire intervals often result in old-growth forests in which abundant gaps are created as old *P. contorta* individuals succumb. Such gaps, in turn, are colonized by seedlings from open-cone trees. Under such conditions, there would be strong selection for the open-cone trait, and stands developing under such conditions would have a low level of serotiny. When such stands are subjected to infrequent but intense wildfire, as occurred in 1988, resulting seedling densities would be very low, as documented by Tinker and others (1994) and this study. Again, there would be strong selection for the open-cone trait to repopulate sparsely colonized areas.

Perry and Lotan (1979) discuss the conditions necessary to maintain the cone polymorphism in lodgepole pine, assuming that cone serotiny is under one-locus, two-allele control. They note that where selection is strong, as in the case of fire selecting for serotiny, fixation of one allele is expected to occur rapidly. In describing potential conditions that would select for open cones, they describe a situation similar to that which we have proposed: "Even in the absence of low intensity fires, second-generation lodgepole pine often become established in small openings within the parent stand." These second-generation trees, established from open cones, would then be the primary source of seed when the stand is destroyed by a severe crown fire. Perry and Lotan note that this situation requires a longer interval between intense crown fires "than is generally thought to be the case." However, such long intervals are the case in the high-elevation forests of the Yellowstone Plateau, and we think it likely that continued strong selection for the open cone phenotype could result in fixation of that allele despite the fact that such forests are occasionally subjected to severe fires.

SUBSTRATE CHARACTERISTICS

Lodgepole pine has been thought to germinate and establish best on mineral soil or thin, disturbed duff (Lotan and Perry 1983, Parker and Parker 1983). The high thermal conductivity, heat capacity, and moisture retention of mineral soil apparently mitigate temperature extremes at the soil surface and increase water available to seedling roots (Cochran 1970, Eis 1970). Organic seedbeds, such as thick duff and litter, tend to dry quickly and expose seedlings to drought, the primary cause of mortality in the first year (Lotan and Perry 1983). Although high light levels generally are most favorable for seedling establishment (Clements 1910, Baker 1949), some shading, for example, from slash or deadfalls, may enhance early survival by increasing available moisture and decreasing soil temperatures (Noble and Alexander 1977, Stuart and others 1989).

In contrast to these previous characterizations of lodgepole seedling site requirements, we found during our 1989 field studies that large numbers of seedlings had become established in moderate-burn plots where the effects of the fire were minimal (duff and understory plants largely undisturbed). Fine-scale sampling confirmed these observations. In moderate-burn plots, most seedlings were found in unburned or scorched litter and in burned litter/duff (characteristics of litter/duff still detectable; could be picked up in chunks); these were the most common substrate categories on moderate-burn sites. However, seedlings generally were not randomly distributed with respect to substrate. At three of the four moderate-burn plots sampled, seedling presence was positively associated with the presence of unburned litter, and, in three cases, seedling presence was negatively associated with burned litter/duff ($P < 0.05$). At one moderate-burn plot, seedling presence was negatively associated with mineral soil ($P < 0.05$). At the three canopy-burn plots inventoried, seedlings showed positive associations with charcoal and moss ($P < 0.05$), but no negative associations were found.

The positive associations of established seedlings with the presence of litter, and the negative or neutral relationships with burned litter/duff and mineral soil, seem contradictory to the traditional view that lodgepole seedlings germinate and survive best on bare soil or disturbed duff. Several factors may have contributed to the patterns that we observed. It is possible that seeds present on the ground, deposited from open cones before the fires, were destroyed in patches of litter that burned but were unharmed in unburned patches. If such seeds germinated along with newly fallen seeds from serotinous cones after the fires, then more seedlings might occur in litter even if germination and survival were poorer there. Furthermore, as suggested earlier, some seeds released from serotinous cones may have been killed by landing on burning or smoldering patches of litter or duff, whereas those landing on unburned or scorched litter would have been unharmed.

Another factor that probably contributed to the patterns observed was abundant moisture during the initial postfire years. The 1989 and 1990 growing seasons were characterized by frequent rainfall throughout the summer months (Romme and others 1997). Had seedlings been subject to more severe seasonal drought during those years, it is possible that fewer would have survived in unburned litter or duff. In fact, casual observations at one of our study sites support this suggestion.

At the interface between unburned and moderately burned forest at site 11 (where mean seedling density on the moderate-burn plot was 190 m^{-2}), there were far more seedlings where the forest floor had been subject to fire than in immediately adjacent unburned areas. It is unlikely that the seed rain from serotinous cones would have differed much over such small distances or that any differences would have corresponded so closely with the surface burn pattern. Thus substrate differences must have been responsible for the observed pattern. Site 11 lies near the lower elevation limit for lodgepole forests in Yellowstone, adjacent to sagebrush *(Artemisia tridentata)* steppe, where the prospect of late-season drought would be relatively high. The unburned area had a substantial litter/duff layer that may have exacerbated drought stress and reduced germination or increased mortality of seedlings. We emphasize, however, that seedlings were present in the unburned area at this site and in the unburned plots at all of our other study sites, but generally at far lower densities than in the burned areas.

Although our studies do not elucidate the ideal safe site (as per Harper 1977) for lodgepole establishment, they do provide evidence against the traditional view that mineral soil is the best substrate. Coates and others (1991) also found that exposure of mineral soil had no positive impact on seedling survival. It is clear that lodgepole pine seeds are able to germinate and survive on a variety of substrates.

INTERSPECIFIC COMPETITORS

Several investigators have reported that competition from herbaceous plants may also inhibit lodgepole seedling establishment (Fowells 1965, Parker and Parker 1983, Lieffers and others 1993). Stermitz and others (1974) reported increased survival of seedlings where herbaceous vegetation had been removed experimentally. In contrast, our data provide little support for the hypothesis that establishment density of lodgepole pine will be inversely related to the abundance of potentially competing herbs and shrubs. Across moderate-burn transects, lodgepole seedling densities were not correlated with cover of other vascular plants ($P > 0.20$). Across canopy-burn transects, lodgepole seedling densities were positively correlated with cover of other plant species ($r = 0.75$, $P = 0.006$). This correlation probably reflects patterns of mortality of both lodgepole pine seeds and vegetative parts of other species in areas subjected to severe canopy burns.

Our fine-scale sampling at four sites produced only one negative interspecific association, that between lodgepole pine seedlings and pine grass *(Calamagrostis rubescens)*. Other studies indicate that pine grass can retard conifer regeneration when it establishes early after a fire (Stahelin 1943, Lotan and Perry 1983). The negative association that we recorded was of little consequence in terms of seedling establishment, however. It occurred at site 11's moderate-burn plot, where seedling densities averaged 190 m^{-2}! We found positive associations between lodgepole seedlings and fireweed *(Epilobium angustifolium)* at five of the seven plots (moderate- and canopy-burn) and between lodgepole seedlings and strawberry *(Fragaria virginiana)* at one plot. Again, these patterns probably reflect effects of fire severity. Where fire intensities were lower, more understory plants as well as lodgepole seeds survived. We conclude that presence of potential interspecific competitors generally did not significantly affect initial establishment of lodgepole pine seedlings on areas occupied by lodgepole pine and burned by the 1988 Yellowstone fires.

CHANGES IN DENSITY OF LODGEPOLE PINE SEEDLINGS DURING THE FIRST SIX POSTFIRE YEARS

Data from the Sleeping Child fire (Lyon 1976, 1984) suggest that we might expect an annual seedling mortality of about 6 percent over the first decade. Seedling attrition on most of our plots has been considerably less than that (Tables 4.4, 4.5). Our sampling design for seedling densities did not enable us to track the fates of individual seedlings. However, repeated samples from the quadrats on the permanent transects provide estimates of the annual rate of density change, which reflects both mortality and recruitment (Table 4.4). Relatively high rates of decrease in seedling density were observed at only three sites, Mount Washburn, Lewis Lake, and Indian Creek. High mortality at Mount Washburn was a consequence of severe erosion that deposited debris on the areas sampled, especially the moderate- and remote canopy-burn transects. Erosion on the steep slope may have contributed to seedling mortality on the moderate-burn plot at Indian Creek as well. Aside from those three sites, rates of density change were generally modest (Table 4.4). Rates were negative at most of the moderate-burn plots, indicating that mortality was greater than recruitment, but at about half of the canopy- and remote canopy-burn plots, recruitment was sufficient to offset mortality. The estimate of a 35 percent annual increase in density at site 1 probably has a large error associated with it because of the very small number of seedlings present within the area sampled. Nevertheless, we have witnessed continued recruitment at that and most of the other canopy-burn plots. Overall, the results indicate that mortality rates have been

Table 4.4. Annual rate of change in density of *P. contorta* seedlings in moderate-, canopy-, and remote canopy-burn plots at the twelve study sites

		Rate of Density Change (%)		
	Site	Moderate	Canopy	Remote Canopy
1	Solfatara Plateau	−3.9	0	+35.0
2	Mount Washburn	−40.8	−9.2	−26.0
3	Lewis River	−5.8	+3.8	0
4	Lewis Lake	−10.2	−4.0	−8.6
5	Norris Junction East	+7.8	0	
6	Indian Creek	−13.0	−4.0	0
7	Swan Lake Flat	−2.6	−3.8	−5.4
8	Whiskey Flat	0	0	−2.2
9	Norris Junction South	−0.8	−3.2	−2.9
10	West Yellowstone	−2.8	0	
11	Madison River	−4.4	−1.9	−2.4
12	Madison Junction	−2.7	−2.0	0

Note: Rates were estimated as the slope of a regression of the natural logarithm of the total number of seedlings versus year, based on data from 1990 to 1994. Positive values indicate net recruitment.

quite low at most of the study sites. This conclusion is supported by casual observations in the field: we encountered very few dead seedlings as we conducted the annual inventories. Turner and others (1997), working in similar burned portions of Yellowstone, also reported that most lodgepole seedling recruitment occurred in the first two years after the fire, after which both mortality and recruitment rates were very low (Chapter 14).

An additional estimate of rates of density change was available for three study sites where plots were established to study the effects of competitors on lodgepole seedling growth (see Factors Influencing Growth of Lodgepole Pine Seedlings, Competitive Neighborhood, below). At each site, the total numbers of lodgepole seedlings were recorded in 1991 and 1993 in each of ten 1-m^2 quadrats of "control" and "interspecific competitors removed" treatments. Counts for the ten quadrats were pooled to estimate rates of density change (Table 4.5a). These rates, ranging from zero to −4.3 percent, are comparable to those from the seedling density data (Table 4.4). That annual mortality rates were only 2–3 percent for plots where lodgepole seedling densities range from 120/m^2 to 150/m^2 (Table 4.5a) is surprising.

Finally, we were able to estimate mortality directly from loss of marked seedlings in the competition study plots. Four seedlings were marked in each of

Table 4.5a. Annual rate of change in density of lodgepole pine seedlings in control and interspecific competitors removed (ICR) plots from 1991 to 1993 at three study areas

Study Site	Treatment	Seedling 1991	Density (M^{-2}) 1993	Rate of Density Change (%)
Madison River Bridge	ICR	135	127	−2.7
	Control	151	141	−3.4
Swan Lake Flat	ICR	27	26	0
	Control	29	27	−4.3
Whiskey Flat	ICR	80	79	−0.8
	Control	80	83	0

Note: The plots were all in areas of moderate burn severity. Rate of change was estimated as the slope of the change in the natural logarithm of seedling totals over the two years.

Table 4.5b. Annual mortality rates for marked target seedlings in control and ICR plots (pooled) from 1991 to 1994

Study Site	Treatment	Seedling 1991	Totals 1993	Mortality Rate (%)
Madison River Bridge	ICR	40	39	
	Control	40	38	
	Total	80	77	1.3
Swan Lake Flat	ICR	40	39	
	Control	40	39	
	Total	80	78	0.8
Whiskey Flat	ICR	40	38	
	Control	40	38	
	Total	80	76	1.7

Note: Four seedlings were marked in each of ten quadrats for both treatments at each area ($4 \times 10 \times 2 = 80$). Mortality rates were estimated as the change in the natural logarithm of seedling totals over the three years. Total seedling densities for these plots are given in Table 4.5a.

ten 1-m^2 quadrats per treatment. Again, we limit our analyses here to those plots in which the densities of lodgepole seedlings were not manipulated, the "control" and "interspecific competitors removed" treatments. Mortality rates for the 80 marked seedlings were less than 2 percent per year at all three study areas (Table 4.5b). Only 9 of 320 seedlings were lost over four growing seasons! These data confirm that mortality rates were low over the first six postfire years, even in extremely dense stands of seedlings.

Lyon (1984) indicated that attrition would be higher in areas having higher seedling densities. Our data, spanning the first six postfire years, provide no support for the hypothesis that seedling mortality is density dependent, despite the fact that two of our sites have seedling densities two and five times as high as the highest densities reported by Lyon. The highest rates of density decrease were observed at sites where seedling densities were quite low. Sites with the highest seedling densities (sites 7, 8, 11, 12) all had modest rates of density change (Table 4.4).

Additional recruitment has been documented in areas where seedling densities are very low. Turner and others (1994) show that areas subjected to severe crown fires in 1988 were generally in close proximity to unburned or moderately burned areas that could serve as a source of lodgepole pine seeds. They point out that even the major "fire runs" of 1988 were long and narrow, so that "even the most extensive patches of crown fire were lined by less severely burned and unburned forests." Fowells (1965) suggests that lodgepole pine can restock clearcut areas that are within about 60 m of uncut forest. Using that appraisal of dispersal limitation, Turner and others (1994) estimate that 30 percent of the crown fire area resulting from the 1988 Yellowstone fires would be within the seed shadow of living lodgepole pine. Fowells's (1965) distance estimate is based on silvicultural restocking requirements, however. Some seeds undoubtedly will be transported longer distances, resulting in the gradual "filling in" of high-elevation areas where seedling densities are very low, as we have suggested earlier. Thus we expect additional recruitment at our high-elevation sites.

We rarely saw evidence of animal damage to seedlings at any of our study sites. Evidence of pocket gophers *(Thomomys bottae)* was common at most sites, but it is clear that they did not cause much seedling mortality. Lotan and Perry (1983) noted that trampling damage by large herbivores such as cattle is a major cause of mortality of young lodgepole pine. Given large populations of bison and elk in Yellowstone, we expected that trampling might cause significant mortality in some areas. However, that was not the case, even at site 7, where use of the area by both bison and elk was heavy. There was no evidence of browsing on seedlings, probably, as Lotan and Perry (1983) pointed out, because seedlings are covered by snow during the season when ungulates feed on conifers.

FACTORS INFLUENCING GROWTH OF LODGEPOLE PINE SEEDLINGS

In this section, we assess the relative importance of fire severity, site characteristics, and intraspecific and interspecific competition on growth rates of lodgepole pine during the first decade after the 1988 fires.

Fire Severity

Lodgepole pine seedlings measured in 1991 were significantly taller, on average, in the canopy-burn plots at five of the six sites (Fig. 4.4). Similarly, mean seedling stem diameter was significantly larger in canopy-burn plots than in moderate-burn plots at four of the six sites sampled (Fig. 4.4). Mean aboveground biomass on canopy-burn plots was 2.9, 1.9, and 3 times as high as that on moderate-burn plots at sites 7, 8, and 11, respectively (Fig. 4.4).

Samples taken at five sites in 1992 confirmed that height and stem diameters of seedlings were greater on canopy-burn plots than on moderate-burn plots (Table 4.6). Moreover, there was more variability in both seedling height and diameter on the canopy-burn plot than on the paired moderate-burn plot at every site (Table 4.6). This trend is clearly apparent in frequency polygons of seedling height and diameter for moderate- and canopy-burn plots (Fig. 4.5). These data indicate that not only are seedlings larger but also that there tend to be relatively more seedlings in the larger size classes on the canopy-burn plots.

At all three sites where seedlings were harvested in 1990, seedlings generally accumulated more aboveground biomass per unit height in the severe-burn plots than in the moderate-burn plots (Fig. 4.6). This indicates that seedlings in the canopy-burn plots were stockier. The difference was particularly conspicuous and seedlings were considerably larger at sites 7 and 11 (Fig. 4.6). There was no difference between moderate- and canopy-burn plots in the slope of the relationship between aboveground biomass and stem cross-sectional area at any of the three sites ($P > 0.2$ in each case, Fig. 4.6). Thus, within a site, the amount of stem and leaf biomass supported per unit cross-sectional area of stem was the same for moderate and canopy-burn plots.

Height data taken at sites 11 and 7 in 1996 and early 1997, respectively, show that not only do size differences between moderate- and canopy-burn plots persist, but they also become exaggerated (Fig. 4.7). On average, seedlings gained 3.3 cm/yr in height on the moderate-burn plot at site 11, compared with 8.9 cm/yr on the canopy-burn plot. At site 7, the seedlings on the canopy-burn plot grew an average of 22 cm taller each year, compared with 14 cm on the moderate-burn plot.

All of these data demonstrate that lodgepole pine seedlings consistently grew faster in areas subjected to severe crown fires than in adjacent areas subjected to fire of moderate severity. In addition, seedlings in the canopy-burn areas were generally more robust, supporting more aboveground biomass per unit height. These differences in seedling growth could reflect differences in abiotic condi-

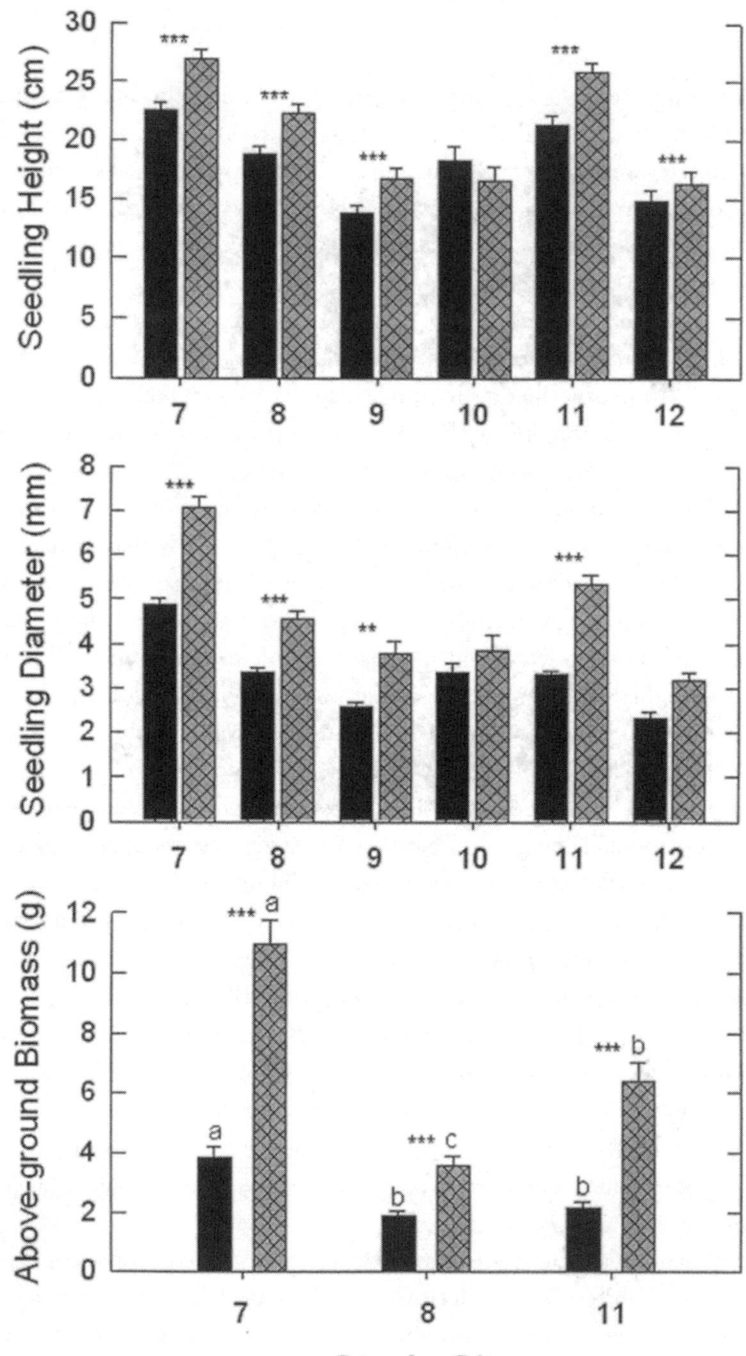
Seedling Height (cm)
30
25
20
15
10
5
0
***
***
***
***
***
7
8
9
10
11
12
Seedling Diameter (mm)
8
7
6
5
4
3
2
1
0
***
***
**
***
7
8
9
10
11
12
Above-ground Biomass (g)
12
10
8
6
4
2
0
*** a
a
*** c
b
*** b
b
7
8
11
Study Site

tions or differences in the competitive environment between moderate- and canopy-burn areas, or both.

The companion study in which we manipulated the competitive environment of target lodgepole pine seedlings (Ellis 1993, see below) demonstrated clearly that intraspecific competition strongly affects seedling growth. Thus faster growth in the canopy-burn areas may, in part, reflect lower densities of seedlings. However, the fact that we did not find consistent relationships between seedling size and various indexes of the immediate competitive environment (see below) suggests that differences in abiotic factors may be largely responsible for the observed size differences between moderate- and canopy-burn areas. There may have been postfire differences in soil nutrient availability between the adjacent plots because more litter and duff on the forest floor were consumed in the canopy-burn plots, which may have released more nutrients from the accumulated organic matter. Ash from the consumption of needles, fine twigs, and coarse woody debris in the canopy-burn areas may have deposited additional nutrients (Chapter 12), but one would also expect higher losses due to volatilization in those areas. We would expect, however, that any differences in nutrient availability would be relatively short-lived and would not account for the persistent differences in rates of seedling growth between moderate- and canopy-burn plots. It seems more likely that the observed differences in seedling size were due to light availability. Quantifying irradiance or temperature regimes in moderate- and canopy-burn plots was beyond the scope of this study, but casual observations indicated that considerably more light reached the forest floor in canopy-burn plots. This was a simple consequence of the fact that needles, fine twigs and branches, and many cones were consumed in crown fires, resulting in a much more open postfire canopy. Thus more light would have been available for photosynthesis, and temperatures probably would have been more favorable for metabolic processes in the canopy-burn plots. Furthermore, it is likely that snow would melt earlier in crown fire areas and that temperatures would become favorable for growth earlier in the season.

Figure 4.4 (facing page). Mean seedling height (top) or stem diameter (center) at six sites, and mean seedling aboveground biomass at three sites (bottom) in moderate- and canopy-burn plots. Error bars represent one standard error. Asterisks indicate a significant difference between the moderate- (solid-colored) and canopy-burn (cross-hatched) plots at a particular site (** = $P < 0.01$; *** = $P < 0.001$). In the bottom graph, results of across-site comparisons using Student-Newman-Keuls multiple range test are indicated with lowercase letters. Means within a burn category associated with different letters are significantly different at $P = 0.05$. Data were taken in 1991 from areas burned in 1988.

Table 4.6a. Results of two-sample *t*-tests for differences in mean stem diameters of lodgepole pine seedlings in moderate and canopy burn areas at five sites in Yellowstone National Park in June 1992

Site	Burn Type	N	Diameter (mm)	S.D.	*t*	P
11	Moderate	200	1.84	0.93	−14.84	<0.001
	Canopy	200	3.64	1.44		
12	Moderate	200	2.01	0.08	−13.66	<0.001
	Canopy	200	3.59	1.42		
9a	Moderate	200	2.67	1.21	−10.17	<0.001
	Canopy	200	4.74	2.62		
9b	Moderate	200	1.38	0.64	−15.58	<0.001
	Canopy	200	3.20	1.50		
7	Moderate	200	4.16	1.70	−10.64	<0.001
	Canopy	200	6.59	2.74		
10	Moderate	200	4.35	1.73	−4.28	<0.001
	Canopy	200	5.28	2.53		

Table 4.6b. Results of a two-sample *t*-test for mean height of lodgepole pine seedlings in moderate and canopy burn areas of five sites in Yellowstone National Park in June 1992

Site	Burn Type	N	Height (cm)	S.D.	*t*	P
11	Moderate	200	12.59	5.59	−12.09	<0.001
	Canopy	200	19.82	6.34		
12	Moderate	200	13.91	5.15	−7.46	<0.001
	Canopy	200	18.40	6.78		
9a	Moderate	200	13.13	6.16	−7.80	<0.001
	Canopy	200	18.76	7.85		
9b	Moderate	200	6.90	3.20	−15.11	<0.001
	Canopy	200	14.52	6.39		
7	Moderate	200	20.67	8.40	−6.51	<0.001
	Canopy	200	26.70	10.05		
10	Moderate	200	20.89	7.31	−10.64	<0.001
	Canopy	200	24.84	9.39		

Site Factors

Across sites, there were significant differences in seedling height, seedling stem diameter, and biomass in both canopy- and moderate-burn plots (Fig. 4.4). Mean aboveground biomass in the canopy-burn plot at site 7 was 3.1 and 1.7 times that at sites 8 and 11, respectively (Fig. 4.4). At site 11, the slope of the relationship be-

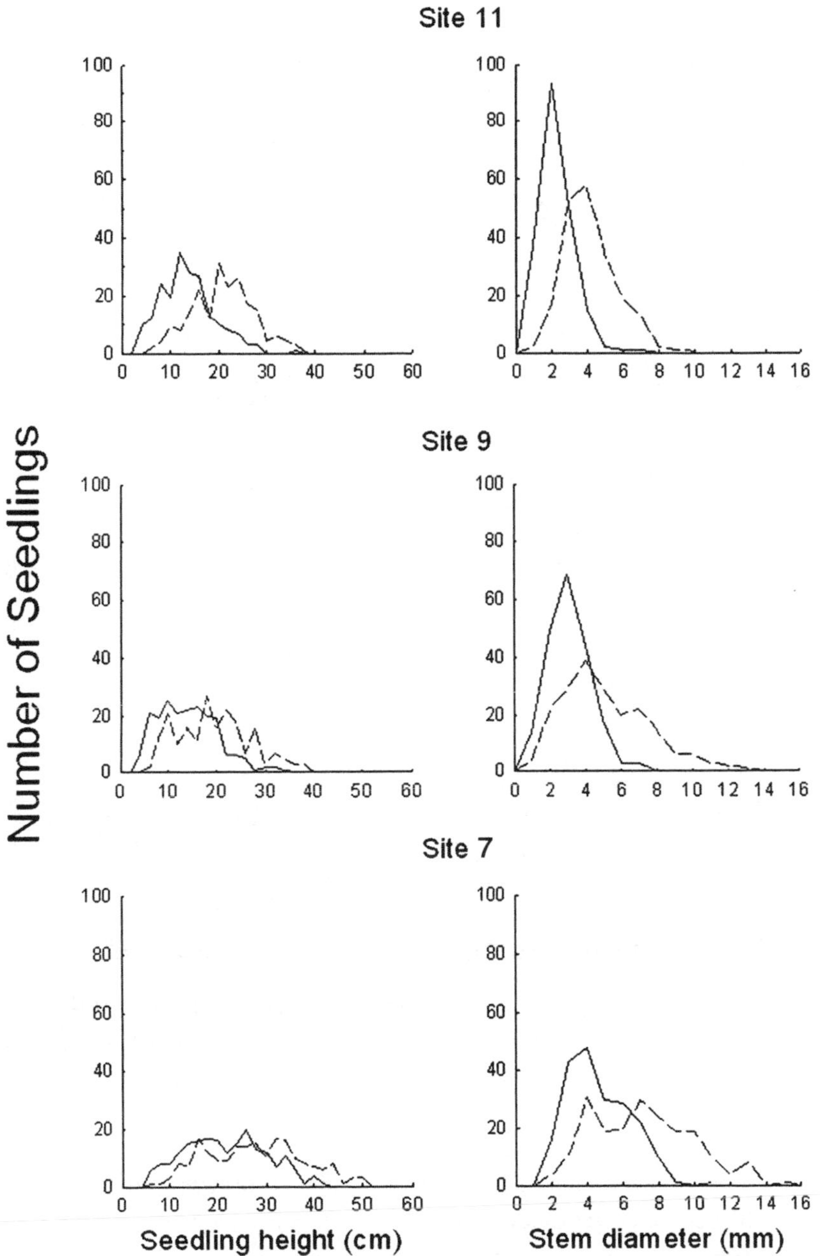

Figure 4.5. Frequency polygons for lodgepole pine seedling height and stem diameter at three study sites in Yellowstone National Park. Solid lines represent moderate burn, broken lines canopy burn. Data were taken in 1992 from areas burned in 1988.

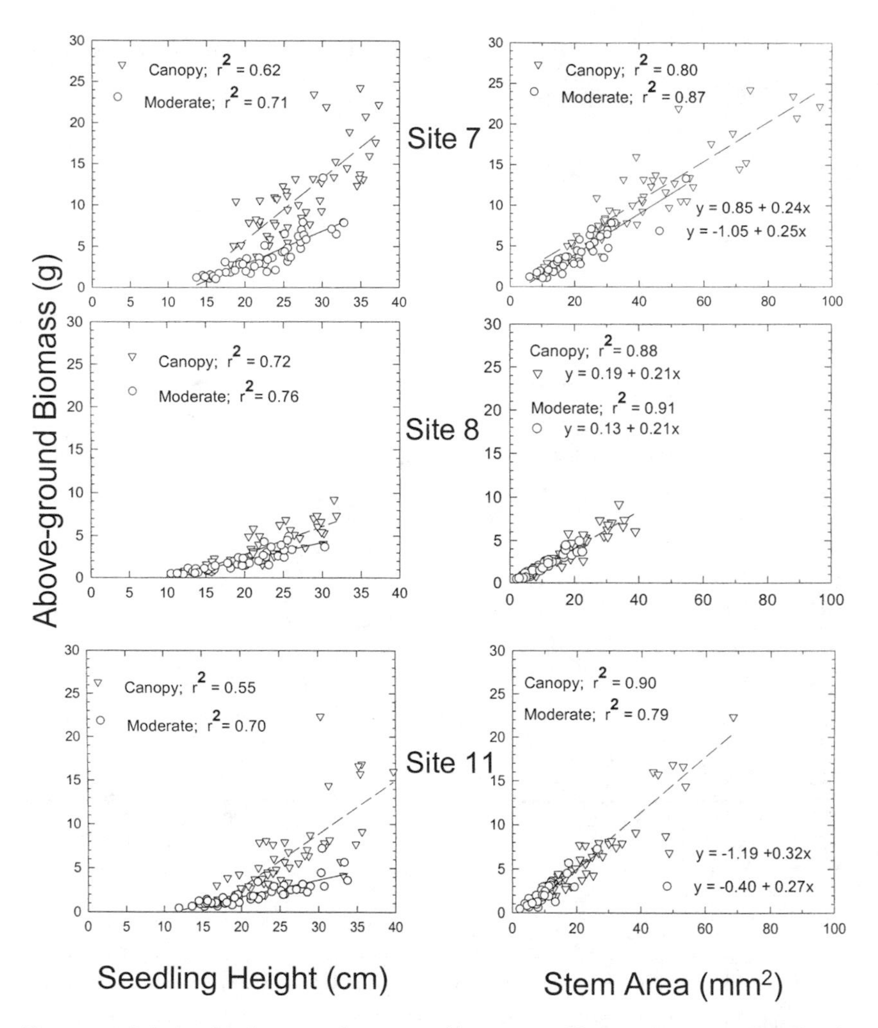

Figure 4.6. Relationship between aboveground biomass and lodgepole pine seedling height or cross-sectional stem area for moderate- and canopy-burn plots at sites 7, 8, and 11. Data were taken in 1991 from areas burned in 1988.

tween aboveground biomass and stem cross-sectional area was significantly ($P < 0.001$) higher than that at sites 7 and 8 (Fig. 4.6), indicating that seedlings at site 11 supported more aboveground biomass per unit cross-sectional area of stem.

Our studies of factors affecting seedling growth were restricted to the six low- and mid-elevation sites that had the highest seedling densities (sites 7–12) in an effort to determine whether seedling growth was dependent upon seedling density. Neither seedling height nor stem diameter was correlated with seedling den-

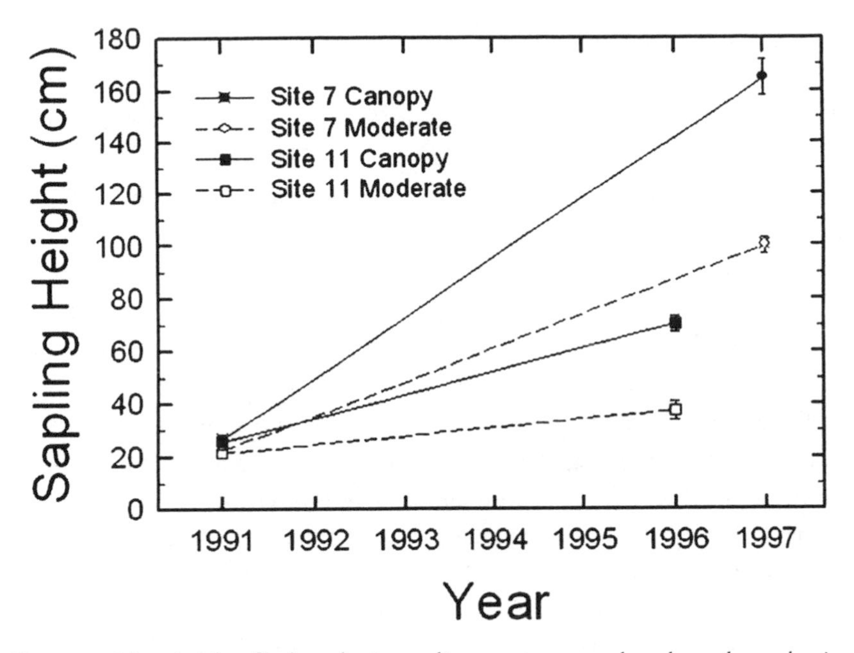

Figure 4.7. Mean height of lodgepole pine saplings on canopy- and moderate-burn plots in 1991 and 1996 (site 11) or 1997 (site 7). Error bars represent standard errors.

sity across the six sites on either moderate- or canopy-burn plots; in fact, site 11, which had by far the highest seedling densities, had the second-tallest seedlings and, among canopy-burn plots, seedlings with the second-largest stem diameters (Fig. 4.4). Clearly, any density-dependent effects on seedling growth are secondary to environmental differences among the sites, such as nutrient and water availability, soil depth and moisture storage capacity, and temperature. Site 7, which consistently produced the largest seedlings of any site in the study (Fig. 4.4), appeared to be the most mesic, supporting numerous mature *Picea engelmannii* trees in the prefire stand. Although the underlying substrate at site 7 is basalt, the soils are derived from glacial till, as evidenced by the presence of large and small rounded boulders. Similarly, site 11, which supports a vigorous growth of understory forbs and perennial grasses, appears to be more fertile than other sites. In contrast, the soils at site 8, which produced relatively small seedlings (Fig. 4.4), were derived from rhyolite, which typically produces coarse textured, infertile soils (Despain 1990). This site has a southwesterly aspect and appears to be among the more xeric of the sites. These results are not surprising; differences in site index among areas supporting stands of lodgepole pine are well known (for example, see Alexander and Edminster 1980).

Competitive Neighborhood

There is little doubt that crowded neighborhoods can affect an individual plant's growth and performance. The competitive effects within a neighborhood are a consequence of the number of neighbors, their size, and their proximity to a target plant. Plants that establish first or that encounter fewer neighbors may monopolize local resources and grow more rapidly than those that establish later or that are found in more crowded neighborhoods (Shainsky and Radosevich 1992). Given these arguments, we predicted that rates of seedling growth would be inversely proportional to both the number and proximity of neighbors.

The mean number of potential competitors within 0.31 m of a target lodgepole pine seedling was significantly greater in the moderate-burn than in the canopy-burn plot at all five sites sampled (Table 4.7). The mean number of potential intraspecific competitors and the mean number of interspecific competitors was higher in the moderate-burn plot at four of the five sites (Table 4.7). Mean distance from a target seedling to the nearest intraspecific competitor was larger in the canopy-burn plot at three of the five sites, and the difference was marginally significant at the other two sites (Table 4.7). In contrast, distance to the nearest interspecific competitor was significantly greater in the canopy-burn plot at only two of the five sites (Table 4.7).

Correlation analyses of lodgepole pine seedling heights or diameters versus numbers, types, or distances of potential competitors revealed few significant relationships, and the results were inconsistent. Significant correlations occurred only for three of the five sites (Table 4.8). At site 7, seedling height and diameter were negatively correlated with mean number of all competitors in the moderate-burn plot. Similarly, at site 10, seedling diameter was negatively correlated with mean number of all competitors and with the mean number of interspecific competitors. In contrast, at site 12, seedling height and diameter were positively correlated with numbers of interspecific or all competitors in the moderate-burn plot, whereas in the canopy-burn plot, those correlations were negative (Table 4.8).

In 1991, we established experimental plots at three sites (7, 8, and 11) in which the competitive environment of target lodgepole seedlings was manipulated (see Ellis 1993 for details). Height and diameter growth of target seedlings on control plots were compared to that on plots in which interspecific, intraspecific, or all vascular plant competitors were removed. Removal of intraspecific competitors resulted in significant increases in height and stem diameter of target lodgepole pine seedlings at all three study sites, whereas removing interspecific com-

Table 4.7. Mean numbers of intraspecific, interspecific, and total competitors within a 0.31-m radius of target lodgepole pine seedlings, and mean distances from target seedlings to the nearest intraspecific or interspecific competitor, in moderate- and canopy-burn plots at five study sites in Yellowstone National Park

	Number of Competitors			Mean distance (cm) to nearest competitor	
	Intraspecific	Interspecific	Total	Intraspecific	Interspecific
Site 7					
Moderate	7.0	12.0	17.0	12.0	10.0
Canopy	1.0	9.0	10.0	21.0	13.0
P	0.001	0.05	0.001	0.001	0.095
Site 8					
Moderate	12.0	7.0	19.0	10.0	17.0
Canopy	3.0	1.0	4.0	20.0	26.0
P	0.001	0.02	0.010	0.001	0.001
Site 9					
Moderate	4.0	7.0	11.0	15.0	15.0
Canopy	1.0	3.0	4.0	27.0	19.0
P	0.001	0.05	0.001	0.001	0.116
Site 10					
Moderate	2.0	15.0	18.0	21.0	14.0
Canopy	1.0	3.0	4.0	26.0	23.0
P	0.119	0.001	0.001	0.06	0.001
Site 12					
Moderate	16.0	35.0	51.0	10.0	6.0
Canopy	5.0	25.0	30.0	13.0	8.0
P	0.001	0.188	0.077	0.09	0.198

Note: P is the probability that the difference between moderate- and canopy-burn plots is due to chance.

petitors had minor effects on seedling growth (Figure 4.8). At sites 8 and 11, and for stem diameter at site 7, seedling growth on plots where interspecific competitors were removed was not statistically different from that of controls ($P > 0.05$). With the exception of stem diameter at site 11 (Fig. 4.5), the effects of removing all potential competitors and of removing only intraspecific competitors were statistically indistinguishable.

Table 4.8. Results of correlation analyses between height and diameter of target lodgepole pine seedlings versus density of intraspecific, interspecific, or all competitors, or versus distance to the nearest intraspecific or interspecific competitor in moderate- and canopy-burn plots at three study sites in Yellowstone National Park

		Moderate		Canopy	
		R	p	r	p
Site 7					
Height	Intraspecific competitors	NS	NS	NS	NS
	Interspecific competitors	NS	NS	NS	NS
	All competitors	−0.39	0.049	NS	NS
	Nearest intraspecific competitor	NS	NS	NS	NS
	Nearest interspecific competitor	NS	NS	NS	NS
Diameter	Intraspecific competitors	NS	NS	NS	NS
	Interspecific competitors	−0.53	0.060	NS	NS
	All competitors	−0.46	0.020	NS	NS
	Nearest intraspecific competitor	NS	NS	NS	NS
	Nearest interspecific competitor	NS	NS	NS	NS
Site 10					
Diameter	Intraspecific competitors	NS	NS	NS	NS
	Interspecific competitors	−0.45	0.022	NS	NS
	All competitors	−0.47	0.016	NS	NS
	Nearest intraspecific competitor	NS	NS	NS	NS
	Nearest interspecific competitor	NS	NS	NS	NS
Site 12					
Height	Intraspecific competitors	NS	NS	0.41	0.040
	Interspecific competitors	0.41	0.040	−0.48	0.014
	All competitors	0.47	0.018	−0.41	0.040
	Nearest intraspecific competitor	−0.54	0.005	NS	NS
	Nearest interspecific competitor	NS	NS	NS	NS
Diameter	Intraspecific competitors	NS	NS	0.41	0.040
	Interspecific competitors	0.62	0.001	0.63	0.001
	All competitors	0.61	0.001	−0.58	0.020
	Nearest intraspecific competitor	−0.41	0.040	NS	NS
	Nearest interspecific competitor	−0.43	0.030	0.42	0.040

Note: NS = no significant correlation.

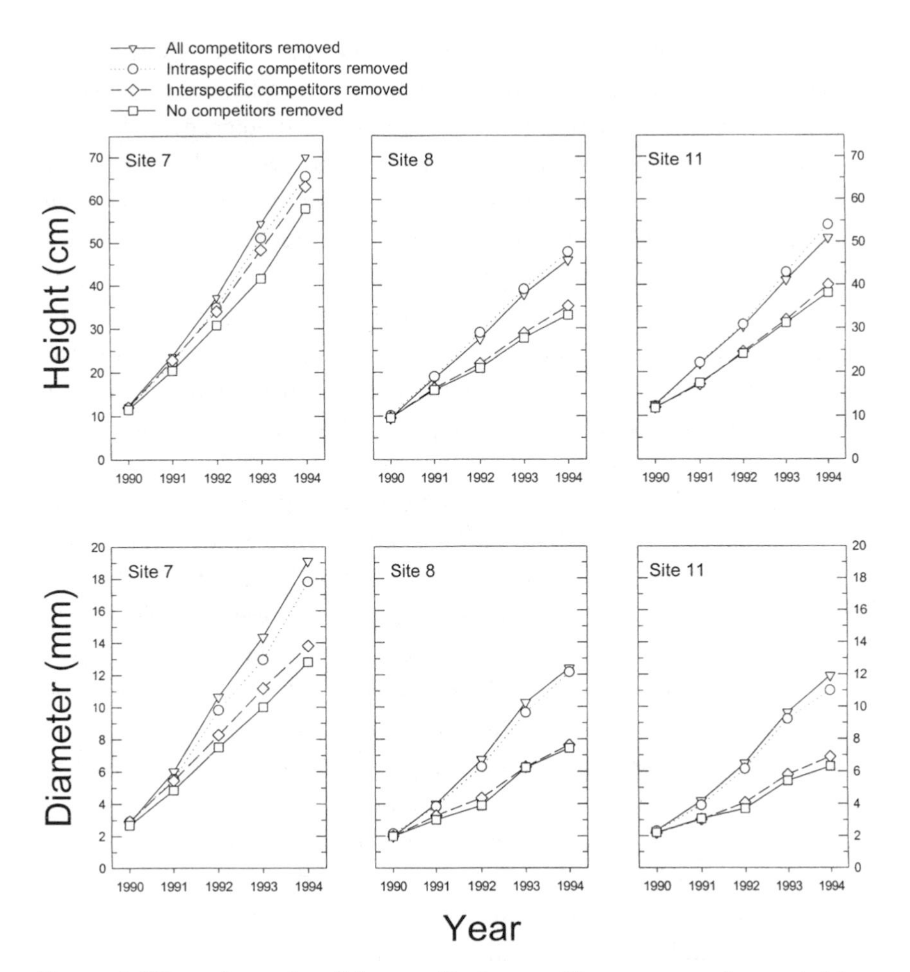

Figure 4.8. Effects of removing all, intraspecific, interspecific, or no competitors on height and stem diameter of target lodgepole pine seedlings at sites 7, 8, and 11. Data shown for 1990 were actually taken at the beginning of the 1991 growing season, before seedling growth was initiated.

It is generally held that intraspecific competition will be more intense than interspecific competition because conspecific neighbors have similar physiology, phenology, and morphology (Barbour and others 1980). Strong effects of intraspecific competition have been demonstrated in crowded stands of pines, including reduced stem diameter (Oren and others 1987), retarded hydraulic and canopy development (Keane and Weetman 1987), and decreased water use efficiency (Drivas and Everett 1988). Thus we anticipated that the intraspecific competition would have stronger effects on seedling growth, but the scarcity of perceptible effects of interspecific competitors was unexpected given the emphasis on the importance of interspecific competitors by other investigators (for example, Brown 1975, Lotan and Perry 1983, Powell and others 1994). The significant, if modest, effects of interspecific removals at sites 7 and 11 (Figure 4.8) are probably attributable to the presence of *Calamagrostis rubescens,* a strongly rhizomatous, robust perennial that is thought to be among the most important interspecific competitors with lodgepole pine seedlings (Lotan and Perry 1983). *C. rubescens* was common at both sites, and given its vigorous growth and height, may have reduced soil resources as well as light availability to neighboring lodgepole pine seedlings. At site 8, the most abundant interspecific competitor was *Carex rossii;* these short-statured sedges apparently had little influence on growth of lodgepole pine seedlings.

Our data show that the abundance of potential competitors had little or no influence on the postfire establishment of lodgepole pine over a broad range of sites in Yellowstone National Park. Lyon's (1984) results indicated that interspecific competitors had little effect on survival of lodgepole pine seedlings following the Sleeping Child fire in Montana. Given the minor effects of interspecific competitors on growth of lodgepole pine seedlings documented here, it seems reasonable to conclude that the interspecific competitors encountered here play a minor role in the recovery of lodgepole pine following stand-replacing fires. Interspecific competitors may have much more significant effects in forest clear-cuts, especially if aggressive aliens are seeded into the sites (Powell and others 1994).

Seedlings grew more rapidly at site 7, both with and without the presence of competitors (Fig. 4.8). This is consistent with the data in figure 4.4, which shows that aboveground biomass of seedlings harvested in 1991 from moderate- and canopy-burn plots was higher for seedlings at site 7 than at sites 8 or 11. That these site differences were apparent when competitors were removed provides additional evidence that the growing conditions were generally more favorable at site 7. Nevertheless, the relative effects of removing competitors were generally consistent with the expectation that competitive effects would be density dependent. Mean density of lodgepole pine seedlings on the manipulation plots prior to removal of intraspecific competitors were 36, 81, and 134 m^{-2} at sites 7, 8, and 11, respectively. Seedling height after four years at these sites was 21 percent, 39 percent, and 34 percent greater, and stem diameters were 48 percent, 66 percent, and 89 percent greater with all competitors removed. Thus the magnitude of competitive effects was roughly proportional to the density of intraspecific competitors.

Of the various indexes of local crowding, available area, as estimated by

Thiessen polygons, has most frequently been found to correlate with plant growth (Mead 1966, Liddle and others 1982, Mithen and others 1984, Matlack and Harper 1986, Kenkel and others 1989). Thiessen polygons are constructed from a map by drawing lines (Mithen and others 1984) between a target plant's point and those for all nearby neighbors. Polygons are formed by the intersections of perpendicular bisectors of these lines. Thus the polygon around a target plant includes all points in the plane that are closer to that plant than to any other, thereby providing an index of the resource area available to that plant. We constructed Thiessen polygons for all target lodgepole pine seedlings in the control and plots with interspecific competitors removed at the three manipulation study sites.

Available area, as estimated by Thiessen polygons, was positively correlated with height ($r = 0.49$, $P = 0.002$) and stem diameter ($r = 0.57$, $P < 0.001$) of lodgepole pine seedlings at site 8 but not at sites 7 or 11. The range of seedling sizes and polygon areas was similar for the three sites (data not shown); thus the differences among sites did not appear to be a consequence of limited variability in polygon area at sites 7 or 11. It is not immediately clear why a strong relationship was found at site 8 but not at sites 7 and 11. Seedling densities at site 8 were intermediate in relation to the other two sites, making it unlikely that the differences in responses were related to seedling densities. Another possibility is that area is somehow more critically related to available resources at site 8. As noted earlier, seedlings were smaller at site 8 than at sites 7 or 11, and our general impression is that site 8 is more xeric (and probably less fertile) as a consequence of coarse-textured soils derived from rhyolite. The soils at that site may be shallower and have lower water holding capacities. Thus available area may be a relatively more important parameter affecting seedling growth at site 8 because soil resources tend to be more limiting.

Seedling aboveground biomass was closely correlated with both seedling height and stem diameter ($r > 0.86$ in all cases), so we used the summation of either heights or stem diameters of neighboring seedlings (those defining the Thiessen polygons) as surrogates for the size of potential conspecific competitors. Contrary to expectations, we found no evidence that size of target seedlings on plots having intraspecific competitors was negatively correlated with the size of potential competitors. In fact, the only significant correlations were positive (target seedling height versus sum of neighbor heights, $P < 0.001$; target seedling diameter versus sum of neighbor diameters, $P = 0.003$), indicating, for those two cases at site 8, that individual seedlings tended to occur in neighborhoods of similar-sized seedlings.

EFFECTS OF MICROSCALE HETEROGENEITY ON GROWTH OF LODGEPOLE PINE SEEDLINGS

Studies that have addressed the importance of local crowding on growth typically have found that the local competitive environment often accounts for only a relatively small fraction of observed variability in size (for example, Mead 1966, Mack and Harper 1977, Watkinson and others 1983, Mithen and others 1984, Eissenstat and Caldwell 1988). If environmental conditions were homogeneous with respect to availabilities of light, water, and nutrients, differences in seedling performance presumably would reflect differences in the competitive environment and genetic variation in growth potential and among individuals. That we did not find close correlations between size of seedlings and various parameters thought to index the intensity of competition suggests that microscale environmental heterogeneity may be a critical factor affecting early seedling growth. Of course, we do not know the extent to which individual seedlings may vary in genetic potential for rapid growth. We cannot rule out the possibility that many seedlings in a local neighborhood may be at least half-siblings with similar genetic potentials. Nevertheless, the fact that size of target seedlings was, in some cases, positively correlated with the size of their closest intraspecific neighbors indicates that local differences in environmental conditions, such as soil fertility or shading by standing dead trees, may be very important.

Other studies also have found that neighbors of large plants tend to be large, and vice versa (Reed and Burkhart 1985, Kenkel and others 1989), but such autocorrelations are thought to result from self-thinning in response to competition for light, which selectively removes smaller trees and leaves the relatively larger trees in a local patch. This results in a positive autocorrelation because the only small trees that survive the self-thinning are those surrounded by other small trees (Kenkel and others 1989). It is clear, however, that competition for light and self-thinning cannot explain the positive correlation between size of target seedlings and their neighbors in this study. Exposure of local groups of seedlings to similar environmental conditions in a heterogeneous environmental matrix appears to be the most plausible explanation.

Although we found that competition among seedlings can be intense and that removal of intraspecific neighbors will result in substantial gains in size of the remaining target seedlings, our data suggest that competitive effects are secondary to the effects of fine-scale heterogeneity in resource availability and, perhaps, to genetic variability among seedlings. Which seedlings become stand dominants is not simply a consequence of available area or proximity and size of neighbors. Instead, it appears that numerous factors interact to determine the fate of individual seedlings. Several of these factors are probably stochastic (for example, the proximity of a seedling to standing dead trees that may cast shade on it for several hours each day). If differential seedling performance and, ultimately, survivorship are largely consequences of local environmental conditions and, at least to some extent, the local competitive environment, both of which probably are highly stochastic, selection for genetically "superior" individuals may be precluded (Kenkel and others 1989). A seed that lands and germinates in a fertile microsite and that happens to have few neighbors may become a stand dominant and survive to contribute progeny to the next generation regardless of its genetic potential. On the other hand, a "superior" seedling may become established on an impoverished microsite or in a highly competitive neighborhood only to have its fate sealed by being overtopped. Thus chance may serve to maintain the genetic diversity within the population (Kenkel and others 1989).

We have documented the first decade of forest reestablishment following the Yellowstone fires of 1988. Our studies had three major components. First, we measured pine seedling densities at twelve sites representing a wide range of conditions of elevation, topography, and geologic substrate within Yellowstone National Park. We correlated pine seedling density with variables of fire severity, seedbed characteristics, and potential herbaceous competitors. Second, we measured the size and growth rates of individual pine seedlings, and correlated size and growth rate with the local density of potential competitors (both intraspecific and interspecific) and with general site conditions such as elevation. Third, we experimentally tested the effects of intra- and interspecific competition by removing potential competitors from the immediate vicinity of selected lodgepole pine seedlings.

Postfire densities of lodgepole pine seedlings (which are now large enough to be called saplings in most areas) vary enormously across the burned landscape, from less than one hundred to more than a million stems per hectare. Seedling density is a function primarily of seed availability immediately after the fire. Seed availability was highest in stands that contained a large proportion of serotinous trees at the time of the fire, and it is in these stands that we see extremely high postfire seedling densities. For any given level of serotiny, stands subjected to crown fires had lower postfire seedling densities than stands in which canopy needles and twigs were not consumed, probably because of greater seed mortality in the more severe burns. Our data suggest that a severe crown fire will destroy 80–88 percent of the seed stored in a lodgepole forest canopy. Despite striking variability in seedling density, every stand that we sampled had adequate seedling densities to restock the stand at densities similar to or higher than the prefire stand density. These results underscore the importance of fire in shaping the structure of lodgepole pine forests of the northern Rocky Mountains (Chapters 2, 3, 12, 14).

Unexpected was our finding that interspecific competitors (herbs and shrubs) had little effect on pine seedling density throughout the first decade after the fires. This is contrary to some previous studies that suggest that competitors, especially grasses, can limit establishment of pine seedlings after fire. We also were surprised to find that lodgepole pine seedlings established readily on litter and duff as well as on mineral soil, contrary to silvicultural literature that emphasizes the importance of a mineral seedbed for optimal establishment of lodgepole pine seedlings.

Although interspecific competition was relatively unimportant during the first decade of reforestation, intraspecific competition had a substantial influence on size and growth rate of individual pine seedlings. The experimental studies, in particular, showed that removal of all other pine seedlings within a 0.3-m radius led to as much as a doubling of seedling height and diameter after five years (Fig. 4.8). Thus even though seedling density was greater in areas of moderate fire severity, the size and growth rates of seedlings were greater in areas of crown fire. Both surveys and experimental studies showed that competition from herbs and shrubs had a minor effect on lodgepole seedling size and growth rate. Lodgepole seedling growth rates probably are also influenced by fine-scale heterogeneity in soil and light conditions, and perhaps also by genetic differences among seedlings. However, additional research is needed to quantify these effects.

Despite the significant influence of intraspecific competition on growth of lodgepole seedlings, competition has not yet been severe enough to cause any substantial density-dependent mortality in the pine seedlings. Mortality rates throughout the first decade after the fire have been uniformly low—even in stands with more than a million pine seedlings per hectare!

Our research has documented some of the major spatial patterns, and has elucidated some of the major ecological processes controlling those patterns during the first decade of forest reestablishment following the extensive Yellowstone fires of 1988. This work provides a basis for continued, long-term study of post-fire forest development within a wilderness landscape that is shaped primarily by natural processes with minimal influence of human activities. One of the important questions to be addressed during the next couple of decades is when substantial density-dependent mortality will begin to occur. It will have to begin soon, because no mature stand in the Yellowstone region supports densities anywhere near the high densities that we have documented (W. Romme, personal observation). Perhaps the small, crowded saplings will hold out until a drought year creates additional stress, or perhaps the extremely dense stands will reach a threshold in total biomass after which self-thinning begins. Related to the question of when and how self-thinning begins is the question of when and how the eventual stand dominants will begin to emerge. It appears that a dominance hierarchy is beginning to emerge in the stands of low to moderate density, where some saplings are somewhat larger than their neighbors. In contrast, no hierarchy may develop in extremely dense stands until substantial self-thinning begins to occur.

The permanent plots that we have established in twelve diverse sites across the Yellowstone Plateau are well marked and should be relatively easy to locate for many decades to come. Ideally, they will continue to be a source of long-term data well into the future—data that will provide new insights into the natural processes of forest reestablishment following large infrequent disturbances like the 1988 Yellowstone fires.

References

Alexander, R. R., and C. B. Edminster. 1980. Lodgepole pine management in the central Rocky Mountains. J. For. 78:196–201.

Anderson, J. E., and W. H. Romme. 1991. Initial floristics in lodgepole pine *(Pinus contorta)* forests following the 1988 Yellowstone fires. Int. J. Wildl. Fire 1:119–124.

Arno, S. F. 1980. Forest fire history in the northern Rockies. J. For. 78:460–465.

Arnott, J. T. 1973. Germination and seedling establishment. Environment Canada, Canadian For. Ser. Pub. no. 1339.

Baker, F. S. 1949. A revised tolerance table. J. For. 47:179–181.

Barbour, M. G., J. H. Burk, and W. D. Pitts. 1980. Terrestrial plant ecology, 2d ed. Benjamin/Cummings, Menlo Park, Calif.

Barrett, S. W. 1994. Fire regimes on andesitic mountain terrain in northeastern Yellowstone National Park, Wyoming. Int. J. Wildl. Fire 4:65–76.

Brown, J. K. 1975. Fire cycles and community dynamics in lodgepole pine forests. Pages 429–456 in D. M. Baumgartner, ed., Management of lodgepole pine ecosystems. Coop. Ext. Serv., Washington State Univ., Pullman.

Christensen, N. L., J. K. Agee, P. F. Brussard, J. Hughes, D. H. Knight, G. W. Minshall, J. M. Peek, S. J. Pyne, F. J. Swanson, J. W. Thomas, S. Wells, S. E. Williams, and H. A. Wright. 1989. Interpreting the Yellowstone fires of 1988. BioScience 39:678–685.

Clements, F. E. 1910. The life history of lodgepole burn forests. USDA For. Serv. Bull. 79. Government Printing Office, Washington, D.C.

Coates, K. D., W. H. Emmingham, and S. R. Radosevich. 1991. Conifer-seedling success and microclimate at different levels of herb and shrub cover in a *Rhododendron-Vaccinium-Menziesia* community of south central British Columbia. Canadian J. For. Res. 21:858–866.

Cochran, P. H. 1970. Seeding ponderosa pine. Pages 28–35 in R. K. Hermann, ed., Regeneration of ponderosa pine. Symposium proceedings, School of Forestry, Oregon State Univ., Corvallis.

Despain, D. G. 1987. The two climates of Yellowstone National Park. Proceedings of the Mont. Acad. Sci. 47:11–20.

———. 1990. Yellowstone vegetation. Roberts Rinehart, Boulder, Colo.

Despain, D. G., D. L. Clark, and J. J. Reardon. 1996. Simulation of crown fire effects on canopy seed bank in lodgepole pine. Int. J. Wildl. Fire 6:45–49.

Drivas, E. P., and R. L. Everett. 1988. Water relations characteristics of competing singleleaf pinyon pine seedlings and sagebrush nurse plants. For. Ecol. Mgmt. 23:27–37.

Eis, S. 1970. Root growth relationships of juvenile white spruce, alpine fir, and lodgepole pine on three soils in the interior of British Columbia. Canadian Dept. Fish and For., Canadian For. Serv. Publication no. 1276.

Eissenstat, D. M., and M. M. Caldwell. 1988. Competitive ability is linked to rates of water extraction: A field study of two aridland tussock grasses. Oecologia 75:1–7.

Ellis, M. 1993. Factors affecting establishment and performance by lodgepole pine following the 1988 fires in Yellowstone National Park. M.S. thesis, Idaho State Univ., Pocatello.

Ellis, M., C. D. von Dohlen, J. E. Anderson, and W. H. Romme. 1994. Some important factors affecting density of lodgepole pine seedlings following the 1988 Yellowstone fires. Pages 139–150 in D. G. Despain, ed., Plants and their environments: Proceedings of the first biennial scientific conference on the Greater Yellowstone Ecosystem. U. S. Dept. Int., Nat. Park Serv., Denver.

Fahey, T. J., and D. H. Knight. 1986. Lodgepole pine ecosystems. BioScience 36:610–616.

Floyd, D. A., and J. E. Anderson. 1982. A new point frame for estimating cover of vegetation. Vegetatio 50:185–186.

Fowells, H. A. 1965. Silvics of forest trees of the U. S. Agric. Hndbk. no. 271, USDA, Washington, D.C.

Habeck, J. R., and R. W. Mutch. 1973. Fire dependent forests in the northern Rocky Mountains. Quaternary Res. 3:408–424.

Harper, J. L. 1977. Population biology of plants. Academic Press, London.

Hellum, A. K., and B. S. P. Wang. 1985. Lodgepole pine seed: Seed characteristics, handling, and use. Pages 187–197 in D. M. Baumgartner, R. G. Krebill, J. T. Arnott, and G. F. Westman, eds., Lodgepole pine: The species and its management. Coop. Ext. Serv., Washington State Univ., Pullman.

Horton, K. W. 1953. Causes of variation in the stocking of lodgepole pine regeneration following fire. Canada Department of Northern Affairs and Natural Resources, For. Res. Div., Silviculture Leaflet no. 95.

Houston, D. B. 1973. Wildfires in northern Yellowstone Park. Ecology 54:1111–1117.

Johnson, E. A. 1992. Fire and vegetation dynamics. Cambridge Univ. Press, Cambridge.

Johnson, E. A., and S. L. Gutsell. 1993. Heat budget and fire behaviour associated with the opening of serotinous cones in 2 *Pinus* species. J. Vege. Sci. 4:745–750.

Johnson, E. A., and D. R. Wowchuk. 1993. Wildfires in the southern Canadian Rocky Mountains and their relationship to mid-tropospheric anomalies. Canadian J. For. Res. 23:1213–1222.

Keane, M. G., and G. F. Weetman. 1987. Leaf area-sapwood cross-sectional relationships in repressed stands of lodgepole pine. Canadian J. For. Res. 17:205–209.

Kenkel, N. L., J. A. Hoskins, and W. D. Hoskins. 1989. Local competition in a naturally established jack pine stand. Canadian J. For. Res. 76:2630–2635.

Knapp, A. K., and J. E. Anderson. 1980. Effect of heat on germination of seeds from serotinous lodgepole pine cones. Am. Midl. Nat. 104:370–372.

Knight, D. H., and L. L. Wallace. 1989. The Yellowstone fires: Issues in landscape ecology. BioScience 39:700–706.

LaMont, B. B., D. C. Le Maitre, R. M. Cowling, and N. J. Enright. 1991. Canopy seed storage on woody plants. Bot. Rev. 57:277–317.

LaRoi, G. H., and R. J. Hnatiuk. 1980. The *Pinus contorta* forest of Banff and Jasper National Parks: A study in comparative synecology and syntaxonomy. Ecol. Monog. 50:1–29.

Liddle, M. J., C. S. J. Budd, and M. J. Hutchings. 1982. Population dynamics and neighbourhood effects in establishing swards of *Festuca rubra.* Oikos 38:52–59.

Lieffers, V. J., S. E. MacDonald, and E. H. Hogg. 1993. Ecology of and control strategies for *Calamagrostis canadensis* in boreal forest sites. Canadian J. For. Res. 23:2070–2077.

Lotan, J. E. 1967. Cone serotiny of lodgepole pine near West Yellowstone, Montana. For. Sci. 13:55–59.

———. 1968. Cone serotiny of lodgepole pine near Island Park, Idaho. USDA For. Serv. Res. Paper INT-52. Intermountain Forest and Range Experiment Station, Ogden, Utah.

———. 1975. The role of cone serotiny in lodgepole pine forests. Pages 471–495 in D. M. Baumgartner, ed., Management of lodgepole pine ecosystems, vol. 1. Coop. Ext. Serv., Washington State Univ., Pullman.

Lotan, J. E., and D. A. Perry. 1983. Ecology and regeneration of lodgepole pine. Agriculture Handbook 606. U.S. Department of Agriculture, Washington, D.C.

Lyon, L. F. 1976. Vegetal development on the Sleeping Child burn in western Montana, 1961–1973. USDA For. Serv. Res. Paper INT-184. Intermountain Forest and Range Experiment Station, Ogden, Utah.

———. 1984. The Sleeping Child burn: 21 years of postfire change. USDA For. Serv. Res. Paper INT-330. Intermountain Forest and Range Experiment Station, Ogden, Utah.

Lyon, L. F., and P. F. Stickney. 1976. Early vegetal succession following large northern Rocky Mountain wildfires. Pages 355–375 in Proceedings, Montana tall timbers fire ecology conference and fire and land management symposium no. 14, vol. 14.

Mack, R. N., and J. L. Harper. 1977. Interference in dune annuals: Spatial pattern and neighbourhood effects. J. Ecol. 65:345–363.

Marston, R. A., and J. E. Anderson. 1991. Watersheds and vegetation of the Greater Yellowstone Ecosystem. Conserv. Biol. 5:338–346.

Matlack, G. R., and J. L. Harper. 1986. Spatial distribution and the performance of individual plants in a natural population of *Silene dioica.* Oecologia 70:121–127.

McCune, B. 1988. Ecological diversity in North American pines. Am. J. Bot. 75:353–368.

Mead, R. 1966. A relationship between individual plant-spacing and yield. Ann. Bot. 30:301–309.

Mithen, R., J. L. Harper, and J. Weiner. 1984. Growth and mortality of individual plants as a function of "available area." Oecologia 62:57–60.

Moir, W. H. 1969. The lodgepole pine zone in Colorado. Am. Midl. Nat. 81:87–98.

Muir, P. S., and J. E. Lotan. 1985a. Disturbance history and serotiny in *Pinus contorta* in western Montana. Ecology 66:1658–1668.

———. 1985b. Serotiny and life history of *Pinus contorta* var. *latifolia.* Canadian J. Bot. 63:938–945.

Noble, D. L., and R. R. Alexander. 1977. Environmental factors affecting natural regeneration of Engelmann spruce in the central Rocky Mountains. For. Sci. 23:420–429.

Oren, R., R. H. Waring, S. G. Stafford, and J. W. Barrett. 1987. Twenty-four years of ponderosa pine growth in relation to canopy leaf area and understory competition. For. Sci. 33:538–547.

Parker, A. J., and K. C. Parker. 1983. Comparative successional roles of trembling aspen and lodgepole pine in the southern Rocky Mountains. Great Basin Nat. 43:447–455.

Peet, R. K. 1981. Forest vegetation of the Colorado Front Range: Composition and dynamics. Vegetatio 45:3–75.

Perry, D. A., and J. E. Lotan. 1977. Opening temperatures in serotinous cones of lodgepole pine. USDA Forest Service Research Note INT-228. Intermountain Forest and Range Experiment Station, Ogden, Utah.

———. 1979. A model of fire selection for serotiny in lodgepole pine. Evolution 33:958–968.

Powell, G. W., M. D. Pitt, and B. M. Wikeem. 1994. Effect of forage seeding on early growth and survival of lodgepole pine. J. Range Mgmt. 47:379–384.

Reed, D. D., and H. E. Burkhart. 1985. Spatial autocorrelation in individual tree characteristics in loblolly pine stands. For. Sci. 31:575–587.

Romme, W. H. 1982. Fire and landscape diversity in subalpine forests of Yellowstone National Park. Ecol. Monog. 52:199–221.

Romme, W. H., and D. G. Despain. 1989. Historical perspectives on the Yellowstone fires of 1988. BioScience 39:695–699.

Romme, W. H., M. G. Turner, R. H. Gardner, W. W. Hargrove, G. Tuskan, D. G. Despain, and R. A. Renkin. 1997. A rare episode of sexual reproduction in aspen *(Populus tremuloides Michx.)* following the 1988 Yellowstone fires. Natural Areas J. 17:17–25.

Shainsky, L. J., and S. A. Radosevich. 1992. Mechanisms of competition between Douglas-fir and red alder seedlings. Ecology 73:30–45.

Stahelin, R. 1943. Factors influencing the natural restocking of high altitude burns by coniferous trees in the central Rocky Mountains. Ecology 24:19–30.

Stermitz, J. E., M. G. Klages, and J. E. Lotan. 1974. Soil characteristics influencing lodgepole pine regeneration near West Yellowstone, Montana. USDA For. Serv. Res. Paper INT-163. Intermountain Forest and Range Experiment Station, Ogden, Utah.

Stuart, J. D., J. K. Agee, and R. I. Gara. 1989. Lodgepole pine regeneration in an old, self-perpetuating forest in south central Oregon. Canadian J. For. Res. 19:1096–1104.

Tinker, D. B., W. H. Romme, W. W. Hargrove, R. H. Gardner, and M. G. Turner. 1994. Landscape-scale heterogeneity in lodgepole pine serotiny. Canadian J. For. Res. 24:897–903.

Turner, M. G., W. W. Hargrove, R. H. Gardner, and W. H. Romme. 1994. Effects of fire on landscape heterogeneity in Yellowstone National Park, Wyoming. J. Vege. Sci. 5:731–742.

Turner, M., W. H. Romme, and R. H. Gardner. 2000. Prefire heterogeneity, fire severity, and early postfire plant reestablishment in subalpine forests of Yellowstone National Park, Wyoming. Int. J. Wildl. Fire 9:21–36.

Turner, M., W. H. Romme, R. H. Gardner, and W. H. Hargrove. 1997. Effects of fire size and pattern on early succession in Yellowstone National Park. Ecol. Monog. 76:411–433.

USGS. 1972. Geologic map of Yellowstone National Park. United States Geological Survey, Denver.

Watkinson, A. R., W. M. Lonsdale, and L. G. Firbank. 1983. A neighbourhood approach to self-thinning. Oecologia 61:334–336.

Wheeler, N. C., and W. B. Critchfield. 1985. The distribution and botanical characteristics of lodgepole pine: Biogeographic and management implications. Pages 1–13 in D. M. Baumgartner, R. G. Krebill, J. T. Arnott, and G. F. Weetman, eds., Lodgepole pine: The species and its management. Coop. Ext. Serv., Washington State Univ., Pullman.

Chapter 5 Fire Effects, Elk, and Ecosystem Resilience in Yellowstone's Sagebrush Grasslands

Benjamin F. Tracy

Both large grazing mammals (Frank and others 1999, McNaughton 1985, McNaughton and others 1989) and fire (Daubenmire 1968, Wright and Bailey 1982) strongly affect the functioning of grassland ecosystems. Most of the grasslands burned in 1988 represented important summer, transitional, and winter range for Yellowstone's elk and bison. Some postfire hypotheses suggested that the 1988 fires might increase forage quantity and quality for Yellowstone's ungulates and possibly increase rates of nutrient cycling (Singer and others 1989, Chapters 6, 14). This study was initiated to help answer some of these questions. Specifically, I wanted to learn whether fire affected aboveground net primary production (ANPP) and nutrient cycling in the grassland areas of the park. In addition, I wanted to see how grazing and waste deposition by ungulates interacted with burning effects.

Sites in summer, transitional, and winter ranges for elk were chosen for study. Data were collected on these locations from 1990 to 1993. The summer range site was located in the central portion of the park near Hayden Valley and was grazed mostly by bison from May to October. The transitional site was located near Swan Lake Flat and was grazed only during the spring by small numbers of elk (twenty to thirty). Located on the northern range near Hellroaring Creek, the winter range site was grazed mostly by elk from late October to May. Results and more detailed descriptions of these sites can be found in Tracy and McNaughton (1996). Because most fire effects had disappeared by 1992, an experimental burn was conducted in 1992 to study how fire affected ecosystem function in the short term. Postexperimental fire sampling was designed to address both hypotheses generated from previous data and observations made immediately after the experimental burn. Data collected in 1991 (Tracy and McNaughton 1996) showed that aboveground net primary production (ANPP), forage consumption by ungulates, and mineral nitrogen availability were higher in burned areas than in unburned areas. Although these effects had largely disappeared by the following year, I hypothesized that the same positive effects could be achieved if the site was burned again. I suspected that the significant fire effects were mostly short-lived. I also hypothesized that waste deposition by ungulates would interact with fire such that dung or urine deposited on recently burned soils would stimulate greater nitrogen mineralization than would dung or urine deposited on soils that were not recently burned. The experimental burn also produced new microsites that I hypothesized might influence soil nutrient availability and, indirectly, ANPP. The new microsites included: (1) "black holes" in the landscape created from intensely burned, mature sagebrush plants (Fig. 5.1) and (2) patches of lupines *(Lupinus sericeus),* which bloomed on the experimentally burned plots in midsummer. I hypothesized that both black holes and the lupine patches would possess greater concentrations of mineral nitrogen and more soil water than areas not associated with these microsites.

STUDY SITES

I conducted the experimental burn on Yellowstone's Northern Range, near Hellroaring Creek. A large inclining plateau characterizes this site at an elevation of ~2000 m. The site provides winter range for ungulates that graze the area usually from October to May. Elk *(Cervus elaphus)* are the most abundant large grazers on this site. Between one hundred and two hundred elk were usually observed grazing the site in the early spring. Small numbers of bison and pronghorn antelope ($n <$ 20) were observed using the winter range site as well. Annual temperature and precipitation for winter range are 3.8°C and 349 mm, respectively (Frank 1990). Cool-season grasses dominated vegetation composition on the site: *Festuca idahoensis, Agropyron spicatum, Agropyron caninum, Danthonia unispicata, Poa pratensis,* and *Stipa occidentalis.* Several sedge species *(Carex spp.)* also grew on the site. Sagebrush

Figure 5.1. Image of a "black hole" at the study site. The black hole is formed by an intense burning of mature sagebrush. Photo by Benjamin F. Tracy.

(Artemisia tridentata) was the dominant woody shrub. Dominant forbs included lupine *(Lupinus sericeus)*, sticky geranium *(Geranium viscosissimum)*, yarrow *(Achillea millefolium)*, and weedy milkvetch *(Astragalus miser)*. Plant nomenclature follows Hitchcock and Cronquist 1973. Soils on the study site were glacial tills derived from volcanic parent material and classified as sandy loams (Tracy 1996).

The winter range site originally burned from a backfire ignited to stop an approaching fire in late summer 1988. The experimental burn, ignited in 1992, burned mostly areas that had previously burned in 1988 (greater than 100 ha), and about 5 ha of previously unburned sagebrush grassland. The burn pattern allowed comparison of four adjacent areas each with a different fire history: (1) an area burned only in 1992 (B92), (2) an area burned in 1988 and 1992 (B8892), (3) an area burned only in 1988 (B88), and (4) an area that had no recent fire history (UB). In each area, sampling was confined within a 0.5-ha plot. All four plots were on the same topographic bench and were within 50 m of each other. Elk had equal access to all plots.

METHODS

Immediately before the experimental burn, standing crop was measured by clipping vegetation within three quadrats (50 × 50 cm) that were randomly placed within the areas to be burned. Net aboveground production and consumption of vegetation on plots were measured using small movable exclosures as described in Tracy and McNaughton 1997. Nitrate and ammonium availability in soils were measured by taking a soil core (3.8 × 10 cm) and then immediately extracting the soil with 1M KCl. I measured soil moisture from these cores gravimetrically after drying for 48 hr at 65°C. Nitrate and ammonium concentrations in KCl extracts were analyzed using a Lachat Quikchem Autoanalyzer (Milwaukee).

I sampled three microsites in this study for soil nutrients, black holes, dung and urine patches, and lupine patches. For each microsite sampled, an adjacent control plot was also sampled. The intense burning of the mature sagebrush on the B92 plot created visually striking "black holes" within the surrounding grassland. A typical black hole is circular, about 1 m in diameter, and consists of a thin layer of black ash covering mineral soil. A charred sagebrush stump near the middle of a black hole usually remains after fire. Black holes created by the 1988 fire were still visible in 1993 because colonization of the black holes by grasses and forbs was slow relative to the vegetation of adjacent burned areas. At each plot, I took soil cores beneath sagebrush. Of course, the black hole sampling took place where sagebrush was previously located. Aboveground vegetation beside each black hole was harvested in 25 × 25 cm plots. The vegetation was dried, weighed, and analyzed for N content using a Carlo Erba CNS autoanalyzer.

Because elk grazed the experimental burn site intensely during the first month after snowmelt (mid-May to mid-June), I decided that this would be the best time to evaluate the effects of fresh dung and ungulate urine deposition on soil nutrient availability. Four fresh elk dung piles were selected at each plot. Visibly moist dung was considered "fresh." I measured net N mineralization below fresh dung piles and artificial urine patches. Mineralization tubes—open-topped tubes made of PVC plastic—were placed below dung piles, and I took care to replace the fresh dung over the min tube. I saved a subsample of the dung for future nutrient analyses. Artificial ungulate urine (equivalent to 40 g N/m^2) was applied to the four plots within PVC rings (506 cm^2) to avoid runoff. Artificial urine patches (4 per plot) were created inside the temporary exclosures so that ungulate activity (for example, grazing, additional urination) would not affect the urine patches. Urine patches were at least 1 m from where ANPP measurements were taken inside the exclosure. I measured net N mineralization within urine patches and beneath dung piles over a 30-d period. Net N mineralization was measured using an *in situ* method (Raison and others 1987) employing min tubes. Briefly, this method involved inserting min tubes (3.8 × 10 cm) flush with the ground, such that the plastic served as a wall to keep out plant roots from the incubating soil. Net mineralization was considered the difference in extractable nitrogen from an initial core and extractable N at the end of the incubation period. At the end of the 30-d period the aboveground vegetation over each urine patch was harvested inside a 25 × 25 cm quadrat. The vegetation was dried, weighed, and analyzed for N content using a Carlo Erba CNS autoanalyzer.

The lupines bloomed in distinct patches (0.5–3 m in diameter) at the Hellroaring site. I took soil samples in the middle of randomly chosen lupine patches (four per plot). Aboveground lupine biomass was measured by clipping lupines in randomly located quadrats (50 × 50 cm). Lupine biomass was then dried and weighed. I clipped eight quadrats in each four plots. Besides soil samples taken inside lupine patches, soil samples were taken in control and sagebrush locations simultaneously. Artificial urine patches created the first month could not be effectively resampled by the third month after snowmelt.

To summarize the sampling, soil water and mineral N availability were measured in a total of five different locations: (1) in control regions that contained no sagebrush, lupine, or ungulate waste, (2) beneath existing or burned-up sagebrush, (3) within patches of lupine, (4) under fresh elk dung, and (5) within artificial urine patches.

RESULTS

Although the experimental burn reduced aboveground plant biomass almost to nothing by May 1993, vegetation rapidly recovered to preburn levels by July (Fig. 5.2). In fact, aboveground production on recently burned plots exceeded production on unburned plots by approximately 25 percent. Consumption of aboveground biomass by elk was low on recently burned sites (Fig. 5.3). Elk consumed the same amount of aboveground production, about 20 percent, from recently burned areas and unburned areas during 1993 (Tracy and McNaughton 1997). Compared with recently burned plots in 1991 (b88), consumption on areas that were burned in the experimental fire was almost 30 percent lower (Tracy and McNaughton 1996).

The fire stimulated production of lupine. Lupine production on burned areas, including the area burned in 1988, was more than 60 percent greater than lupine production on the unburned area (Tracy and McNaughton 1997). Lupine patches also had significant effects on soil microenvironments. Over the growing season, soil moisture beneath lupine patches was 15 percent greater than

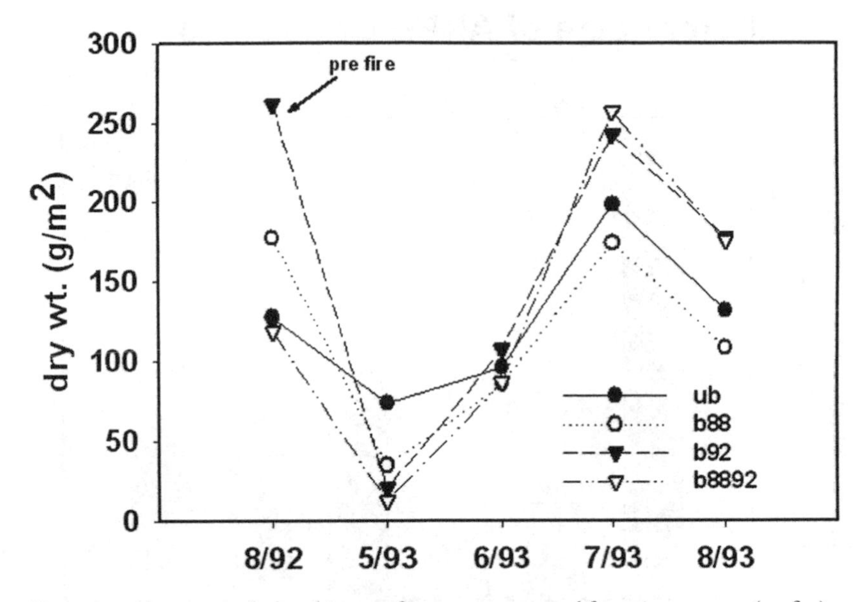

Figure 5.2. Aboveground plant biomass for grasses measured from August 1992 (prefire) to August 1993 at the Hellroaring site. Abbreviations for the study plots are unburned (ub), burned in 1988 (b88), burned in 1992 (b92), and burned both in 1992 and 1988 (b8892).

control areas and 25 percent greater than soil moisture associated with sagebrush patches. The same trend was found for extractable ammonium (Tracy 1996).

The burning of sagebrush and creation of black holes may have affected the distribution of soil nutrients, particularly in the area burned once in 1992. Soil moisture was between 15 and 25 percent higher in control areas than in black holes (Fig. 5.4a). In the unburned area (ub) that still contained large sagebrush, mean soil moisture over the growing season was the same under sagebrush and control areas (Fig. 5.4a). Mineral nitrogen accumulated in new black holes, compared with control areas (Fig. 5.4b). On the area that burned in 1992 (b92), black holes contained almost 70 percent more soil nitrate than control areas. Older black holes, on the b88 and b8892 plots, did not exhibit this pattern of nutrient accumulation. The variation in nitrate concentration within b92 black holes was also higher than microsites sampled on the other three areas (Fig. 5.4b). The nitrogen content of plants growing next to these patches showed the same pattern as soil nitrate values (Fig. 5.4c).

Artificial urine application to burned soils stimulated aboveground production more than urine application to unburned soil (Fig. 5.5). Net nitrification in urine patches showed the same pattern, with rates in burned areas 50–60 per-

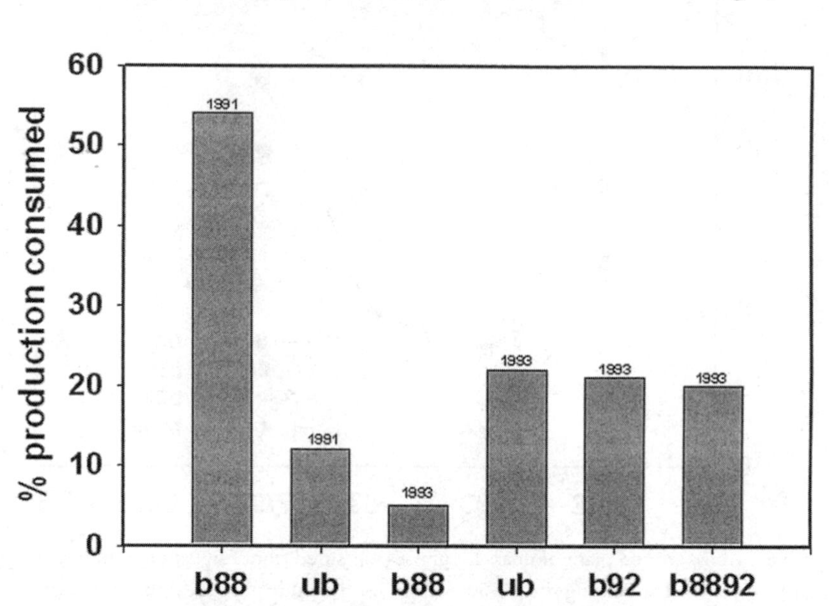

Figure 5.3. Percentage of total annual net primary production (ANPP) consumed by elk at the Hellroaring winter range site. Data were taken from the 1991 growing season and the 1993 growing season following the experimental burn. No differences between burned and unburned sites were noted for 1992. Plot abbreviations are as in Figure 5.2.

cent higher than rates in unburned areas. Net N mineralization exhibited no distinct differences when I compared rates beneath dung piles with control samples.

DISCUSSION

Herbaceous aboveground production rapidly returned to preburn levels after the experimental burn. Burning stimulated grass growth and increased sprouting of lupine from adventitious buds along horizontal roots (Tracy and McNaughton 1997). The result was greater herbaceous aboveground biomass on burned areas than on the unburned areas. Peak production occurred later in the growing season compared with the years before 1993. The later peak in biomass production was probably related to the cold spring of 1993 and, possibly, fire damage to the bunchgrasses growing on the burned plots (Blaisdell 1953, Conrad and Poultan 1966). The bunchgrasses, such as Idaho fescue, may have

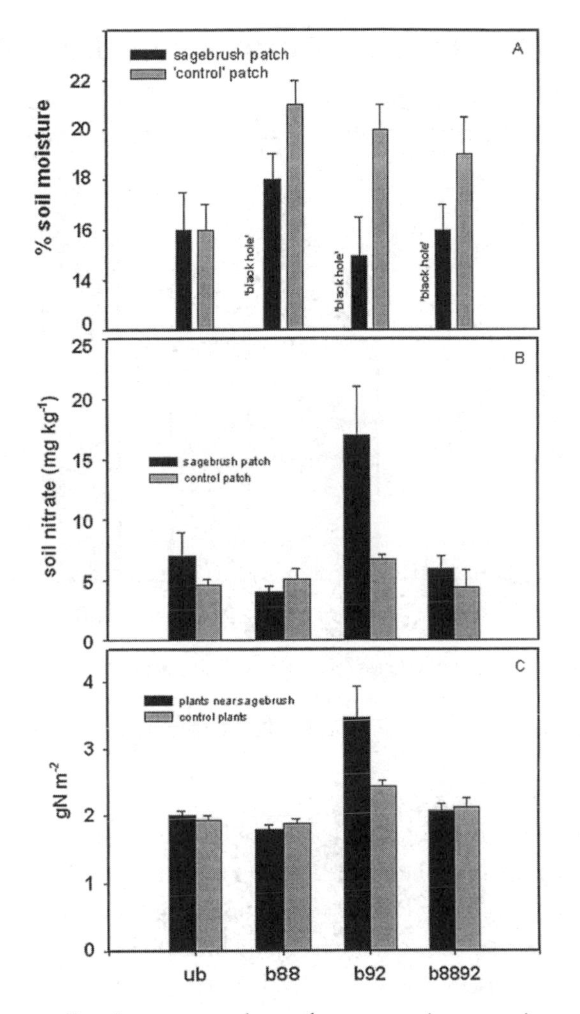

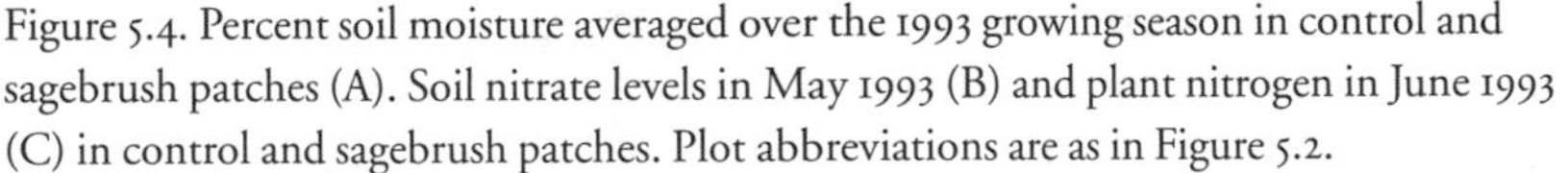
Figure 5.4. Percent soil moisture averaged over the 1993 growing season in control and sagebrush patches (A). Soil nitrate levels in May 1993 (B) and plant nitrogen in June 1993 (C) in control and sagebrush patches. Plot abbreviations are as in Figure 5.2.

needed four to six weeks in the early spring to recover from fire damage before they restored active growth.

Grazing by large herbivores can increase aboveground production of Yellowstone grasses (Frank and McNaughton 1992, Frank and McNaughton 1993). In other semiarid grasslands, grazing sometimes reduces grass growth, particularly after a fire (Wright and Bailey 1982:170). For example, Jirik and Bunting (1994) found that grazing reduced vigor and caused mortality of *Agropyron spicatum* the first year after a burn. Such negative responses to burning are the reason it is

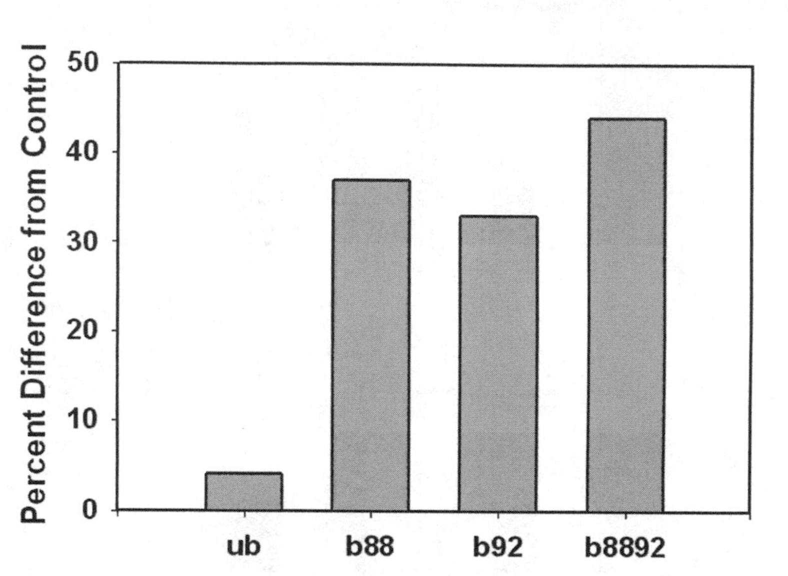

Figure 5.5. Aboveground plant biomass measured in artificial urine patches one month after urine application in 1993. Values are percent difference in plant biomass between urine patches and control areas next to urine patches. Plot abbreviations are as in Figure 5.2.

often recommended that rangelands be "rested" for one or two years before grazing is resumed. Elk grazing, however, did not affect the production of the burned regrowth in 1993. If grazing had any effect on the production of grasses in the experimentally burned areas, the effect was probably positive. Although the grass regrowth on burned plots was highly concentrated in nutrients (Tracy and McNaughton 1997), elk consumed only about 20 percent of the aboveground production and did not discriminate between burned and unburned areas. Compared with 1993, elk consumed 40–50 percent more of the aboveground production on burned plots in 1991 and 1992 (Tracy 1996, Tracy and McNaughton 1996). One reason elk may have avoided the nutritious regrowth in 1993 is the large lupine production on burned plots. It is well known that lupines accumulate alkaloids in their aboveground tissues and that herbivores will avoid eating such plants after smelling or tasting alkaloids (Waller and Nowacki 1978). Although elk movements are dependent on multiple factors, avoidance of unpalatable forbs may be one reason elk failed to eat more of the nutritious aboveground production than previous years.

The high precipitation during the 1993 growing season undoubtedly contributed to the rapid recovery of aboveground production on the experimentally burned plots. Although burning can increase the productivity of semiarid grasslands, such increases are usually evident only during years of adequate precipitation (Wright 1974). The 1993 growing season was definitely adequate. In fact, the 1993 growing season was one of the wettest summers on record in the Yellowstone area. Burning can be harmful to plant growth in semiarid grasslands when precipitation is low (Redmann 1978, Redmann and others 1993). Low precipitation probably explains why most studies in western semiarid grasslands report negative or neutral responses of ANPP after fire (Conrad and Poultan 1966, Merrill and others 1994, Wilson and Shay 1990).

Other smaller patch scale processes may have influenced the recovery of aboveground production on burned plots. These patch scale processes may involve the presence of lupine patches, black holes, areas where large sagebrush once stood, and urine patches from ungulates.

LUPINE PATCHES

Lupine biomass increased after the experimental burn. Abundant lupine blooms also were noted after the 1988 fires (D. Despain, personal communication). Lupines can affect ecosystem function because most species fix nitrogen (Paul and Clark 1989:174). Not all plants in the Fabaceae, however, fix nitrogen in arid regions; many legumes exhibit little N fixation (Paul and Clark 1989). Lupines growing at the experimental burn site did not possess nodules (B. Tracy, personal observation), so their N-fixing rates may have been limited. Though no N fixation may have occurred through the lupine biomass, the lupine patches may have had other effects on nutrient cycling, and possibly ANPP.

The dense canopy created by the lupine patches likely helped increase soil moisture (Fig. 5.4a). Compared with control areas, the shade of the lupine canopy probably reduced evaporative losses from the soil surface, and this effect allowed more water to infiltrate. Plants growing within and next to lupine patches could benefit from this higher soil moisture. For example, after the eruption of Mount St. Helens, Morris and Wood (1989) found that two perennial forbs growing within lupine patches exhibited higher growth rates than did forbs in control treatments. Whether such positive interactions occurred in Yellowstone is unknown. However, accumulating evidence suggests that positive interactions among plants are fundamental characteristics of plant communities in harsher environments (Callaway 1995). Harsh conditions are commonplace during the hot, dry summers in Yellowstone's northern range, and, possibly, plants growing near lupine patches may benefit from an association with lupine. If this is the case, lupines might be considered "autogenic engineers" (Jones and others 1994) in this grassland ecosystem because their physical structure likely alters the environment and ecological processes.

"BLACK HOLES"

When a fire burns through sagebrush grassland, the sagebrush burns more intensely and more completely than the surrounding grasses. This action creates a black hole in the grassland. A typical black hole is circular, about 1 m in diameter, and consists of a thin layer of black ash covering mineral soil. A charred sagebrush stump near the middle of a black hole usually remains after fire. Black holes created by the 1988 fire were still visible in 1993 because grasses and forbs did not colonize the black holes. Organic acids probably accumulated under the burned sagebrush, and these chemicals may have retarded seed germination and colonization by grasses and forbs (Blank and others 1994).

Black holes contained less soil moisture than control regions over the 1993 growing season. Water infiltration in the black holes may have been retarded by hydrophobic materials produced when soils are intensely heated by the burning of shrubs (DeBano and others 1976). It is also possible that temporary removal of the large sagebrush plants by fire may have freed up available water for adjacent plants. If herbaceous plants were in competition with sagebrush for soil and water (Wright and Bailey 1982:170, Harniss and Murray 1973), then the burn may have benefited herbaceous plants growing near sagebrush. The burn killed almost all mature sagebrush in the b92 plot and damaged the sagebrush seedlings growing on the b8892 plot. In a related example, a fire completely removed sagebrush from a grassland in the Bridger-Teton National Forest near Yellowstone (Gruell 1980). Herbaceous plants produced more biomass following this fire. The higher production was attributed to the greater availability of soil moisture and soil nutrients that mature sagebrush was using (Gruell 1980).

Nitrate, and other nutrients (Tracy 1996), accumulated in new black holes created by burning of large sagebrush plants. Why nitrate accumulated in the newly formed black holes is open to speculation. Plants growing next to the black holes, however, may have taken advantage of the high-concentration nitrate. Based on data from the older black holes, it appears that nutrient accumulation inside black holes does not persist, probably because of uptake from surrounding herbaceous vegetation. Regardless of the specific mechanisms, black holes certainly increase the spatial heterogeneity of soil nutrients horizontally. This fire effect may have important repercussions for influencing variation in aboveground production across the landscape.

WASTE DEPOSITION BY UNGULATES

When ungulates consume forage, they convert nonlabile nutrients to more labile forms that they deposit in their dung and urine. Ungulates can excrete 40–90 percent of the nutrients they ingest (VanSoest 1982), but little is known about how dung and urine deposition interacts with fire. The deposition of dung to burned soils did not affect net N mineralization or net immobilization. I did observe small increases in soil nitrate associated with dung application, but compared with the effects of urine addition, such increases are probably small. Overall, our data suggest that elk dung does not provide any substantial mineral nutrients to growing plants, at least over the short term.

On the other hand, deposition of urine supplies plants with much mineral N, and the supply of mineral N appears to be affected by burning. In this study, net nitrification was higher within urine patches on burned sites than on the unburned site. Artificial urine deposited on burned soils also increased the standing crop of herbaceous plants, while urine deposited on unburned soils had no effect. An interaction between nitrogen, phosphorus, and water availability (Cole and others 1977) in the soils of the experimentally burned plots may explain these results. Our results generally support some of those from the tallgrass prairie (Seastedt and others 1991). Their study showed that aboveground production on burned areas is more sensitive to N addition compared with unburned areas. The authors argued that burning makes soil more N limited compared with unburned areas, and this makes burned soils more sensitive to N addition. The same situation may exist in Yellowstone, but the issue remains equivocal. Whatever the mechanism, our data suggest that urine deposition on burned soils may be an important way in which aboveground production can recover rapidly from burning.

The interaction between ungulate grazing, fire, and soil nutrient content may be a mechanism that can yield longer-term changes in community structure and increase grassland heterogeneity. The addition of urine to burned soils may make the plants that grow on or adjacent to burned patches much more attractive to ungulate foragers, as was seen in tallgrass prairie (Steinauer and Collins 1995). This could be one mechanism whereby grazing lawns (McNaughton 1984) could become established in semiarid grasslands.

I found that effects from the 1988 fires on grassland ecosystem function, when measurable, were usually positive (for example, higher net primary production). These positive findings differ from most fire studies in semiarid systems, which generally report negative short-term effects from fire. The wet growing seasons following the 1988 fires may explain our overall positive results. The same results were found following an experimental fire initiated in 1992. The experimental fire created many patches of burned sagebrush (black holes) and lupine *(Lupinus sericeus),* which, indirectly, may influence how aboveground production responds to fire. For example, herbaceous plants may benefit from the high concentrations of soil nutrients (for example, nitrate) that accumulate in black holes. Additionally, lupine patches appear to improve soil moisture for plants associated with these patches. The lupine patches may have an additional benefit because they may deter elk from overusing burned areas and possibly damaging new grass shoots. Beyond these findings, I also found that ungulate urine deposited on burned soil stimulates greater aboveground production. The patchiness created by burned shrubs, forbs unpalatable to large herbivores, and urine inputs may contribute to the speed at which primary production in sagebrush grasslands recovers from fire. The importance of such small-scale processes, and the interaction between these processes and climatic effects, represent fertile ground for future fire-related research.

References

Blaisdell, J. P. 1953. Ecological effects of planned burning of sagebrush-grass range on the upper Snake River Plains. USDA Tech. Bull. 1075.

Blank, R. R., F. Allen, and J. A. Young. 1994. Extractable anions in soils following wildfire in a sagebrush-grass community. Soil Sci. Soc. Am. J. 58:564–570.

Callaway, R. M. 1995. Positive interactions among plants. Bot. Rev. 61:306–337.

Cole, C. V., G. S. Innis, and J. W. B. Stewart. 1977. Stimulation of phosphorus cycling in semiarid grasslands. Ecology 58:1–15.

Conrad, C. E., and C. E. Poultan. 1966. Effect of a wildfire on Idaho fescue and blue-bunch wheatgrass. J. Range Manage. 19:138–141.

Daubenmire, R. 1968. The ecology of fire in grasslands. Adv. Ecol. Res. 5:209–266.

DeBano, L. F., S. M. Savage, and D. A. Hamilton. 1976. The transfer of heat and hydrophobic substances during burning. Soil Sci. Soc. Am. J. 40:779–782.

Frank, D. A. 1990. Interactive ecology of plants, large mammalian herbivores, and drought in Yellowstone National Park. Ph.D. diss., Syracuse University, Syracuse, N.Y.

Frank, D. A., and S. J. McNaughton. 1992. The ecology of plants, large mammalian herbivores and drought in Yellowstone National Park. Ecology 73:2043–2058.

———. 1993. Evidence for the promotion of aboveground grassland production in Yellowstone National Park. Oecologia 96:157–161.

Frank, D. A., S. J. McNaughton, and B. F. Tracy. 1999. Functional similarities among the

Earth's grazing ecosystems: Comparing the Serengeti and Yellowstone National Park. Bioscience 48:513–521.

Gruell, G. E. 1980. Fire's influence on wildlife habitat on the Bridger-Teton National Forest, Wyoming. Vol. 2, Changes and causes, management implications. USDA For. Serv. Res. Pap. INT-252:35.

Harniss, R. O., and R. B. Murray. 1973. Thirty years of vegetal change following burning of sagebrush-grass range. J. Range Manage. 26:322–325.

Hitchcock, C. L., and A. Cronquist. 1973. Flora of the Pacific Northwest. University of Washington Press, Seattle.

Jirik, S. J., and S. C. Bunting. 1994. Post-fire defoliation response of *Agropyron spicatum* and *Sitanion hystrix.* Int. J. Wildland Fire 4:77–82.

Jones, C. G., J. H. Lawton, and M. Shachak. 1994. Organisms as ecosystem engineers. Oikos 69:373–386.

McNaughton, S. J. 1984. Grazing lawns: Animals in erds, plant form, and coevolution. Am. Nat. 124:863–886.

———. 1985. Ecology of a grazing ecosystem. Ecol. Monogr. 55:259–294.

McNaughton, S. J., M. Oesterheld, D. A. Frank, and K. J. Williams. 1989. Ecosystem-level patterns of primary productivity and herbivory in terrestrial habitats. Nature 341:142–144.

Merrill, E. H., N. L. Stanton, and J. C. Hak. 1994. Responses of bluebunch wheatgrass, Idaho fescue, and nematodes to ungulate grazing in Yellowstone National Park. Oikos 69:231–240.

Morris, W. F., and D. M. Wood. 1989. The role of lupine in succession on Mount St. Helens: Facilitation or inhibition? Ecology 70:697–703.

Paul, E. A., and F. E. Clark. 1989. Soil microbiology and biochemistry. Academic Press, San Diego.

Raison, R. J., M. J. Connell, and P. K. Khanna. 1987. Methodology for studying fluxes of soil mineral-N in situ. Plant Soil 51:73–108.

Redmann, R. E. 1978. Plant and soil water potentials following fire in a northern-mixed grassland. J. Range Manage. 31:443–445.

Redmann, R. E., J. T. Romo, and B. Pylypec. 1993. Impacts of burning on primary productivity of *Festuca* and *Stipa-Agropryron* grasslands in central Saskatchewan. Am. Mid. Nat. 130:262–273.

Seastedt, T. R., J. M. Briggs, and D. J. Gibson. 1991. Controls of nitrogen limitation in tallgrass prairie. Oecologia 87:72–79.

Singer, F. J., W. Schreier, J. Oppenheim, and E. O. Garton. 1989. Drought, fires, and large mammals. Bioscience 39:716–722.

Steinauer, E. M., and S. L. Collins. 1995. Effects of urine deposition on small-scale patch structure in prairie vegetation. Ecology 76:1195–1205.

Tracy, B. F. 1996. Fire effects in the grasslands of Yellowstone National Park. Ph.D. diss., Syracuse University, Syracuse, N.Y.

Tracy, B. F., and S. J. McNaughton. 1996. Comparative ecosystem processes in summer and winter ungulate ranges following the 1988 fires in Yellowstone National Park. Pages 181–191 in J. M. Greenlee, ed., The ecological implications of fire in Greater Yellowstone: Proceedings. International Association of Wildland Fire, Fairfield, Wash.

———. 1997. Elk grazing and vegetation responses following a late season fire in Yellowstone National Park. Plant Ecology 130:111–119.

VanSoest, P. J. 1982. Nutritional ecology of the ruminant. O and B, Corvallis, Ore.

Waller, G. R., and E. K. Nowacki. 1978. Alkaloid biology and metabolism in plants. Plenum, New York.

Wilson, S. D., and J. M. Shay. 1990. Competition, fire, and nutrients in a mixed grass prairie. Ecology 71:1959–1967.

Wright, H. A. 1974. Range burning. J. Range Manage. 27:5–11.

Wright, H. A., and A. W. Bailey. 1982. Fire ecology, United States and southern Canada. Wiley, New York.

Chapter 6 Elk Biology and Ecology Before and After the Yellowstone Fires of 1988

Francis J. Singer, Michael B. Coughenour, and Jack E. Norland

Elk occupy an intermediate position in postfire vegetal succession. Neither recent burns nor unburned continuous forests support the highest density of elk (Lyon 1966, Leege 1969, Martinka 1976). Martinka (1976) concluded that the highest densities of elk were found in a complex of multiaged conifer stands used for thermal and predator avoidance, intermixed across the landscape with previously burned bunchgrass and seral shrub communities that were used primarily for feeding. The fires of 1988 in Yellowstone National Park (YNP) provided a unique opportunity to study the effects of large-scale wildfire on one of the largest migratory elk populations in North America. Several large, hot, fall fires burned about one-third of the 590,000 ha summer range and about one-quarter of the 134,000 ha winter range for elk (Singer and others 1989).

Fire may benefit elk through increased forage quality (DeWitt and Derby 1955, Hobbs and Spowart 1984, Leege 1969, Rowland and others 1983, Chapter 5), increased forage quantity (West and Hassan 1985, Chapter 5), earlier green-up and later plant senescence (Daubenmire 1968, Peet and others 1975, Skovlin and others 1983, Hobbs and Spowart 1984), and increased foraging efficiency (Canon and others 1987). Elk strongly prefer forages from previously burned sites (Leege 1969, Nelson 1976, Davis 1977, Roppe and Hein 1978). Most of these prior studies of elk-fire relations were of relatively small, prescribed burns (tens of ha), and most were spring burns (Hobbs and Spowart 1984, Canon and others 1987). Other than Turner and others (1994) and Martinka (1976), few studies documented responses of elk to large-scale wildfires ignited during the more typical lightning season in later summer or autumn.

The generalization that fire benefits elk (Peek 1980) is supported by some but not all empirical studies. For example, herbaceous protein concentrations (and thus nitrogen concentrations) are sometimes enhanced following fire (Rowland and others 1983, Seip and Bunnell 1985, Singer and Mack 1993, Pearson and others 1995, Singer and Harter 1996) and sometimes are increased by only a few percentage points (Bendel 1974, Lyon and Stickney 1976, Hobbs and Spowart 1984). Protein enhancement may also disappear within only one to two years postfire (Dills 1970, Lloyd 1971, Lyon and Stickney 1976, Wood 1988).

The purpose of our investigation was to document the effects of the fires on elk habitat, survival, productivity, and dietary responses. We predicted that elk cow:calf ratios, recruitment, and elk population size would increase following the fires (Leege 1969, Christensen and others 1989, Boyce and Merrill 1991, Singer and others 1989, Chapters 13, 14) although increased mortality might occur immediately after the fires. We also predicted forage quality (in other words, concentrations of nitrogen, macronutrients, and forage digestibility) would increase in burned nonwoody plants (DeWitt and Derby 1955, Daubenmire 1968, Old 1969, Dills 1970, Lloyd 1971, Grelen and Whitaker 1973, Wilms and others 1981, Umoh and others 1982, Rowland and others 1983, Ohr and Bragg 1985, Chapter 5). We predicted that the 1988 fires would alter elk distribution and habitat preferences, due to these changes in forages and forest cover. For example, elk should avoid burned forests in the winter because these burned forests accumulate more snow than unburned forests and the snow will be more crusted from wind and diurnal thawing (Chapter 13). Thermal cover in burned forests will be reduced resulting in greater heat loss by elk (Meiman 1968, Jones 1974, Davis 1977). Burned grasslands may green up earlier in the spring than unburned grasslands due to blackened surfaces (Daubenmire 1968, Hobbs and Spowart 1984).

METHODS

Herbaceous Biomass Responses to the Fires

We estimated herbaceous biomass (in other words, all nonwoody growth) in paired burned and unburned key winter range plant communities on the northern range of YNP (Fig. 1.2). The winter range communities sampled included big sagebrush, Douglas fir, and dry bunchgrass. We separated our sampling of burned Douglas fir into those stands burned by cool fires (mostly understory) and those stands burned by hotter soil heating (canopy fires, residual burning of larger down fuels), following the soil heating mapping of Despain and others (1989). Two summer range communities were also sampled: lodgepole forest and wet meadow. In the grassland and shrub communities, we sampled fifteen randomly located 0.5 m^2 (circular) plots in both burned and unburned sites. Plots were sampled for peak herbaceous biomass in 1990 and 1991 using the point intercept method and regressions of biomass to hits established for the area by Frank and McNaughton (1990). We separated live from dead material. In forested communities, we clipped all herbaceous vegetation in 0.5 m^2 plots to ground level, then dried and weighed it. We clipped vegetation in the forest since the point intercept method was not extensively tested there (Frank and McNaughton 1990). At each random point, the placement of the plot frame was chosen by randomly selecting a direction and number of steps from the point. The sampling design included a two-stage stratified random sampling with four replicate burned/unburned sites for each vegetation type (primary units) and fifteen replicate biomass plots (secondary units in each burned treatment). Plots at each study site and burn treatment were compared with a blocked ANOVA (blocked by site).

Effects of the Fire on Plant Nutrient Concentration

Nutrient concentrations of samples of the live biomass of the three most common winter range grasses (*Pseudoroegnaria spicata, Festuca idahoensis,* and *Koeleria cristata* [bluebunch wheatgrass, Idaho fescue, and Junegrass, respectively]) were sampled in unburned and burned grassland (n = 12 bunchgrass clumps/species/year) in the fall season just prior to elk arriving onto the winter range in 1989 and 1990. Nitrogen concentration, fat fiber, Van Soest fiber, total ash, gross energy, Ca, P, K, Mn, and Mg concentrations were analyzed (Rangeland Ecosystem Science Department, Nutritional Analysis Laboratory, Colorado State University) according to the Association Office of Analytical Chemists (AOAC 1970). *In-vitro* dry matter digestibility (DMD, hereafter referred to as digestibility) was determined by the method of Tilley and Terry (1963) as modified by Pearson (1970). We additionally collected midwinter samples for forage quality (percent N and digestibility) in elk feeding craters (see also Chapter 13) in burned and unburned sites (Norland and others 1996) to document forage quality of what animals were actually feeding on. We also sampled summer range forages from the Central Plateau of the park (lodgepole, wet meadow types) in midsummer to coincide with the time elk were using these forages.

Samples of burned bark from aspen and lodgepole pine and burned needles from Engelmann spruce *(Picea engelmannii)*, lodgepole pine, and Douglas fir *(Pseudotsuga menziezii)* were collected from trees where elk were observed feeding on them. Similar samples from unburned trees were collected for comparison from as near as possible to the burned trees. Mean nutrients were compared with the t-test (Proc t-test, SAS), except for nutrient concentrations of spruce tree needles which were not normally distributed. Percent N was transformed with x/log(x), and digestibility was transformed with log(log(x)) for spruce.

Elk Diet Composition Before and After the Fires

Winter elk diets were determined by fecal analysis. Fresh fecal samples were collected at the same approximate fifteen locations from within the park during each of the six consecutive winters from 1985–1986 through 1990–1991, including three prefire winters and three postfire winters. Each sample was a composite of 5 g of fresh fecal material from 10–12-pellet groups. Botanical composition of each sample was expressed as a percentage of relative cover of identifiable plant fragments in two hundred random microscope fields (Washington State Univ., Wildlife Habitat Lab, Pullman). Elk diet samples were compared among years with a nonparametric test (SAS 1990) because neither the raw data nor a log transformation was normal. Only forage species that comprised > 5 percent of the elk diets in any single year were included in the analysis. Each plant species was analyzed separately. Tukey's procedure was used for all pairwise comparisons ($p \leq 0.05$).

Elk Population After the Fires

Aerial counts of elk were made from fixed-wing aircraft (Super Cub) and helicopter surveys during the winters of 1986–1993. During each fixed-wing survey, the northern elk range (about 1,340 km^2) was surveyed both inside and outside the park boundary. To facilitate counting, the entire winter range area was divided into sixty-six count units of 20.3 km^2 ± 1.3 km^2 (mean ± SE, range = 4–45 km^2). The boundaries of the count units consisted of rivers, creeks, and other readily recognizable topographic features. Four different aircraft teams (pilot and observer) surveyed about one-quarter each of the northern range during each survey (approx. seventeen count units per team). Aircraft teams always surveyed the entire area in one day. Each count unit was surveyed in elevation contours approximately 0.5 km apart.

To evaluate the efficiency of the fixed-wing surveys, we fitted forty-seven elk (thirty-four adult cows, twelve adult bulls, and one yearling bull) with white radiocollars (Samuel and others 1987). Radiocollared elk missed during the survey were located by a fifth aircraft, and the sighting conditions (pilot/observer experience, elk density strata [high, medium, low], location in and out of the park, snow and forest cover variables) were recorded.

The dichotomous classification of elk groups seen or missed was treated as the dependent variable in a logistic regression analysis (Samuel and others 1987, Singer and others 1989, Singer and Garton 1994). A logistic regression model following Samuel and others (1987) was used for predicting visibility y, where visibility probability was:

$$y = \frac{e^u}{1 + e^u} \qquad (1)$$

where u was the equation describing sightability and e was the base of the natural log. Elk population estimates were then corrected for each count unit by logistic regression of visibility factors found to be significant (density strata, location in or out of the park, snow and forest cover variables). Stepwise logistic regression suggested that visibility for elk groups was best explained by group size, vegetation cover, and activity according to:

$$u = 0.97 + 0.37\,G - 0.54\,C - 1.70\,A \qquad (2)$$

where G = group size, C = percent conifer canopy cover, and A = elk activity (Singer and Garton 1994, Unsworth and others 1994). Percent conifer canopy cover was defined as the percentage of cover that could obscure an elk over the area the group observed as described in the technique manual (Unsworth and others 1994).

Habitat Selection by Elk

We sampled elk habitat selection during ten fixed-wing surveys of the entire northern winter range conducted between December 1986 and March 1992 (one to three surveys per winter). The locations of all elk groups observed were plotted on a 1:62,500 topographic map, and group size, percent forest canopy, percent snow cover, and elk behavior (standing, bedded, moving) were recorded. Forest canopy that would obscure an elk from observation and snow cover were estimated visually for the area connecting the outermost elk in the group following Samuel and others (1987) and Unsworth and others (1994).

The aerial observations also sampled winter feeding habitats. Nearly all elk groups were feeding or bedded near feeding sites at the time of the surveys. We observed little tendency for elk within the park to move from feeding sites into tree cover until afternoon. Therefore, most surveys were conducted in the morning (before noon local time). We later generated elevation, slope, aspect, vegetative cover type, habitat type, and burn category (burned, unburned, mosaic burn) (Mattson and Despain 1984, Despain and others 1989) using the park's GIS. Availability of each habitat category was also generated from the GIS. We also analyzed habitat use from the helicopter surveys in order to document any differential habitat and burn area preference by elk sex and age class.

Confidence intervals on proportional elk use of all habitat and burn categories were compared to availability using Bonferroni confidence intervals (Neu and others 1974, Miller 1981, Byers and others 1984). The terms *selected, avoided,* and *expected* imply elk habitat use greater than (+), less than (−), or equal to (0) availability at $p < 0.10$ level, respectively. We conducted habitat preference tests on groups because individuals within a group were not independent, which would favor Type I errors (White and Garrott 1990). Ten habitat types recognized on the northern range were aggregated into four types for purposes of the preference test because each type must contain less than 5 percent of the total elk groups (Neu and others 1974). The five categories were: lodgepole pine forests, Douglas fir forests, other forests, sagebrush/grasslands, and wet grasslands dominated by *Deschampsia* and *Carex* spp. Differences in selection of habitat and burn categories amongst years and between sexes of elk were tested with a log likelihood chi-square procedure (SAS 1990). Fisher exact tests are reported for comparisons with small sample sizes (< 10).

RESULTS

Fire Effects on Herbaceous Biomass

In winter range communities, burning increased forb biomass in sagebrush the first growing season postfire (1989) and in Douglas fir forest and bunchgrass communities the second growing season (1990) after the fire (Fig. 6.1). Grass biomass was reduced by burning both years in big sagebrush communities, but grass biomass increased in the bunchgrass type. Considerably less biomass was

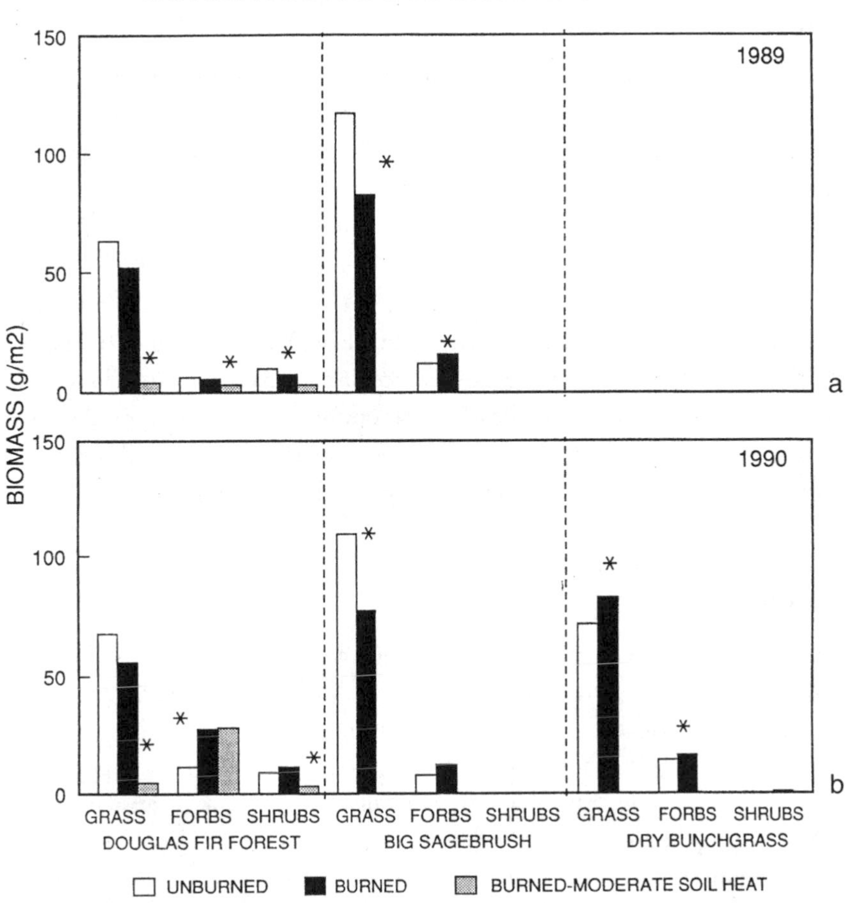

Figure 6.1. Aboveground dry matter plant biomass in unburned and burned (a) winter and (b) summer range plant communities in 1989 and 1990 on Yellowstone's northern elk winter range. Lodgepole 3 = a mature lodgepole forest and other communities as described in Mattson and Despain (1984). Differences between burned and unburned sites were tested with ANOVA * = $p < 0.05$.

produced in burned lodgepole pine forests on summer range the first and second postfire growing season, except for biomass of grasses and one forb, fireweed *(Epilobium augustifolium),* which were increased by burning ($p < 0.05$, Fig. 6.1a). Grass biomass production increased about 20 percent on burned wet meadows on summer range both the first and second postfire growing seasons. Other than fireweed, forb biomass in wet meadows was not influenced by burning (Fig. 6.1b).

Effects of the Fires on Snow Depth and Density

Snow depths did not differ between burned and unburned sagebrush or Douglas fir stands in either early or late winter ($p > 0.05$). But snow was 5 percent denser in burned compared with unburned sagebrush communities late during the second postfire winter (1989–1990), while snow was 6–7 percent denser early in both the second and third postfire winters in burned versus unburned ($p < 0.05$, Norland and others 1996). These minor differences in snow density did not affect elk feeding, since the density of elk feeding craters did not differ between either burned and unburned sagebrush or Douglas fir stands ($p > 0.05$, Norland and others 1996). But the differences in snowpack in burned lodgepole pine stands on the northern range where canopy burns were more typical in 1988 were apparently substantial enough to cause an avoidance of burned compared with unburned lodgepole pine stands (Singer and Harter 1996).

Effects of the Fires on Forage Quality in Elk Winter Range

Forage quality was generally higher on burned sites, in both the first and second postfire winter. Nitrogen concentrations of the three most common grass species averaged 32 percent higher in burned versus unburned habitats both winters ($p < 0.05$). Dry matter digestibility (DMD) was higher for all three species the first postfire growing season (1989), but only Idaho fescue possessed higher DMD during the second postfire growing season (1990).

Macronutrients were not higher on burned sites the first and second growing seasons following the fires. Percent K was reduced in Idaho fescue during the first growing season and percent Mn was reduced in Idaho fescue and Junegrass during the second growing season. We found no fire effects on Ca, P, or Mg. However, fibrous constituents (percent cellulose, lignin, fiber, and ash) were higher in forages on burned sites by the second growing season.

Effects of the Fires on Forage Quality in Elk Summer Range

Nitrogen concentration was higher on burned sites for all forages in lodgepole pine forests on summer range the first summer postfire ($p < 0.05$). No differences in nitrogen concentration were noted for forages in burned and unburned wet meadow communities. Differences in nitrogen concentration had completely disappeared by the third summer postfire. Very few differences were noted in digestibility with increases found in burned fireweed and other forbs in the lodgepole pine forests the first summer postfire. Again, all differences had disappeared by the third postfire summer (1991).

Effects of the Fires on Burned Bark Forages

Burned aspen bark had a lower nitrogen concentration ($p = 0.05$) and digestibility ($p = 0.07$) than unburned bark. Nitrogen concentration was not higher in burned Douglas fir needles ($p = 0.98$), but digestibility was marginally higher ($p = 0.07$) in the first postfire winter. Digestibility was lower in burned aspen bark ($p = 0.04$).

Elk Mortality During the Fires

A total of 243 elk were found dead from the fires within the park and another 84 elk within burned portions of the remainder of the Greater Yellowstone Area outside of the park. This represented about 0.7 percent of the elk present in the area at the time of the fires, based upon population estimates for eight elk herds that used the park each summer (Singer 1991).

Elk Mortality the First Winter Following the Fires

A total of 1,004 elk carcasses were counted during April of 1989 on the random sample (22 of 66) count units. Based on this sample, we estimated 3,021–5,757 elk (95 percent confidence interval) died over the entire winter range during the winter of 1988–1989. These elk died primarily from malnutrition (DelGiudice and Singer 1996, DelGiudice and others 1991). Another 2,773 elk were harvested outside of the park. The elk population was estimated to be 23,237 ± 834 at the beginning of the winter. Thus an estimated 24–37 percent of the northern Yellowstone elk population was lost during the first postfire winter due to the combination of winter malnutrition and harvest (Fig. 6.2).

Survival of adult cows and bulls averaged 0.91–1.00 prior to the fires (Fig. 6.2). Fifty-seven percent of all adult cow mortalities and 88 percent of all bull mortalities that were recorded during four years of monitoring occurred during the winter of 1988–1989. Adult bull survival was only half (0.46) the survival of adult cows (0.84) during the first postfire year. Adult cow survival rate did not vary statistically among years ($p > 0.05$), although numerically cow survival dropped to 0.87 during the first postfire winter (1988–1989). Adult bull survival

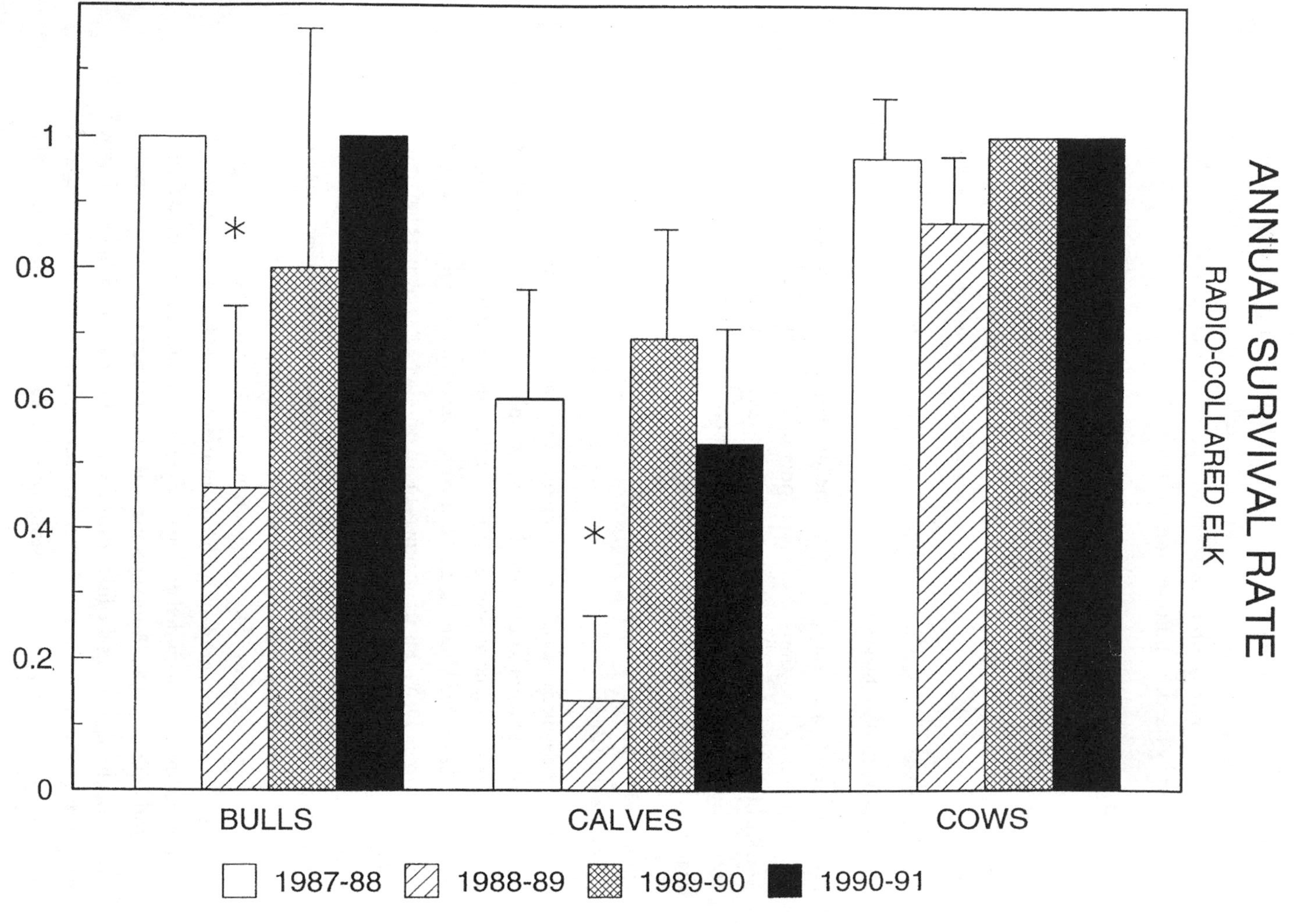

Figure 6.2. Annual survival rates (± SE) for adult cow ($n = 34$), adult bull ($n = 13$), and calf elk ($n = 127$) before and after the fires of 1988 in Yellowstone

was significantly lower over the winter of 1988–1989 (0.46) compared with the three other winters ($p < 0.05$). Calf survival was also lowest the first postfire winter (0.16) compared with the other three winters (0.51) ($p < 0.05$). Winter undernutrition or closely related causes (DelGiudice and others 1991) accounted for most of the adult elk mortalities during four years (87 percent of all adult mortalities). Winter malnutrition was the second leading cause of mortality for radiocollared calves during four years (22.7 percent of all calf mortalities). The fires and climatic events of 1988–1989 reduced survival of elk calves. Calves were born lighter and later in 1989 the first spring following the fires (Singer and others 1997).

Elk Diets Before and after the Fires

Elk consumption of all grasses other than *Calamagrostis rubescens* declined dramatically the first winter following the fires, but steadily increased the second and third winters (Fig. 6.3, $p < 0.0001$). Consumption of *Carex* and *Juncus* increased the first and second postfire winters ($p = 0.001$).

Winter severity was high the first winter after the fire. Snow densities were higher and extensively crusted due to high winds and severe cold (the winter was rated severe at -2.5 on a scale of $+4$ = mildest and -4 = most severe (Farnes 1996). The crusted snows and burned herbaceous forages forced elk to eat forage which protruded above the snow, such as the tall big sagebrush, tree bark, tree needles, aspen bark, and mosses growing on trees. Consumption of *Artemesia tridentata* increased the winter after the fires ($p = 0.006$), but decreased to prefire levels in the second postfire winter (Fig. 6.3). Consumption of pine bark and needles from both lodgepole pine and Douglas fir continued into the second winter (Fig. 6.3).

Elk Population Responses to the Fires

The northern Yellowstone elk population grew steadily following cessation of artificial controls from 1968 to 1988. In 1988, peak numbers of 23,240 ± 830 elk were estimated for the northern Yellowstone winter range (16–17 elk/km^{-2}). High elk numbers prefire in early 1988 were associated with a series of mild winters (nine of the ten winters prior to 1988 had less than long-term-average precipitation [Chapter 10]). During this time period, elk expanded their range and increased their use of areas north of the park. Both the mild winters and the range expansion (see also Coughenour and Singer 1995, Lemke and others 1998) resulted in a higher potential carrying capacity for elk, primarily due to the larger effective foraging area during the winter for elk (Chapter 13). During the first

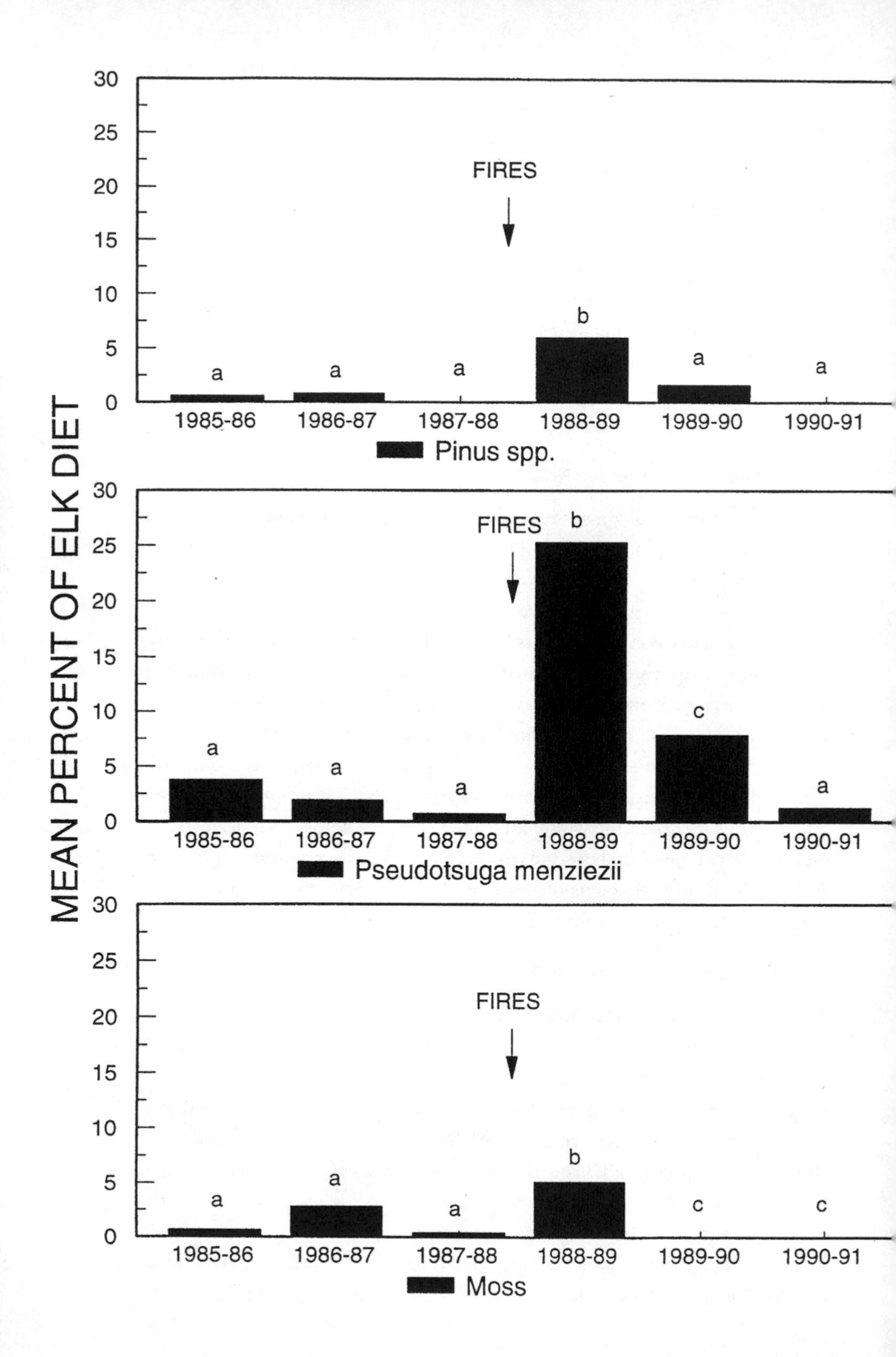

MEAN PERCENT OF ELK DIET
FIRES
30
25
20
15
10
5
0
a
a
a
b
a
a
1985-86
1986-87
1987-88
1988-89
1989-90
1990-91
Pinus spp.
FIRES
b
a
a
a
c
a
Pseudotsuga menziezii
FIRES
a
a
a
b
c
c
Moss

postfire winter, we estimated that the elk herd declined by 24–37 percent. By 1994, the elk herd had recovered to prefire levels.

Elk movement patterns were also significantly altered by the fires and associated events of the 1988–1989 season. Elk migrated in large numbers from the park during the winter of 1988–1989 due to the combined effects of drought, burning of 22 percent of the winter range in the park, and deep winter snows in the park. Fifty-four percent of the elk population migrated and were counted north of the park by late winter, an area that comprised only 18 percent of the winter range area but where snows were not as deep and no fires had occurred. This was the third-largest migration out of the park since 1916 (Houston 1982). Some young elk even remained outside of the park during the following summer, as shown by three radiocollared yearling elk observed north of the park boundaries during the summer, a phenomenon never observed with marked elk prior to the fires.

Elk Habitat Selection

Radiocollared elk cows were quite mobile, migrating out of the park and away from burned areas. Bull elk were less migratory that winter, staying in the park and staying in burned areas where they consumed (1) the few sedges that sprouted on burned wet meadows following the fires (no fall green-up occurred in any other plant communities), (2) tree bark from felled, burned aspen trees, and (3) lodgepole pine bark and needles from the mosaic of burned, unburned, and partially burned trees. Apparently these were all submaintenance foods because bull elk died at a higher rate than cows the first winter.

Bull and cow groups differed not only in their use of burned areas ($p = 0.013$), but also in the aspect on which they were sighted ($p < 0.001$) and in the plant communities they frequented ($p = 0.019$). But there was no difference between the sexes in their use of slopes ($p = 0.790$). Bulls used northeast aspects more and northwest aspects less than cows (Fisher exact test, $p = 0.036$). Both sexes preferred moderate slopes (< 21 percent), both pre- and postfire. Cows used dry grass and shrub habitats more and moist grass and unburned shrubs less than bull groups (Fisher exact test, $p = 0.019$, 0.015, respectively). Differential use of burned areas between genders was found only immediately postfire.

Elk use of burned areas was influenced by slope ($p = 0.011$), aspect ($p =$

Figure 6.3 (facing page). Elk diets on Yellowstone's northern winter range three winters before and three winters after the fires of 1988. Different letters denote statistical differences in mean percent composition of diets between years using ANOVA for the species in each individual graph.

0.016), and elevation ($p < 0.001$). Elk were more likely to use burned areas with gentle slopes (0–6 percent), with southwest aspects, and on middle elevation areas of the winter range. Groups of elk seen during aerial counts (n = 3,831 groups) used southwest slopes more than expected, but no other consistent selection for aspects was demonstrated. Elk utilized forests (either burned or unburned) less than expected during all surveys (Table 6.1). Elk groups consistently used burned dry grassland/dry shrubs (the burned type was used 52 percent more than the unburned patches; selection index of observed/expected = 1.858 ± 0.107 versus 1.22 ± 0.120), and burned moist grasslands/moist shrublands greater than expected following the fires (Table 6.1).

Selection for burned moist grasslands/moist shrubland (selection index = 1.407 ± 0.052) was 18 percent greater than for unburned patches of the same type (index = 1.195 ± 0.138). Elk group sizes were larger in grassland than in forests, and elk groups were larger in unburned than in burned grasslands ($p < 0.05$).

DISCUSSION

Effects of the Fires on Forage Production and Quality

The modest increase (20 percent) in forage production in dry grasslands and wet meadows that was attributable to the fires of 1988 was biologically important because these are two very important elk habitats. Small to no increases in production were observed in sagebrush, Douglas fir, and lodgepole pine forests. Part of the variation among and within habitats could be attributable to the different fire intensities experienced. Hotter fires and higher soil heating were typical in habitats that exhibited short-term declines in forage production (Despain and others 1989). Hotter fires in forest or tall shrub communities with high fuel loadings killed seeds and many plants (Despain and others 1989, Knight and Wallace 1989, Bartos and others 1994). Plant recovery in these types required several years. For example, recovery of plant biomass in burned summer range lodgepole forest required three growing seasons, similar to the three- to five-year span noted for forested communities in other areas (Leege 1969, Lyon 1971, Peek and others 1979, Merrill and others 1994, Brown and DeByle 1989, Bartos and others 1994, Chapters 4, 14).

The Yellowstone grasslands did not experience the large increase in biomass in response to fire seen in tallgrass prairie (Knapp and Seastedt 1986, Seastedt

Table 6.1. Winter habitat preferences by elk in burned and unburned habitats before and after the fires of 1988, northern Yellowstone National Park, Wyoming

		Use: Availability Ratio[a]			
		1 Cell		9 Cells	
Habitat Category	% of Area in Habitat Category	1st Winter After Fires	2d and 3d Postfire Winters	1st Winter After Fires	2d and 3d Postfire Winters
Unburned in 1988					
Lodgepole Pine/ Subalpine Fir	22.63	0.438⁻	0.389⁻	0.451⁻	0.367⁻
Whitebark Pine/ Douglas Fir	16.08	0.888	0.720	0.845	0.696⁻
Sagebrush/Grassland	37.82	1.833⁺	1.463⁺	1.796⁺	1.492⁺
Wetlands	2.54	0.520⁻	1.920⁺	0.560⁻	1.800⁺
Burned in 1988					
Lodgepole Pine/ Subalpine Fir	6.30	0.048⁻	0.079⁻	0.063⁻	0.111⁻
Whitebark Pine/ Douglas Fir	4.10	0.341⁻	0.488⁻	0.390⁻	0.512⁻
Sagebrush/Grassland	9.30	0.516⁻	1.785⁺	0.503⁻	1.710⁺
Wetlands	0.50	0.400⁻	0.800⁻	0.400⁻	1.200⁺

Notes: Percentage of northern winter range covered by habitat types (availability), and the ratio of observed herd fraction to availability fraction on: (a) single 50 × 50-m or, (b) nine contiguous 50 × 50-m pixels sampled: the winter just after the fires (1989), and subsequent postfire winters (1990, 1991). Minus signs indicate that the pixel or pixel group was used less than expected by chance. Plus signs indicate usage more than expected by chance.

[a]The ratio of the percent of all group observations of elk (n = 3831 groups) to the percent of the total area in that habitat category. A "+" implies use greater than, "−" less than, no sign equal to availability of the habitat category with Bonferroni confidence intervals (Neu and others 1974, Miller 1981, Byers and others 1984).

and others 1991) and other more productive grassland systems (Moomaw 1957, Ribinski 1968, Wright and others 1979, West and Hassan 1985). The fires in Yellowstone's grasslands were cool and spotty because there was not the heavy litter and fuel accumulation seen in more productive grasslands. Typically, Yellowstone's grasslands have only 85 g/m^{-2} annual standing crop biomass compared to several hundred grams per m^{-2} in more productive grassland ecosystems.

In contrast to our predictions and to other reports, spring regrowth did not occur earlier, nor was senescence later on burned grasslands (Old 1969, Peet and others 1975, Skovlin and others 1983, Hobbs and Spowart 1984, Seip and Bunnell 1985; however, see Chapter 5).

Percent nitrogen and digestibility were enhanced in four of the six winter forages that we sampled from the winter range. On the summer range, there were few differences in forage quality attributable to the fires. Percent nitrogen was enhanced in only five of the eighteen species studied (see Chapter 5). Forage quality improvement was most notable in YNP in those communities with higher fuel loadings (big sagebrush, lodgepole forest, Douglas fir forest), where fires burned hotter and longer, thus releasing more nutrients. A number of other authors have also found more release of nutrients where fuels are higher and fires are hotter (DeWitt and Derby 1955, Daubenmire 1968, Old 1969, DeBano 1991, Seastedt and others 1991, Chapters 4, 5).

Elk foraging efficiency and elk condition were likely enhanced following the fires:

1. Biomass and quality were increased in two important elk habitats and several key forage species.
2. Leaf sizes of forbs and willows doubled and shoots were approximately 1.5 times larger on burned sites (Norland and others 1996).
3. Elk fed more on sedges and rushes following the fires possibly because accumulated dead litter in these types was removed by the fires. Foraging was probably far more efficient in sedge and rush communities once the dead litter was removed by burning.
4. Bunchgrasses were taller and more erect on the northern range (Singer and Harter 1996), also likely enhancing foraging efficiency in winter.

Higher urea nitrogen:creatinine ratios (DelGiudice and Singer 1996) and higher elk:calf ratios indicated northern Yellowstone elk were in better condition by the second postfire winter. Elk foraging efficiency (Canon and others 1987) and improvements in elk body weight and condition were observed due to burning (Rowland and others 1983).

We observed increased fiber, cellulose, and lignin concentrations in dry grasslands, rather than decreased levels as shown by Wilms and others (1981) and Smith (1960). We also observed little difference in macronutrient concentrations in these same forages in contrast to most studies that reported increased concentrations in regrowth following fires (Old 1969, Lloyd 1971, Wilms and others 1981, Umoh and others 1982, Ohr and Bragg 1985).

As predicted, elk consumed previously unpalatable forage species following burning. Burned bark and needles were consumed in substantial amounts during the first postfire winter. Burned Douglas fir had higher nitrogen and digestibility values, but nitrogen and digestibility in aspen were lower than in unburned forages, and these items were likely inadequate forages for elk. Radiocollared elk that were observed consuming burned aspen and conifer bark and conifer needles died that first winter. In contrast to our findings, Jakubas and others (1994) found higher forage quality in burned lodgepole pine bark in the Madison-Firehole winter range of YNP. They reported doubled digestibility and protein contents in burned lodgepole bark. Their analyses were conducted four years following the fires of 1988 and in a deep snow environment where elk continued to consume bark. They looked only at lodgepole pine bark, a species we did not sample. In our study area, consumption of burned bark by elk, which included Douglas fir and spruce, was restricted entirely to the first winter after the fires. We did not observe consumption of burned bark in the following winters. Regrowth of sedges and rushes following burning was also consumed at higher rates postfire.

Elk Distribution and Habitat Use

Elk distributions and habitat use were greatly altered by the fires of 1988. More elk migrated north of the park each winter following the fires. Prior to the fires (1971–1987), migrations north of the park averaged 15 percent of the population (Houston 1982, Singer 1991). Following the fires, more than 50 percent of the herd migrated north of the park most winters (Lemke and others 1998). After the fires, 3,000–4,500 elk consistently spent the winters at the northernmost extremity of the northern Yellowstone elk winter range on the Dome Mountain Wildlife Management Area, where, prior to the fires, only 700 elk spent the winter (Lemke and others 1998).

Elk groups demonstrated a strong preference for burned over unburned dry and moist bunchgrass/shrublands areas from within the park. Elk groups in our study selected burned grassland habitats 52 percent more often than unburned grasslands, similar to the 40 percent greater preference of burned habitats that Pearson and others (1995) reported for the same area using different methods. Davis (1977), Roppe and Hein (1978), and Canon and others (1987) also reported a high preference by elk for burned habitats in other study areas, although Skovlin and others (1983) observed no preference by elk for burned grasslands in Oregon. We observed a decline in use of burned lodgepole forest, where total canopies were burned by fires, but equivalent use by elk of burned and unburned Douglas fir forests where canopies were largely or mostly intact. Surface fires tended to occur only in these forests. Davis (1977) also reported that elk preferred burned forests with standing dead timber to burned forests where no snags were left.

In spite of the short-term forage increases due to the fires within the park, more elk continued to migrate north of the park all winters since the fires. This area is substantially lower in elevation, snowpacks are lighter, and elk survival rates, as suggested by higher calf:cow ratios, were also substantially higher each winter (Coughenour and Singer 1995).

In burned forests, snow is deeper, snow cover duration is longer, and there is more crusting due to a combination of increased albedo, lower humidity, and greater wind velocity on burned areas (Lyon and Stickney 1976). Elk prefer closed-canopy forests in late winter, when snows become deep and dense in open areas (Bergen 1971, Beall 1976). Norland and others (1996) observed slightly deeper snows in burned forests in northern YNP. We observed slightly higher snow density on burned forest sites. These values (11 percent increase in snow density and 2.5 cm deeper snow) were apparently too minor to influence elk feeding crater densities in burned versus unburned Douglas fir forests (see Chapter 13). However, Singer and Harter (1996) reported that elk avoided burned lodgepole pine forests compared with unburned lodgepole pine forests.

Elk Population Responses to the Fires

Following immediate postfire declines, elk population subsequently increased rapidly, recovering to prefire levels by 1995, only five years postfire (YNP 1997, Lemke and others 1998). Summer survival of elk calves was low both the first and second years postfire (1989 and 1990), thus slowing the initial herd recovery. But higher calf:cow ratios were observed by 1991–1994 (YNP 1997, Lemke and others 1998). Subsequent elk populations have not continued to increase due to a combination of doubling in hunter harvests north of the park, more severe winters (Chapter 13), and possibly also the effects of wolf restoration in 1995 (YNP 1997, Lemke and others 1998, J. Mack, personal communication). The calculations of Singer and Mack (1999) predict that wolves would limit the northern Yellowstone elk population in conjunction with continued levels of antlerless harvest north of the park. All model scenarios with wolf predation predicted declining elk populations if antlerless elk harvests are not reduced (Singer and Mack 1999), even with the improved short-term habitat conditions for elk due to the fires. This has apparently proven to be the case.

Acknowledgments
We wish to thank staff of Yellowstone National Park, Montana Department of Fish, Wildlife and Parks, U.S. Forest Service, numerous Student Conservation Association Volunteers, Sagebrush Aero, Inc., Gallatin Flying Service, Yellowstone Air Service, Wyoming Game and Fish, Yellowstone Grizzly Foundation, and Montana Wildlife Diagnostic Lab. Nutrient and digestibility tests were conducted by the Rangeland Ecosystem Lab, Colorado State University, and diet analyses by the Wildlife Habitat Lab, Washington State University. Suggestions from two anonymous reviewers are also gratefully acknowledged.

References
Association of Official Analytical Chemists. 1970. Official methods of analysis, 11th ed. Association of Official Analytic Chemists, Washington, D.C.
Bartos, D. L., J. K. Brown, and G. D. Booth. 1994. Twelve years biomass response in aspen communities following fire. Journal of Range Management 47:79–83.
Beall, R. C. 1976. Elk habitat selection in relation to thermal radiation. Pages 97–100 in S. R. Hieb, ed., Proceedings of Elk-Logging Symposium, University of Idaho, Moscow.
Bendel, J. E. 1974. Effects of fire on birds and animals. Pages 73–138 in T. T. Kozlowski and C. E. Ahlgren, eds., Fires and ecosystems. Academic Press, New York.
Bergen, J. D. 1971. Vertical profiles of windspeed in a lodgepole pine stand. Forest Science 17:314–324.
Boyce, M. S., and E. H. Merrill. 1991. Effects of the 1988 fires on ungulates in Yellowstone National Park. Proceedings of Tall Timbers Fire Conference 29:121–132.
Brown, J. K., and N. V. DeByle. 1989. Effects of prescribed fire on biomass and plant succession in western aspen. United States Department of Agriculture, Forest Service Research Paper INT-412.
Byers, C. R., R. K. Steinhorst, and P. R. Krausman. 1984. Clarification of a technique for analysis of utilization-availability data. Journal of Wildlife Management 45:1050–1053.
Canon, S. K., P. J. Urness, and N. V. DeByle. 1987. Habitat selection, foraging behavior, and dietary nutrition of elk in burned aspen forest. Journal of Range Management 40:433–438.
Christensen, N. L., J. K. Agee, P. F. Brussard, J. Hughes, D. H. Knight, G. W. Minshall, J. M. Peek, S. J. Pyne, F. J. Swanson, J. W. Thomas, S. Wells, S. E. Williams, and H. A. Wright. 1989. Interpreting the Yellowstone fires of 1988. BioScience 39:678–685.
Coughenour, M. B., and F. J. Singer. 1995. Elk population processes in Yellowstone National Park under the policy of natural regulation. Ecological Applications 6:573–593.
Daubenmire, R. 1968. Ecology of fire in grasslands. Advances in Ecological Research 5:209–266.

Davis, P. R. 1977. Cervid response to forest fire and clearcutting in southeastern Wyoming. Journal of Wildlife Management 41:785–788.

DeBano, L. F. 1991. The effect of fire on soil properties. Pages 151–156 in A. E. Harvey and L. F. Neuenshwander, eds., Proceedings: Management and productivity of western-montane forest soils. Forest Service General Technical Report INT-280.

DeByle, N. V., P. J. Urness, and D. L. Blank. 1989. Forage quality in burned and unburned aspen communities. Forest Service Research Paper INT-404.

DelGiudice, G. D., and F. J. Singer. 1996. Physiological responses of Yellowstone elk to winter nutritional deprivation and the 1988 fires: A preliminary examination. Pages 133–136 in J. M. Greenlee, ed., The ecological implications of fire in Greater Yellowstone: Second Biennial Scientific Conference on the Greater Yellowstone Ecosystem, 1993. International Association of Wildland Fire, Fairfield, Wash.

DelGiudice, G. D., F. J. Singer, and U. S. Seal. 1991. Physiological assessment of winter deprivation in elk of Yellowstone National Park. Journal of Wildlife Management 55:653–664.

Despain, D., A. Rodman, P. Schullery, and H. Shovic. 1989. Burned area survey of Yellowstone National Park: The fires of 1988. Yellowstone National Park, Wyoming, 14 pp.

DeWitt, J. B., and J. V. Derby. 1955. Changes in nutritive value of browse plants following forest fires. Journal of Wildlife Management 19:65–70.

Dills, G. G. 1970. Effects of prescribed burning on deer browse. J. of Wildlife Management 34:540–545.

Farnes, P. E. 1996. An index of winter severity for elk. In F. J. Singer, ed., Effects of grazing response by wild ungulates in Yellowstone National Park. Technical Report NPS/96-01, Natural Resource Information Division, Denver.

Frank, D. A., and S. J. McNaughton. 1990. Aboveground biomass estimation with the canopy intercept methods: A plant growth form caveat. Oikos 57:57–60.

Gasaway, W. C., R. D. Boertje, D. V. Grangaard, D. G. Kellyhouse, R. G. Stephenson, and D. G. Larsen. 1992. The role of predation in limiting moose at low densities in Alaska and Yukon and implications for conservation. Wildlife Monographs 120:1–59.

Geist, V. 1971. Mountain sheep, a study in behavior and evolution. University of Chicago Press, Chicago.

Grelen, H. E., and L. B. Whitaker. 1973. Prescribed burning rotations on pine-bluestem range. Journal of Range Management 26:152–153.

Haugen, A. O., and D. W. Speake. 1958. Determining age of young fawn white-tailed deer. Journal of Wildlife Management 22:319–321.

Hobbs, N. T., and R. A. Spowart. 1984. Effects of prescribed fire on nutrition of mountain sheep and mule deer during winter and spring. Journal of Wildlife Management 48:551–560.

Houston, D. B. 1973. Wildfires in northern Yellowstone National Park. Ecology 54:1111–1117.

———. 1982. The northern Yellowstone elk: Ecology and management. Macmillan, New York.

Jakubas, W. J., R. A. Garrott, P. J. White, and D. R. Mertens. 1994. Fire-induced changes in the nutritional quality of lodgepole pine bark. Journal of Wildlife Management 58:35–45.

Jones, J. R. 1974. Silviculture of southwestern mixed conifers and aspen: The status of our knowledge. United States Department of Agriculture, Forest Service RM-122.

Knapp, A. K., and T. R. Seastedt. 1986. Detritus accumulation limits the productivity of tall-grass prairie. BioScience 36:662–668.

Knight, D. H., and L. L. Wallace. 1989. The Yellowstone fires: Issues in landscape ecology. BioScience 39:700–706.

Leege, T. A. 1969. Burning seral brush ranges for big game in northern Idaho. Transactions of the North American Wildlife and Natural Resources Council 34:429–438.

Lemke, T. O., J. A. Mack, and D. B. Houston. 1998. Winter range expansion by the northern Yellowstone elk herd. Intermountain Journal of Sciences 4:1–9.

Lloyd, P. S. 1971. Effects of fire on the chemical status of herbaceous communities of the Derbyshire Dales. J. of Ecology 59:261–273.

Lyon, L. J. 1966. Initial vegetation development following prescribed burning in south-central Idaho. United States Department of Agriculture, Forest Service Research Paper INT-29. Ogden, Utah.

———. 1971. Vegetal development following prescribed burning of Douglas fir in south-central Idaho. United States Department of Agriculture, Forest Service Research Paper INT-24. Ogden, Utah.

Lyon, L. J., and P. F. Stickney. 1976. Early vegetal succession following large northern Rocky Mountain wildfires. Proceedings of the Tall Timbers Fire Ecology Conference 16:355–375.

Martinka, C. J. 1976. Fire and elk in Glacier National Park. Proceedings of the Tall Timbers Fire Ecology Conference 14:377–389.

Mattson, D. J., and D. G. Despain. 1984. Grizzly bear habitat component mapping handbook for the Yellowstone ecosystem. Interagency Grizzly Bear Study Team, National Park Service and United States Forest Service, Bozeman, Mont.

Meiman, J. R. 1968. Snow accumulation relative to elevation, aspect and forest canopy. Pages 35–47 in Snow Hydrology Proceedings, University of New Brunswick.

Merrill, E. H., N. L. Stanton, and J. C. Halk. 1994. Response of bluebunch wheatgrass, Idaho fescue, and nematodes to ungulate grazing in Yellowstone National Park. Oikos 69:231–240.

Miller, R. G., Jr. 1981. Simultaneous statistical inference, 2d ed. Springer-Verlag, New York.

Moomaw, J. C. 1957. Some effects of grazing and fire on vegetation of the Columbia Basin region, Washington. Dissertation Abstracts 17:733.

Nelson, J. R. 1976. Forest fire and big game in the Pacific Northwest. Proceedings of the Tall Timbers Fire Ecology Conference 15:85–102.

Neu, C. W., C. R. Byers, and J. M. Peek. 1974. A technique for analysis of utilization-availability data. Journal of Wildlife Management 38:541–545.

Norland, J. N., F. J. Singer, and L. Mack. 1996. Effects of the fires of 1988 on elk habitats. Pages 223–232 in J. M. Greenlee, ed., The ecological implications of fire in Greater Yellowstone: Second Biennial Science Conference on the Greater Yellowstone Ecosystem, 1993. International Association of Wildland Fire, Fairfield, Wash.

Ohr, K. M., and T. B. Bragg. 1985. Effects of fire on nutrient and energy concentration of fire prairie grass species. Prairie Naturalist 17:113–126.

Old, S. M. 1969. Microclimate, fire, and plant production in an Illinois prairie. Ecological Monographs 39:355–384.

Pearson, H. A. 1970. Digestibility trials: *In vitro* techniques. Pages 85–92 in Range and

wildlife habitat: A research symposium. United States Department of Agriculture, Miscellaneous Publication 1147.
Pearson, S. M., M. G. Turner, L. L. Wallace, and W. H. Romme. 1995. Winter habitat use by large ungulates following fire in northern Yellowstone National Park. Ecological Applications 5:744–755.
Peek, J. M. 1980. Natural regulation of ungulates. Wildlife Society Bulletin 8:217–227.
Peek, J. M., R. A. Riggs, and J. L. Lauer. 1979. Evaluation of fall burning on bighorn sheep winter range. J. of Range Management 32:430–432.
Peet, M., R. Anderson, and M. S. Adams. 1975. Effects of fire on big bluestem production. American Midland Naturalist 94:15–26.
Ribinski, R. F. 1968. Effects of forest fires on quantity and quality of deer browse production in mixed oak forests of central Pennsylvania. Quarterly Report of Pennsylvania Cooperative Wildlife Research Unit, Pennsylvania State University 30:13–15.
Roppe, J. A., and D. Hein. 1978. Effects of fire on wildlife in a lodgepole pine forest. Southwestern Naturalist 23:279–288.
Rowland, M. M., A. W. Allredge, J. E. Ellis, B. J. Weber, and G. C. White. 1983. Comparative winter diets of elk in New Mexico. Journal of Wildlife Management 47:924–932.
Samuel, M. D., E. O. Garton, M. W. Schlegel, and R. G. Carson. 1987. Visibility bias during aerial surveys of elk in north-central Idaho. Journal of Wildlife Management 51:622–630.
SAS Institute. 1990. SAS user's guide, 4th ed., version 6. SAS Institute, Cary, N.C.
Seastedt, T. R., J. M. Briggs, and D. J. Gibson. 1991. Controls of nitrogen limitation in tallgrass prairie. Oecologia 87:72–79.
Seip, D. R., and F. L. Bunnell. 1985. Nutrition of Stone's sheep on burned and unburned ranges. J. of Wildlife Management 49:397–405.
Singer, F. J. 1991. The ungulate prey base for wolves in Yellowstone National Park. Pages 323–348 in R. B. Keiter and M. S. Boyce, eds., The Greater Yellowstone Ecosystem: Man and nature in America's wildlands. Yale University Press, New Haven.
Singer, F. J., and E. O. Garton. 1994. Elk sightability model for the Super Cub. Pages 47–48 in J. W. Usworth, F. A. Leban, D. J. Leptich, E. O. Garton, and P. Zager, eds., Aerial survey: User's manual, 2d ed. Idaho Fish and Game, Boise.
Singer, F. J., and M. K. Harter. 1996. Comparative effects of elk herbivory and the 1988 fires on northern range Yellowstone National Park grasslands. Ecological Applications 6:185–199.
Singer, F. J., A. Harting, K. K. Symonds, and M. B. Coughenour. 1997. Elk calf mortality in Yellowstone National Park: The evidence for density dependence, compensation, and environmental effects. Journal of Wildlife Management 61:12–25.
Singer, F. J., and J. Mack. 1993. Potential ungulate prey for wolves. Pages 75–117 in R. S. Cook, ed., Ecological issues on reintroducing wolves in Yellowstone National Park, National Park Service Scientific Monograph no. 22. National Park Service, Washington, D.C.
———. 1999. Predicting the effects of wildfire and carnivore predation on ungulates. Pages 189–237 in T. W. Clark, A. P. Curlee, S. C. Minta, and P. M. Karleva, eds., Carnivores in ecosystems: The Yellowstone experience. Yale University Press, New Haven.
Singer, F. J., W. Schreier, J. Oppenheim, and E. O. Garton. 1989. Drought, fires, and large mammals. BioScience 39:716–722.

Skovlin, J. M., P. J. Edgerton, and B. R. McConnell. 1983. Elk use of winter range as affected by cattle grazing, fertilization, and burning in southeastern Washington. Journal of Range Management 36:184–189.

Smith, D. R. 1960. Description and responses to elk use of two mesic grassland and shrub communities in the Jackson Hole region of Wyoming. Northwest Science 34:25–36.

Steele, R., S. V. Cooper, D. V. Ondov, D. W. Roberts, and R. D. Pfisterer. 1983. Forest habitat types of eastern Idaho–western Wyoming. United States Department of Agriculture, Forest Service, General Technical Report INT-44. Intermountain Forest and Range Experiment Station, Ogden, Utah.

Tilley, J. M. A., and R. A. Terry. 1963. A two-stage technique for the *in vitro* digestion of forage crops. Journal of the British Grassland Society 18:104–111.

Turner, M. G., Y. Wu, L. L. Wallace, W. H. Romme, and A. Brenkert. 1994. Simulating winter interactions among ungulates, vegetation, and fire in non Yellowstone Park. Ecological Applications 4:472–496.

Umoh, J. E., L. H. Harbers, and E. F. Smith. 1982. The effects of burning on mineral contents of Flint Hill forages. Journal of Range Management 35:231–234.

Unsworth, J. W., L. Kuck, M. D. Scott, and E. O. Garton. 1993. Elk mortality in the Clearwater drainage of north central Idaho. Journal of Wildlife Management 57:495–502.

Unsworth, J. W., F. A. Leban, D. J. Leptich, E. O. Garton, and P. Zagar. 1994. Aerial surveys: User's manual, 2d ed. Idaho Department of Fish and Game, Boise.

Ursek, D. W., J. F. Cline, and W. H. Rickard. 1975. Impact of wildfire on three perennial grasses in south-central Washington. J. of Range Management 29:309–310.

West, N. E., and M. A. Hassan. 1985. Recovery of sagebrush-grass vegetation following wildfire. Journal of Range Management 38:131–134.

White, G. C., and R. A. Garrott. 1990. Analysis of wildlife radiotracking data. Academic Press.

Wilms, W., A. W. Bailey, A. McLean, and C. Kalnin. 1981. Effects of fall clipping or burning on the distribution of chemical constituents of bluebunch wheatgrass in spring. Journal of Range Management 34:267–269.

Wood, G. W. 1988. Effects of prescribed fire on deer forage and nutrients. Wildlife Society Bulletin 16:180–186.

Wright, H. A., L. F. Neunschwander, and C. M. Britton. 1979. The use of fire in sagebrush-grass and pinyon-juniper plant communities: A state of the art review. United States Department of Agriculture, Forest Service, General Technical Report INT-58, Intermountain Forest and Range Experiment Station, Ogden, Utah.

Yellowstone National Park. 1997. Yellowstone's Northern Range: Complexity and change in a wildland ecosystem. Yellowstone National Park, Wyoming.

Part III **Effects on Aquatic Systems**

We tend to think that a stream runs through a valley, but the valley also runs through the stream.
—*O. J. Reichman,* Konza Prairie

Chapter 7 Effects of Wildfire on Growth of Cutthroat Trout in Yellowstone Lake

Robert E. Gresswell

Although fire is an important agent of disturbance and ecological change, research prior to the late 1980s was generally limited to the effects of prescribed fires (including slash burns following clear-cut logging) on terrestrial site productivity, vegetation regeneration, and succession (Gresswell 1999). Studies that examined relationships between fire and aquatic ecosystems were generally focused on the loss of soil through erosion and changes in water yield and quality. A few studies considered the effects of fire on aquatic biota, but the emphasis was on primary production and invertebrates (Albin 1979, Bradbury 1986, Tarapchak and Wright 1986).

Since fires burned through the Greater Yellowstone Ecosystem in 1988 and central Idaho in 1992 and 1994, there has been a substantial increase in research evaluating the biological consequences of fire on stream systems (Minshall and others 1997, Rieman and others 1997, Minshall and others 2001). In contrast, postfire evaluation of fire effects on lakes has continued to concentrate on trends in water quality, nutrient status, and primary production (Theriot and others 1997, Lamontagne and others 2000, Planas and others 2000). Despite concern in many regions about the negative effects of nutrient enrichment on aquatic biota (McEachern and others 2000), the relationship between nutrient changes associated with fire and the vertebrate community is poorly understood. For example, it is reasonable to expect that fish populations would be especially sensitive to changes in top-down and bottom-up trophic interactions related to nutrient dynamics following fire, but thus far only one published study has specifically examined these linkages (St-Onge and Magnan 2000).

Yellowstone Lake is the largest high-altitude lake (elevation > 2000 m) in North America, and it is the most important remaining enclave for the Yellowstone cutthroat trout *Oncorhynchus clarki bouvieri* (Gresswell 1995). All wildfires in Yellowstone National Park were actively suppressed from 1885 to 1972, but in 1988, free-ranging fires burned nearly 39 percent of the Yellowstone Lake drainage (U.S. National Park Service, Yellowstone National Park, Wyoming). Although the influence of these fires on lake productivity was not measured directly, a long-term monitoring program yielded data describing demographic structure of cutthroat trout in Yellowstone Lake from 1978 to 1992. This time series provided an opportunity to explore changes in population structure of cutthroat trout before, during, and after the 1988 fires. Previous studies in lakes have demonstrated that nutrients such as nitrogen and phosphorus are directly related to salmonid growth (Warren 1971), and therefore, if nutrients increased substantially in Yellowstone Lake following the fires of 1988, changes in primary productivity might be reflected by increased growth of Yellowstone cutthroat trout.

The purpose of this chapter is to evaluate annual growth of Yellowstone cutthroat trout in relation to the 1988 fire. Trends in year-to-year variation in growth for individual age groups and 25-mm-length groups of cutthroat trout in Yellowstone Lake were examined in an attempt to detect shifts in growth patterns associated with the fires. Answers were sought to the following questions: To what extent does growth of cutthroat trout in Yellowstone Lake vary among years? Are there discernable patterns of cutthroat trout growth through time? Did patterns of cutthroat trout growth in the lake shift after 1988?

STUDY AREA

Yellowstone Lake lies at an elevation of 2,357 m and has a surface area of about 34,000 ha (Kaplinski 1991). Vegetation in the drainage basin (261,590 ha, Benson 1961) is predominately lodgepole pine *Pinus contorta* forest and subalpine meadows. A total of 124 tributaries to Yellowstone Lake (including ephemeral streams) have been identified (Hoskins 1974, 1975, Jones and others 1986, 1987).

Spawning cutthroat trout have been observed in sixty-eight tributaries, but sixteen of these are used only in years when discharge is adequate for upstream passage (Jones and others 1987).

Yellowstone Lake is covered with ice for approximately 160 days annually, so the growing season for cutthroat trout is short (Gresswell and others 1994). Diatoms dominate the phytoplankton community, but blooms of a blue-green alga *(Anabaena flos-aquae)* occur during periods of thermal stratification (Benson 1961, Garrett and Knight 1973). Zooplankton (primarily *Diaptomus shoshone, Daphnia schøedleri,* and *Conochilus unicornis;* Benson 1961) are the primary food of immature Yellowstone cutthroat trout inhabiting the pelagic areas of the lake. Mature cutthroat trout feed on crustacean zooplankton, two amphipod taxa *(Gammarus lacustris* and *Hyallela azteca),* and aquatic insect larvae (Benson 1961, Jones and others 1990), and these larger fish are most abundant in the littoral zone.

The only other fish species native to Yellowstone Lake is the longnose dace *(Rhinichthys cataractae)* (Simon 1962). Redside shiner *(Richardsonius balteatus),* lake chub *(Couesius plumbeus),* and longnose sucker *(Catostomus catostomus)* became established during the twentieth century following nefarious introductions (Gresswell and Varley 1988). In 1994, lake trout *(Salvelinus namaycush)* were found in Yellowstone Lake for the first time (Kaeding and others 1996, Mahony and Ruzycki 1997). The presence of lake trout is a major concern because of potential competitive interactions between the two salmonid species (for example, Marrin and Erman 1982, Griffith 1988, McIntyre 1995), and Yellowstone cutthroat trout have become the major prey of lake trout (Ruzycki and Beauchamp 1997). Although the presence of this aggressive predator may substantially alter the abundance and demography of Yellowstone cutthroat trout in the lake (McIntyre 1995, Ruzycki and Beauchamp 1997, Gresswell and others 1997), data used in the present analysis were collected before lake trout numbers had increased to a point of detection. Therefore it is assumed that the influence of lake trout on the results and interpretation of this study is minimal.

The Yellowstone Lake watershed has never been subjected to large-scale environmental degradation, but anthropogenic activities have affected the dynamics of the Yellowstone cutthroat trout population. For instance, hatchery supplementation and high levels of angler harvest during the first half of the twentieth century were associated with alterations in the population structure of native trout and changes in species composition in the lake (Gresswell and Varley 1988, Gresswell and others 1994). Hatchery operations ended in the mid-1950s, and harvest has been restricted since the early 1970s.

A shift in population structure toward a greater proportion of older and larger fish was documented following the implementation of more restrictive angling regulations in the Yellowstone Lake basin (Gresswell and Varley 1988, Gresswell and others 1994). Mean length of spawning cutthroat trout at Clear Creek increased from approximately 365 mm in the late 1960s to almost 400 mm in the early 1990s; average age changed from about 3.9 to 5.8 years. Mean length of fish captured by anglers increased from about 345 mm to > 370 mm during this time period, and similarly, mature cutthroat trout captured in experimental gill nets increased from about 350 mm to about 390 mm.

Approximately 100,000 ha of the Yellowstone Lake watershed burned between early July and late August 1988. The majority of area burned was in the southern part of the drainage, including the Yellowstone River, the largest tributary to the lake. Pelican Creek, the second largest tributary to the lake (drainage area = 17,656 ha), was the only major tributary entering the lake from the north that was affected by the fire.

METHODS

Information for the current study was obtained from gill-net samples collected annually from 1978 to 1992. Each year during the third week of September, gill nets were set at eleven sites in Yellowstone Lake (Fig. 7.1). Each site was sampled once during a five-day period, and five nets were set at each site. The distance between nets at individual sites was approximately 100 m. Monofilament nets were 38 m long and 1.5 m deep with 7.6-m-long panels of 19-, 25-, 32-, 38-, and 51-mm mesh netting (bar measure). Nets were set perpendicular to shore in water 1.5 to 5 m deep.

During processing, cutthroat trout from each net were arranged from longest to shortest, and total length of each individual was measured to the nearest millimeter. Scales were collected from the longest fish and then from every third fish in length order. Age was backcalculated from scales (Everhart and others 1975); sample sizes ranged from 96 to 341 fish per year. Scales that lacked a first-year annulus were identified using criteria developed by Laakso and Cope (1956), and age was adjusted upward when necessary. To ensure that estimates from fish in individual age or length groups were independent of those from other groups, only the growth increment from the last complete year of growth prior to capture (growth year) was estimated for each fish (Ricker 1975, Gutreuter 1987). In order to further reduce the influence of fish size on observed growth, growth increments were converted to relative growth rates by dividing the growth increment by the length of fish at the beginning of the growth year:

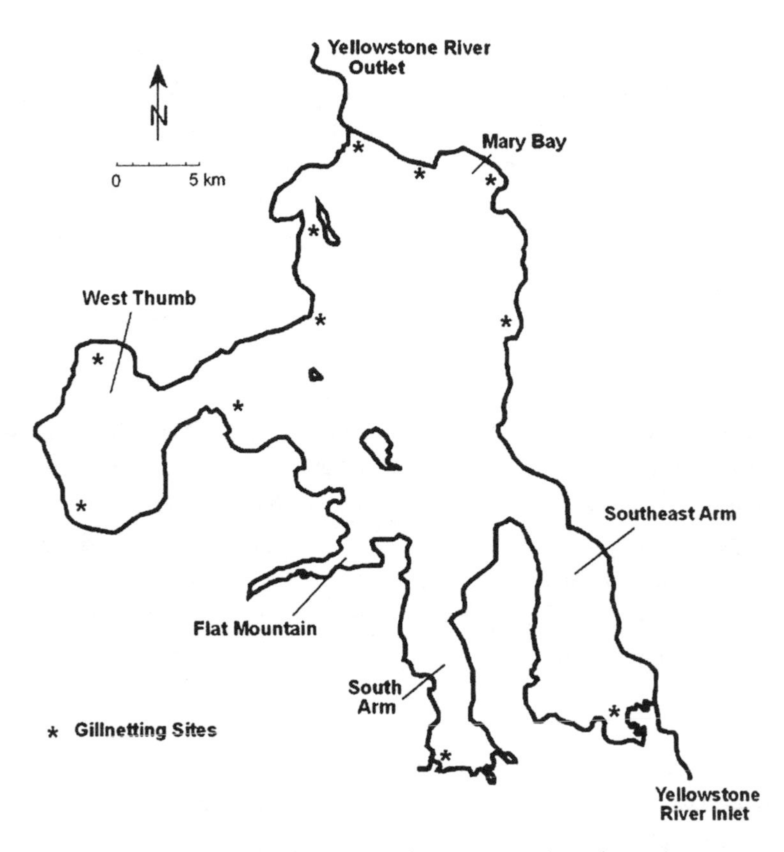

Figure 7.1. Gill netting sites (*), Yellowstone Lake, 1978–1992 (sample years).

$$RG_i = (L_{ie} - L_{ib}) / L_{ib}$$

where:

RG_i = relative growth rate (mm/mm/year),
L_{ib} = length of cutthroat trout$_i$ (mm) at the beginning of the growth year, and
L_{ie} = length of cutthroat trout$_i$ at the end of the growth year.

Evaluating effects (for example, changes in fish growth) related to a disturbance event occurring at a single point in time is complicated by the number of factors and interactions of those factors. Interannual variation in food availability, physical characteristics of the environment, and fish density makes it difficult to combine years in a traditional before-and-after analysis of growth, especially when a suitable control site is not available. Unequal sample sizes among years and among age groups and 25-mm length groups exacerbate these interannual fluctuations, and differences in age interpretation related to inaccuracies in the assignment of annuli also confound analyses. For this reason, statistical procedures focused both on evaluating differences among groups (for example, growth years and age groups and length groups) using analysis of variance (ANOVA) and investigating trends of age and length groups through time using regression techniques. This approach can be useful for explaining observed changes and interpreting subtle gradients associated with complex observational data (Tabachnick and Fidell 1989).

Statistical analyses were accomplished using NCSS (Hintze 1999). To compare growth through time, relative growth rates for age and length groups were organized by growth year (sample year—1). Because analysis and interpretation of time-series data were susceptible to confounding by serial autocorrelation (Montgomery and Reckhow 1984), data (growth increments and relative growth rates) were initially examined for temporal correlation. Each age group and length group was examined separately using the original data series with the series average removed. Autocorrelations and partial autocorrelations were calculated without seasonal differencing (one season of growth was assumed annually).

Two-factor analysis of variance (ANOVA) was used for comparisons of mean relative growth rates. Growth year was the primary factor of interest. Initial comparisons included age group as a second factor that reflected among-group variation in relative growth rate. Both factors (growth year and age group) were fixed. Growth can sometimes be more closely related to length of fish at the beginning of a growth period than the age of the fish (Gutreuter 1987), and therefore, a second comparison was conducted using growth year and 25-mm-length group as fixed factors. Because factorial analysis requires observations in all treatment combinations, age-group comparisons were limited to ages 3, 4, 5, and 6 from 1978 through 1992 (growth years 1977 through 1991). Comparisons for 25-mm-length groups ranged from 100 mm to 350 mm for the same time periods. Multiple mean comparisons were conducted *a posteriori* for significant analyses using Tukey-Kramer procedures (Zar 1984). Emphasis was placed on identifying differences in growth patterns among individual years and age groups and length groups; these comparisons provided insight into among-group patterns occurring before and after the 1988 fire.

In order to aid interpretation of trends through time, relative growth rates of cutthroat trout in individual age and length groups were evaluated using linear regression. Simple linear regressions used relative growth rate as the response variable and growth year as the predictor. Subsequently, multiple regression analysis was used to investigate possible changes in trends after the fire (1988 to 1991 growth years), and a dummy variable that influenced the response only in postfire years was incorporated using the following model:

$$y_i = \beta_0 + \beta_1 t_i + \beta_2 d_i + \varepsilon_i$$

where:

t_i = growth year,
y_i = mean relative growth rate for growth year t_i,
$d_i = 0$ if $t_i < 1988$; $d_i = t_i$ if $t_i \geq 1988$,
ε_i = random error.

To further reduce confounding associated with demographic changes resulting from angling restrictions (Gresswell and Varley 1988), regressions were assessed for two different time periods. Initial analyses included all of the sample years (1978–1992), and subsequently relationships were reevaluated using data collected for four years immediately preceding the fires (1984–1987 growth years) and four years following the fires (1988–1991 growth years). Using an equal number of years before and after the fire ensured equal weighting for prefire and postfire periods.

RESULTS

Serial correlation was not evident in the individual age groups or length groups for either growth increments or relative growth rates, and therefore no transformations were performed prior to statistical analysis. Although relative growth in individual age groups and length groups varied among growth years (Tables 7.1 and 7.2), growth patterns among groups remained relatively stable through time. Younger and smaller trout grew faster than older and larger individuals (Fig. 7.2), a pattern that is characteristic of fishes and other organisms that exhibit indeterminate growth (Warren 1971, Everhart and others 1975). Differences in relative growth rate for individual age groups were evident between the prefire and postfire periods (Fig. 7.2a), but the relationships for relative growth rate and length groups were virtually identical before and after the fire (Fig. 7.2b).

Statistical comparisons of relative growth rates among age groups and growth years (1977 to 1991) yielded statistically significant differences ($P < 0.01$) for the main effects and the interaction of the main effects. Growth was greatest in the

Table 7.1. Mean annual relative growth rates (mm/mm/year) for age groups of cutthroat trout captured in gill nets from eleven sites in Yellowstone Lake, 1978–1992

	Age Group							
Growth Year	Age 3	Age 4	Age 5	Age 6	Mean	n	SD	CV (%)
Prefire								
1977	0.68	0.22	0.16	0.09	0.35	160	0.34	96
1978	0.74	0.22	0.16	0.12	0.43	167	0.35	81
1979	0.83	0.26	0.13	0.17	0.44	146	0.41	94
1980	0.57	0.27	0.18	0.12	0.34	340	0.23	70
1981	0.71	0.33	0.19	0.17	0.40	124	0.32	81
1982	0.60	0.38	0.19	0.12	0.35	169	0.22	62
1983	0.59	0.27	0.15	0.11	0.33	169	0.24	73
1984	0.70	0.35	0.16	0.17	0.52	144	0.35	67
1985	0.64	0.32	0.16	0.10	0.36	173	0.27	76
1986	0.57	0.29	0.16	0.10	0.36	247	0.22	62
1987	0.62	0.35	0.21	0.14	0.39	188	0.22	56
Postfire								
1988	0.67	0.34	0.19	0.14	0.39	187	0.26	66
1989	0.63	0.36	0.20	0.14	0.33	154	0.22	67
1990	0.68	0.39	0.20	0.14	0.37	135	0.27	73
1991	0.74	0.39	0.19	0.13	0.37	120	0.28	76
Mean	0.65	0.30	0.18	0.13	0.38			
n	942	730	681	270		2623		
SD	0.28	0.12	0.06	0.04	0.28			
CV (%)	43	40	34	35				

Note: Mean, sample size (n), standard deviation (SD), and coefficient of variation (as a percent of the mean; CV [%]) were calculated for each year and each age group. Relative growth increments were estimated for each fish from the last complete year of growth (growth year) prior to capture (prefire: 1977–1987, postfire: 1988–1991).

younger age groups and least in the older age groups (Table 7.1), and Tukey-Kramer multiple means comparisons yielded statistically significant differences among all age groups ($P < 0.05$). Among-year comparisons indicated that the mean annual relative growth rate (0.52 mm/mm/year) was greatest in 1984; this estimate was statistically different from all other years. Mean annual relative growth rate was lowest for the 1989 growth year, and it differed statistically from estimates for 1981, 1978, 1979, and 1984 growth years (in order from the lowest mean annual relative growth rate). Relative growth rate estimates for the other

Table 7.2. Mean annual relative growth rates (mm/mm/year) for 25-mm length groups of cutthroat trout captured in gill nets from eleven sites in Yellowstone Lake, 1978–1992

Growth	25-mm Length Group													
Year	100	125	150	175	200	225	250	275	300	325	350	Mean	n	SD	CV(%)
Prefire															
1977	0.83	0.50	0.62	0.50	0.34	0.31	0.26	0.20	0.17	0.13	0.11	0.30	148	0.23	77
1978	0.99	0.73	0.57	0.45	0.36	0.33	0.25	0.22	0.17	0.14	0.10	0.40	157	0.32	80
1979	1.20	0.95	0.66	0.58	0.44	0.32	0.25	0.20	0.16	0.14	0.10	0.39	136	0.34	87
1980	0.76	0.56	0.43	0.37	0.32	0.26	0.22	0.18	0.15	0.11	0.09	0.32	332	0.19	61
1981	1.06	0.82	0.58	0.47	0.41	0.31	0.25	0.27	0.17	0.15	0.14	0.36	118	0.25	68
1982	0.72	0.61	0.52	0.51	0.39	0.30	0.26	0.19	0.15	0.14	0.11	0.35	168	0.22	62
1983	0.87	0.65	0.47	0.46	0.33	0.30	0.23	0.17	0.16	0.13	0.11	0.33	164	0.22	66
1984	1.04	0.76	0.67	0.53	0.50	0.36	0.27	0.22	0.19	0.15	0.15	0.48	134	0.27	55
1985	0.82	0.67	0.48	0.38	0.43	0.32	0.25	0.19	0.15	0.11	0.09	0.34	171	0.24	72
1986	0.89	0.62	0.48	0.37	0.36	0.27	0.18	0.20	0.15	0.12	0.09	0.36	247	0.22	62
1987	0.77	0.60	0.49	0.44	0.34	0.30	0.26	0.22	0.16	0.15	0.11	0.38	194	0.22	58
Postfire															
1988	0.82	0.62	0.47	0.51	0.34	0.28	0.26	0.20	0.16	0.14	0.11	0.38	188	0.25	66
1989	0.82	0.59	0.52	0.39	0.38	0.31	0.25	0.20	0.17	0.15	0.11	0.32	158	0.22	67
1990	0.83	0.61	0.45	0.48	0.41	0.28	0.26	0.21	0.18	0.14	0.11	0.35	127	0.23	67
1991	0.87	0.67	0.46	0.47	0.33	0.20	0.28	0.21	0.18	0.17	0.12	0.37	114	0.27	73
Mean	0.86	0.64	0.51	0.46	0.37	0.29	0.25	0.20	0.16	0.14	0.11	0.36			
n	175	326	242	154	211	230	231	294	325	194	174		255		
SD	0.23	0.16	0.14	0.14	0.10	0.08	0.06	0.05	0.04	0.04	0.03	0.25			
CV (%)	27	26	28	31	28	26	24	26	24	26	26				

Note: Mean, sample size (n), standard deviation (SD), and coefficient of variation (as a percent of the mean; CV[%]) were calculated for each year and each length group. Relative growth increments were estimated for each fish from the last complete year of growth (growth year) prior to capture (prefire: 1977–1987, postfire: 1988–1991).

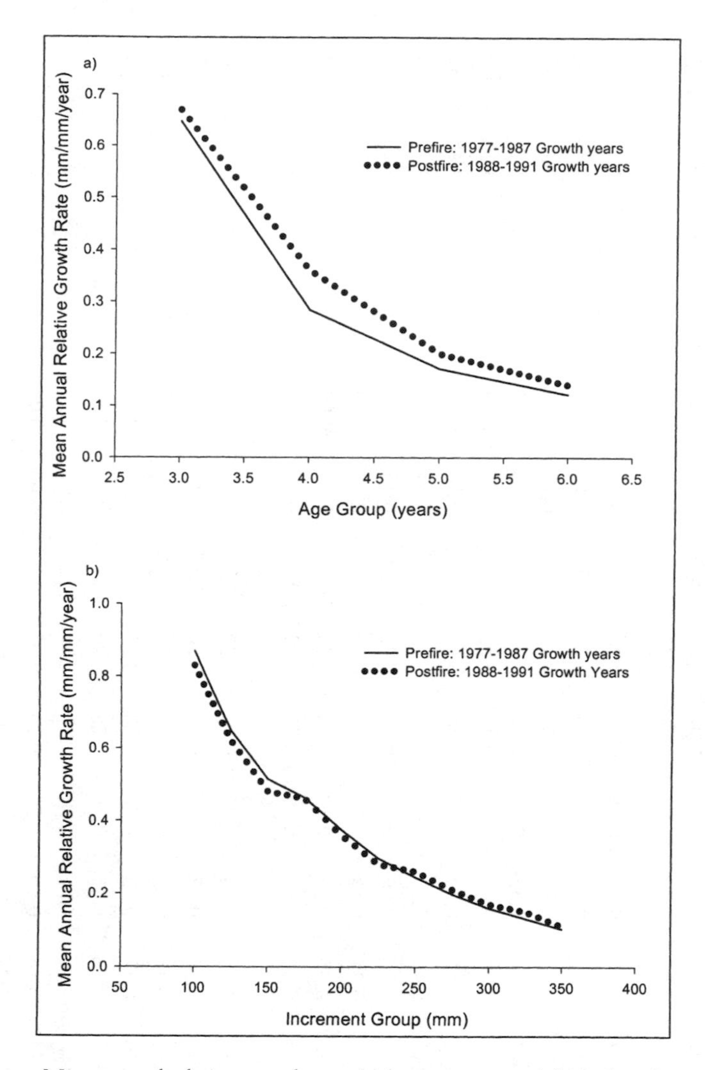

Figure 7.2. Mean annual relative growth rate, (a) by age group and (b) by length group, for Yellowstone cutthroat trout captured in Yellowstone Lake, 1977–1991 growth years.

postfire growth years (1991, 1990, and 1988) were moderate (0.37–0.39 mm/mm/year), but they did not differ significantly from any other growth year except 1984 ($P > 0.05$).

Main effects (growth year and length group) and the interaction of the main effects were statistically significant ($P < 0.01$) when relative growth rates for 25-mm length groups were compared among 1977–1991 growth years. Smaller Yellowstone cutthroat trout exhibited greater relative growth rates than larger fish

(Table 7.2), and rates differed significantly among all size groups ($P \leq 0.05$). When length groups were combined and compared among years, relative growth rate was highest for 1984 (0.48 mm/mm/year), and this estimate was statistically different from rates in all other years. Lowest relative growth rates were estimated for the 1977, 1980, 1989, and 1983 growth years (differences among these years were not significant; $P > 0.05$). Relative growth estimates for the remaining postfire years (1990, 1991, and 1988) were again moderate (0.35–0.38 mm/mm/year), differing significantly only from growth years with the highest (1979, 1978, and 1984) and lowest estimates.

Individual age and length groups exhibited temporal relationships in relative growth that were generally consistent throughout the period of record (1977–1991 growth years), and trends did not appear to change following the fires (Fig. 7.3). Mean annual relative growth rates of age-4 cutthroat trout increased persistently from 1977 through 1991 (growth years). Rates for age-5 and age-6 fish also increased, but trends were more gradual. Only age-3 cutthroat trout exhibited a trend of declining relative growth during this period. Relative growth rates for length groups appeared to decline slightly (100-mm and 150-mm) or remain relatively constant (Fig. 7.3).

Linear regressions for the entire period of record yielded statistically significant ($P \leq 0.05$) positive slopes for age groups 4 and 5 (Table 7.3). These simple linear models explained about 65 percent and 36 percent of the variation in relative growth for age-4 and age-5 cutthroat trout ($r^2 = 0.65$ and $r^2 = 0.36$, respectively). Adding a second predictor variable (a dummy variable that affected only postfire growth years) to the multiple regression model did not substantially change these relationships, and only the regression for age-4 cutthroat trout was significant ($P < 0.05$). The slope (β_2) for the dummy variable was < 0.000 in all cases, and it was not statistically significant ($P > 0.05$). The linear regression for the 150-mm length group yielded a statistically significant intercept and negative slope ($P < 0.05$), but the coefficient of determination was low ($r^2 = 0.32$). None of the other regressions (either simple linear or multiple regression) using length groups as a response variable produced statistically significant intercepts or slopes (Table 7.4).

When relationships were reevaluated for the temporally restricted data set (1984–1991 growth years), comparisons of relative growth rates for age groups and growth years were again statistically significant ($P < 0.01$), but the interaction of the main effects was not. Multiple means comparisons for this time period yielded significant differences among age groups, and relative growth decreased with age (Table 7.1). Mean annual relative growth rate by age groups was highest in 1984; this estimate was significantly different from all other years (P

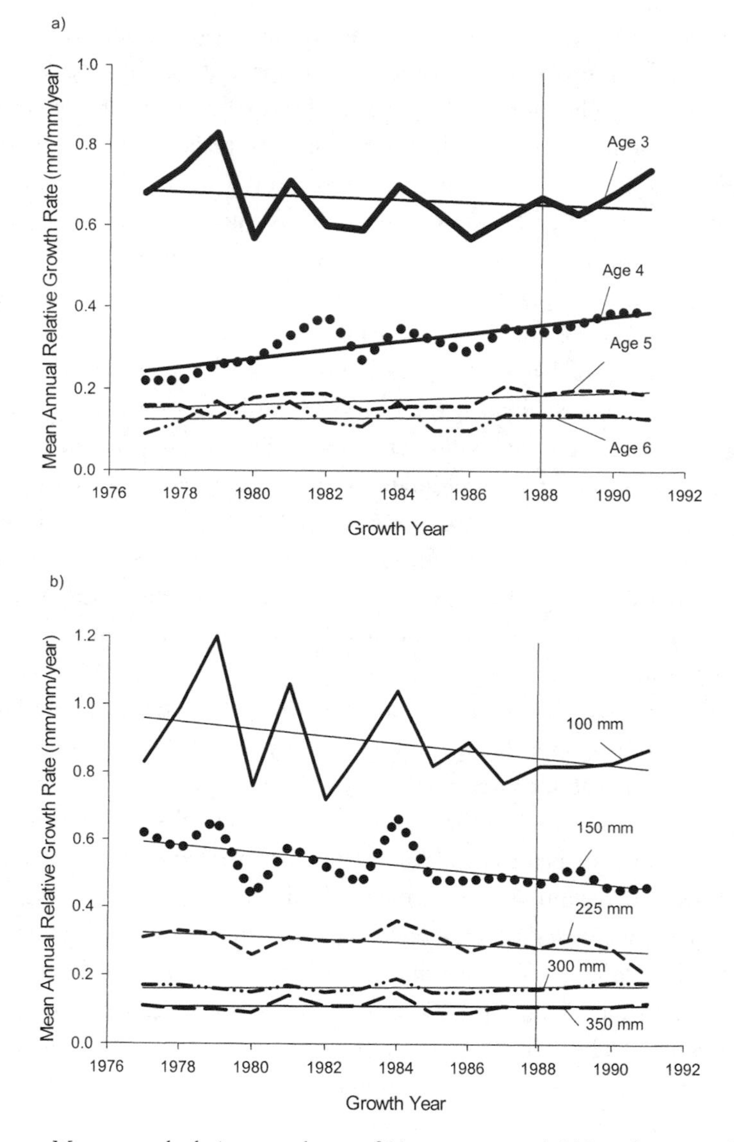

Figure 7.3. Mean annual relative growth rate of (a) age groups and (b) length groups, for Yellowstone cutthroat trout captured in Yellowstone Lake, 1977–1991 growth years.

< 0.05). Relative growth rate was lowest in 1989, and this estimate was statistically different from the estimates for 1988, 1987, and 1984. Remaining postfire years (1991, 1990, and 1988) had rates between the highest and lowest values.

Analysis of relative growth for 25-mm length groups in the temporally restricted data set (1984–1991 growth years) yielded significant differences in main

Table 7.3. Regression coefficients and coefficients of determination for simple linear and two-variable multiple regressions for cutthroat trout captured in gill nets from eleven sites in Yellowstone Lake, 1978–1992

	Age Group			
	Age 3	Age 4	Age 5	Age 6
Coefficient	1977–1991 Growth years (linear regression)			
β_0	6.47	−20.37[a]	−5.85[a]	−1.22
β_1	−0.003	0.010[a]	0.003[a]	0.001
r^2	0.03	0.65	0.36	0.01
	1977–1991 Growth years (multiple regression)			
β_0	22.74	−22.11[a]	−4.31	0.39
β_1	−0.011	0.011[a]	0.002	0.000
β_2	0.000	0.000	0.000	0.000
R^2	0.21	0.66	0.38	0.03
	1984–1991 Growth years (linear regression)			
β_0	−16.14	−19.29[a]	−11.88[a]	−0.81
β_1	0.008	0.010[a]	0.006[a]	0.000
r^2	0.15	0.52	0.52	0.00
	1984–1991 Growth years (multiple regression)			
β_0	5.68	−14.54	−14.74	12.06
β_1	−0.003	0.007	0.008	−0.006
β_2	0.000	0.000	0.000	0.000
R^2	0.24	0.53	0.53	0.15

Notes: Mean annual relative growth rate (mm/mm/year) for individual age groups was the response variable, and growth year (β_1) and a dummy variable valued to emphasize postfire years (β_2) were predictor variables. Regressions were evaluated for all years and a restricted set of growth years 1984–1991.

[a] $P < 0.05$.

effects (growth year and length group) and the interaction of the main effects ($P < 0.01$). Rates ranged from 0.83 mm/mm/year for the 100-mm length group to 0.11 mm/mm/year for the 350-mm group (Table 7.2). The 100-mm, 125-mm, 200-mm, and 275-mm length groups differed significantly from all other groups ($P < 0.05$). Mean growth rate of cutthroat trout was highest in 1984 and lowest in 1989; relative growth in 1984 was significantly different from growth in all other years ($P < 0.05$). Two of the four years with the lowest relative growth estimates (1989, 1985, 1990, and 1986) occurred following the fires.

Table 7.4. Regression coefficients and coefficients of determination for simple linear and two-variable multiple regressions for cutthroat trout captured in gill nets from eleven sites in Yellowstone Lake, 1978–1992

	25-mm Length Group										
	100	125	150	175	200	225	250	275	300	325	350
Coefficient	1977–1991 Growth years (linear regression)										
β_0	21.58	12.78	19.80[a]	6.63	1.51	8.16	−1.52	0.35	−1.11	−2.63	−0.67
β_1	−0.010	−0.006	−0.010[a]	−0.003	−0.001	−0.004	0.001	0.000	0.001	0.001	0.000
r^2	0.13	0.06	0.32	0.05	0.00	0.23	0.03	0.00	0.05	0.15	0.01
	1977–1991 Growth years (multiple regression)										
β_0	24.87	8.70	21.74	16.40	−3.93	4.35	3.09	0.38	1.11	−0.38	−0.49
β_1	−0.012	−0.004	−0.011	−0.008	0.002	−0.002	−0.001	0.000	0.000	0.000	0.000
β_2	0.000	0.000	0.000	0.000	0.000	0.000	0.000	0.000	0.000	0.000	0.000
R^2	0.13	0.06	0.32	0.14	0.04	0.27	0.16	0.00	0.17	0.22	0.01
	1984–1991 Growth years (linear regression)										
β_0	31.62	24.30	36.47	−4.52	29.49	29.16[a]	−7.56	−0.03	−3.15	−8.61	1.29
β_1	−0.015	−0.012	−0.018	0.003	−0.015	−0.015[a]	0.004	0.000	0.002	0.004	−0.001
r^2	0.21	0.27	0.39	0.01	0.38	0.59	0.10	0.00	0.08	0.33	0.01
	1984–1991 Growth years (multiple regression)										
β_0	58.58	36.49	64.15	31.28	55.10	49.97[a]	3.25	−4.76	2.17	−8.80	9.07
β_1	−0.029	−0.018	−0.032	−0.155	−0.028	−0.025[a]	−0.002	0.003	−0.001	0.004	−0.005
β_2	0.000	0.000	0.000	0.000	0.000	0.000	0.000	0.000	0.000	0.000	0.000
R^2	0.26	0.29	0.46	0.17	0.47	0.69	0.16	0.10	0.14	0.33	0.09

Notes: Mean annual relative growth rate (mm/mm/year) for 25-mm length group was the response variable, and growth year (β_1) and a dummy variable valued to emphasize postfire years (β_2) were predictor variables. Regressions were evaluated for all years and a restricted set of growth years 1984–1991.

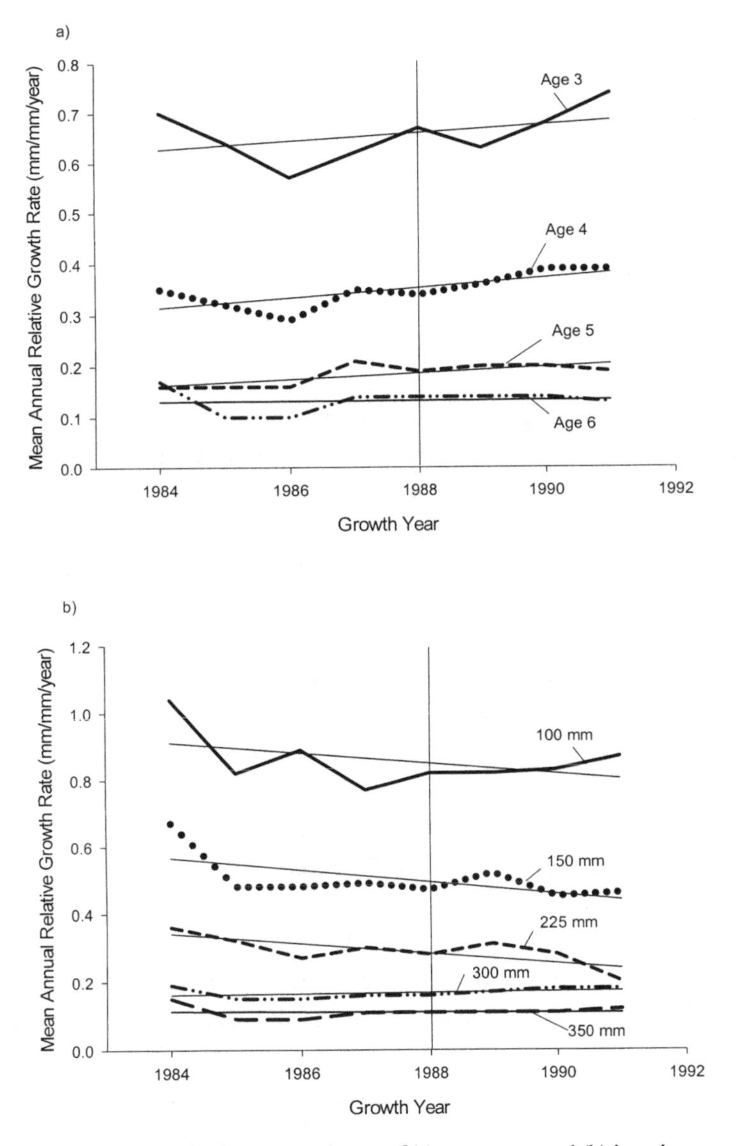

Figure 7.4. Mean annual relative growth rate of (a) age groups and (b) length groups, for Yellowstone cutthroat trout captured in Yellowstone Lake, 1984–1991 growth years.

Most trends in relative growth for individual age or length groups for the 1984–1991 time period did not change substantially from those observed with the long-term data set (Fig. 7.4). Linear regressions for the restricted time period yielded statistically significant ($P < 0.05$) positive slopes for age groups 4 and 5 ($r^2 = 0.52$ and $r^2 = 0.52$, respectively; Table 7.3); however, none of the equations developed using the multiple regression model had statistically significant intercepts or slopes ($P > 0.05$). The coefficient (β_2) derived from the dummy variable was < 0.000 for all of the multiple regressions, and it was never statistically significant ($P > 0.05$). Linear regressions with length group as a response variable had significant negative slopes for the 225-mm length group ($P < 0.05$, $r^2 = 0.69$; Table 7.4). None of the multiple regressions yielded statistically significant coefficients ($P > 0.05$).

DISCUSSION

Analysis of annual relative growth rates from cutthroat trout in Yellowstone Lake suggests that growth varied among years during the study period (1978–1992), but there was no dominant pattern of increased or decreased growth associated with the 1988 fire. Where prefire trends, either increasing or decreasing, were observed, only minor changes in the long-term relationships occurred in postfire years (Figs. 7.3 and 7.4). Although it is possible that other characteristics of the aquatic ecosystem (for example, nutrient dynamics, phytoplankton production, and zooplankton density) were altered by the fire, changes in fish productivity (as measured by fish growth) were not apparent.

Other studies from forested biomes of North America have demonstrated that nutrient input into aquatic systems following fire is related to such factors as fire intensity and size, short- and long-term weather patterns, and the physical, chemical, and biological properties of the watershed, and therefore local responses vary substantially (Gresswell 1999). Immediately following a fire, streams adjacent to burned areas may exhibit rapid postfire peaks of nitrogen and phosphorus that can last for a few weeks (Fredriksen 1971, Brown and others 1973). Nitrate-nitrogen is one of the most mobile chemical components of soil-water systems (Tiedemann and others 1979, Minshall and others 1997), and after the 1988 fires, nitrate increased substantially in streams of burned catchments throughout Yellowstone National Park (Brass and others 1996, Robinson and Minshall 1996). In a study of lakes from eastern Canada, Lamontagne and others (2000) found that increased nitrogen and phosphorus exported from burned landscapes provided significant inputs to lakes. Recent investigation of other small lakes (17–64 ha) in eastern Canada documented a higher biomass of rotifers and total phosphorus, total nitrogen, and chlorophyll *a* concentrations during the first two years following fire (Patoine and others 2000); however, St-Onge and Magnan (2000) failed to detect changes in growth of yellow perch *Perca flavescens* in these lakes that could be attributed to fire.

Other evidence suggests that postfire responses in lakes are quite complex. Bayley and others (1992) reported that larger lakes with extended renewal times might be buffered from the consequences of periodic increases in nutrient availability. In northeastern Minnesota, for example, nutrient export in tributary streams increased following fire, but very little of the phosphorus mobilized by the fire reached the study lakes (McColl and Grigal 1975, Wright 1976). Neither nutrient concentrations nor phytoplankton abundance in the lakes changed as a result of the fire (Wright 1976, Bradbury 1986, Tarapchak and Wright 1986). Precipitation falling directly on the lakes was identified as the principal source for ions and phosphorus in the study lakes (Wright 1976).

The influence of atmospherically derived nutrients may be even more important as lake surface area increases. For instance, annual trends of water quality in Yellowstone Lake and Lewis Lake (a 1,100-ha lake in the Snake River drainage, southwest of Yellowstone Lake) appear to reflect changes in atmospheric inputs before and after the 1988 fires (Lathrop 1994, Theriot and others 1997). Similarly, following a fire (1979) in the Heart Lake watershed (an 870-ha lake located in the Snake River drainage, south of Yellowstone Lake) changes in chemical concentrations were attributed to variations in precipitation inputs (Stottlemyer 1987).

Dissolved silica was the only chemical constituent in Yellowstone Lake that exhibited changes following the 1988 fire that were not directly related to atmospheric inputs (Lathrop 1994, Lathrop and others 1994, Theriot and others 1997). Similar increases in silica were reported after a small fire burned several tributary drainages of the Southeast Arm of Yellowstone Lake in 1976 (Jones and others 1979). Although increases in silica in the lake may be partially attributable to increased postfire erosion in the catchment, Theriot and others (1997) argued that silica was influenced by climatic controls independent of fire. Apparently, silica concentrations rise during periods of drought when there is a reduced biological demand because of declines in nitrogen and diatom productivity (Kilham and others 1996, Theriot and others 1997). Furthermore, Theriot and others (1997) found no evidence that the 1988 fires influenced phytoplankton biomass.

It has been suggested that nitrate-nitrogen limits production in Yellowstone Lake during the annual stratification (Gresswell and Varley 1988, Kilham and others 1996), and there is some indirect evidence of increased nitrate levels in the lake for five years following the fires of 1988. Although diatoms generally predominate the phytoplankton of Yellowstone Lake, *Anabaena flos-aquae,* a nitrogen fixer, becomes dominant as nitrogen declines in the epilimnion during summer stratification (Benson 1961, Garrett and Knight 1973). From 1989 through 1993, however, the bloom did not occur (T. Hansen, TW Services, Yellowstone National Park, personal communication). If nitrate levels in Yellowstone Lake were elevated in the years immediately following the fires, it is possible that the ability of *Anabaena* to fix nitrogen under nutrient-poor conditions was not an important competitive factor during this period.

Growth of cutthroat trout in Yellowstone Lake appears to be the result of complex interactions among food abundance, water temperature, and cutthroat trout density. Cutthroat trout in Yellowstone Lake feed primarily on crustacean zooplankton and benthic invertebrates (Benson 1961). This may be an important factor influencing food-web dynamics in the lake, because in systems where there are no vertebrate piscivores (such as Yellowstone Lake prior to the introduction of lake trout), zooplankton may be limited by predation from fishes rather than phytoplankton production (trophic cascades; Carpenter and others 1985, Proulx and others 1996, Järvinen and Salonen 1998). Increases in nutrient inputs in such systems can escalate primary production, but these changes in production may not be reflected at higher trophic levels. If trophic cascading does occur in Yellowstone Lake, annual variations in growth of cutthroat trout may be more closely related to changes in population density than nutrient input.

Although growth rates of cutthroat trout in Yellowstone Lake apparently did not change as a consequence of the 1988 fires, results of this study suggest that growth of younger and smaller cutthroat trout declined after the implementation of a 330-mm maximum size limit in 1975 (Gresswell and Varley 1988, Gresswell and others 1994). It is possible that decreases were associated with reduced mortality of mature cutthroat trout and subsequent increases in density of younger and smaller individuals (related to greater reproductive potential) after the regulation was implemented. Paradoxical increases observed in relative growth of age-4 and age-5 cutthroat trout suggest that prey items that support the diet of these older and larger fish are not limited, or alternatively, changes may be related to reduced mortality of faster growing fish under the current angling regulations that focus harvest on fish < 330 mm in length. Regardless of the specific mechanisms, results underscore the relative effects of natural and anthropogenic perturbations on the cutthroat trout of Yellowstone Lake. Observed changes in growth rates are consistent with observed alterations of population structure and life-history variation that have been associated with such activities as angler harvest, hatchery operations, and introduction of nonnative fishes (Gresswell and Varley 1988, Gresswell and others 1994), and it appears that the 1988 fires had little effect on growth of cutthroat trout in Yellowstone Lake.

Acknowledgments

This essay incorporates data that were collected with the help of a variety of U.S. Fish and Wildlife Service and U.S. National Park Service, Student Conservation Association, and Young Adult Conservation Corps employees. D. L. Mahony and D. G. Carty supervised field crews, and G. Boltz was responsible for a majority of the cutthroat trout age estimates. R. D. Jones and L. R. Kaeding facilitated use of USFWS equipment and housing. L. Ganio and G. Weaver provided statistical advice. D. S. Bateman, W. P. Dwyer, G. L. Larson, W. J. Liss, G. Weaver, and C. Whitlock reviewed preliminary drafts of the essay, and their comments were incorporated into the final version. The U.S. Fish and Wildlife Service Fishery Assistance Office in Yellowstone National Park and the U.S. National Park Service supported the research on which this essay is based, and the U.S. Geological Survey, Forest and Rangeland Ecosystem Science Center supported manuscript preparation. The opinions expressed are those of the author and do not necessarily reflect those of the acknowledged individuals or agencies.

References

Albin, D. P. 1979. Fire and stream ecology in some Yellowstone tributaries. California Fish and Game 65:216–238.

Bayley, S. E., D. W. Schindler, K. G. Beaty, B. R. Parker, and M. P. Stainton. 1992. Effects of multiple fires on nutrient yields from streams draining boreal forest and fen watersheds: nitrogen and phosphorus. Canadian Journal of Fisheries and Aquatic Sciences 49:584–596.

Benson, N. G. 1961. Limnology of Yellowstone Lake in relation to the cutthroat trout. U.S. Fish and Wildlife Service, Research Report 56, Washington, D.C.

Bradbury, J. P. 1986. Effects of forest fire and other disturbances on wilderness lakes in northeastern Minnesota, part 2. Paleolimnology. Archiv fur Hydrobiologie 106:203–217.

Brass, J. A., V. G. Ambrosia, P. J. Riggan, and P. D. Sebesta. 1996. Consequences of fire on aquatic nitrate and phosphate dynamics in Yellowstone National Park. Pages 53–57 in J. Greenlee, ed., Proceedings of the Second Biennial Conference on the Greater Yellowstone Ecosystem: The ecological implications of fire in Greater Yellowstone. International Association of Wildland Fire, Fairfield, Wash.

Brown, G. W., A. R. Gahler, and R. B. Marston. 1973. Nutrient losses after clear-cut logging and slash burning in the Oregon Coast Range. Water Resources Research 9:1450–1453.

Carpenter, S. R., J. F. Kitchell, and J. R. Hodgson. 1985. Cascading trophic interactions and lake productivity. BioScience 35:634–638.

Everhart, W. H., A. W. Eipper, and W. D. Youngs. 1975. Principles of fishery science. Cornell University Press, Ithaca, N.Y.

Fredriksen, R. L. 1971. Comparative chemical water quality: Natural and disturbed streams following logging and slash burning. Pages 125–137 in J. T. Krygier and J. D. Hall, eds., Forest land uses and stream environment. Oregon State University Press, Corvallis.

Garrett, P. A., and J. C. Knight. 1973. Limnology of the West Thumb of Yellowstone Lake, Yellowstone National Park. Montana State University, Final Report (Contract 2-101-0387), Bozeman.

Gresswell, R. E. 1995. Yellowstone cutthroat trout. Pages 36–54 in M. Young, ed., Conservation assessment for inland cutthroat trout. General Technical Report RM-GTR-256, USDA Forest Service, Rocky Mountain Forest and Range Experiment Station, Fort Collins, Colo.

———. 1999. Fire and aquatic ecosystems in forested biomes of North America. Transactions of the American Fisheries Society 128:193–221.

Gresswell, R. E., H. Li, and P. Rossignol. 1997. Ecological risk analysis of a piscivorous fish introduction into the Yellowstone Lake ecosystem. Pages 122–126 in R. E. Gresswell, P. Dwyer, and R. H. Hamre, eds., Wild Trout VI: Putting the native back in wild trout. Trout Unlimited and Federation of Fly Fishers, Vienna, Va.

Gresswell, R. E., W. J. Liss, and G. L. Larson. 1994. Life-history organization of Yellowstone cutthroat trout *(Oncorhynchus clarki bouvieri)* in Yellowstone Lake. Canadian Journal of Fisheries and Aquatic Sciences 51 (supplement 1):298–309.

Gresswell, R. E., and J. D. Varley. 1988. Effects of a century of human influence on the cutthroat trout of Yellowstone Lake. American Fisheries Society Symposium 4:45–52.

Griffith, J. S., Jr. 1988. Review of competition between cutthroat trout and other salmonids. American Fisheries Society Symposium 4:134–140.

Gutreuter, S. 1987. Considerations for estimation and interpretation of annual growth rates. Pages 115–126 in R. C. Summerfelt and G. E. Hall, eds., The age and growth of fish. Iowa State University Press, Ames.

Hintze, J. L. 1999. Number cruncher statistical system 2000. Jerry L. Hintze, Kaysville, Utah.

Hoskins, W. P. 1974. Yellowstone Lake tributary survey project. Interagency Grizzly Bear Study Team, Bozeman, Mont.

Hoskins, W. P. 1975. Yellowstone Lake tributary study. Interagency Grizzly Bear Study Team, Bozeman, Mont.

Järvinen, M., and K. Salonen. 1998. Influence of changing food web structure on nutrient limitation of phytoplankton in a highly humic lake. Canadian Journal of Fisheries and Aquatic Sciences 55:2562–2571.

Jones, R. D., R. Andresek, D. G. Carty, R. E. Gresswell, D. L. Mahony, and S. Relya. 1990. Fishery and aquatic management program in Yellowstone National Park. U.S. Fish and Wildlife Service, Technical Report for 1989, Yellowstone National Park, Wyo.

Jones, R. D., D. G. Carty, R. E. Gresswell, C. J. Hudson, L. D. Lentsch, and D. L. Mahony. 1986. Fishery and aquatic management program in Yellowstone National Park. U.S. Fish and Wildlife Service, Technical Report for 1985, Yellowstone National Park, Wyo.

Jones, R. D., D. G. Carty, R. E. Gresswell, C. J. Hudson, and D. L. Mahony. 1987. Fishery and aquatic management program in Yellowstone National Park. U.S. Fish and Wildlife Service, Technical Report for 1986, Yellowstone National Park, Wyo.

Jones, R. D., J. D. Varley, D. E. Jennings, S. M. Rubrecht, and R. E. Gresswell. 1979. Fishery and aquatic management program in Yellowstone National Park. U.S. Fish and Wildlife Service, Technical Report for 1978, Yellowstone National Park, Wyo.

Kaeding, L. R., G. D. Boltz, and D. G. Carty. 1996. Lake trout discovered in Yellowstone Lake threaten native cutthroat trout. Fisheries (Bethesda) 21(3):16–20.

Kaplinski, M. A. 1991. Geomorphology and geology of Yellowstone Lake, Yellowstone National Park, Wyo. M.S. thesis, Northern Arizona University, Flagstaff.

Kilham, S. S., E. C. Theriot, and S. C. Fritz. 1996. Linking planktonic diatoms and climate change using resource theory in the large lakes of the Yellowstone Ecosystem. Limnology and Oceanography 41:1052–1062.

Laakso, M., and O. B. Cope. 1956. Age determination in Yellowstone cutthroat trout by the scale method. Journal of Wildlife Management 20:138–153.

Lamontagne, S., R. Carignan, P. D'Arcy, Y. T. Prairie, and D. Pare. 2000. Element export in runoff from eastern Canadian Boreal Shield drainage basins following forest harvesting and fires. Canadian Journal of Fisheries and Aquatic Sciences 57:118–128.

Lathrop, R. G., Jr. 1994. Impacts of the 1988 wildfires on the water quality of Yellowstone and Lewis lakes, Wyoming. International Journal of Wildland Fire 4:169–175.

Lathrop, R. G., Jr., J. D. Vande Castle, and J. A. Brass. 1994. Monitoring changes in Greater Yellowstone lake water quality following the 1988 wildfires. Geocarto International 3:49–57.

Mahony, D. L., and J. R. Ruzycki. 1997. Initial investigations towards the development of a lake trout removal program in Yellowstone Lake. Pages 153–162 in R. E. Gresswell, W. P. Dwyer, and R. Hamre, eds., Wild Trout VI. Trout Unlimited and Federation of Fly Fishers, Vienna, Va.

Marrin, D. L., and D. C. Erman. 1982. Evidence against competition between trout and nongame fishes in Stampede Reservoir, California. North American Journal of Fisheries Management 2:262–269.

McColl, J. G., and D. F. Grigal. 1975. Forest fire: Effects on phosphorus movement to lakes. Science 188:1109–1111.

McEachern, P., E. E. Prepas, J. J. Gibson, and W. P. Dinsmore. 2000. Forest fire induced impacts on phosphorus, nitrogen, and chlorophyll *a* concentrations in boreal subarctic lakes of northern Alberta. Canadian Journal of Fisheries and Aquatic Sciences 57:73–81.

McIntyre, J. D. 1995. Review and assessment of possibilities for protecting the cutthroat trout of Yellowstone Lake from introduced lake trout. Pages 28–33 in J. D. Varley and P. Schullery, eds., The Yellowstone Lake crisis: Confronting a lake trout invasion. U.S. National Park Service, Yellowstone Center for Resources, Yellowstone National Park, Wyo.

Minshall, G. W., C. T. Robinson, and D. E. Lawrence. 1997. Postfire responses of lotic ecosystems in Yellowstone National Park, U.S.A. Canadian Journal of Fisheries and Aquatic Sciences 54:2509–2525.

Minshall, G. W., T. V. Royer, and C. T. Robinson. 2001. Response of the Cache Creek macroinvertebrates during the first 10 years following disturbance by the 1988 Yellowstone wildfires. Canadian Journal of Fisheries and Aquatic Sciences 58:1077.

Montgomery, R. H., and K. H. Reckhow. 1984. Techniques for detecting trends in lake water quality. Water Resources Bulletin 20:43–52.

Patoine, A., B. Pinel-Alloul, E. E. Prepas, and R. Carignan. 2000. Do logging and forest fires influence zooplankton biomass in Canadian Boreal Shield lakes? Canadian Journal of Fisheries and Aquatic Sciences 57:155–164.

Planas, D., M. Desrosiers, S. R. Groulx, S. Paquet, and R. Carignan. 2000. Pelagic and benthic algal responses in eastern Canadian Boreal Shield lakes following harvesting and wildfires. Canadian Journal of Fisheries and Aquatic Sciences 57:136–145.

Proulx, M., F. R. Pick, A. Mazumder, P. B. Hamilton, and D. R. S. Lean. 1996. Effects of nutrients and planktivorous fish on the phytoplankton of shallow and deep aquatic systems. Ecology 77:1556–1572.

Ricker, W. E. 1975. Computation and interpretation of biological statistics of fish populations. Fisheries Research Board of Canada, Bulletin 191.

Rieman, B. E., D. Lee, G. Chandler, and D. Myers. 1997. Does wildfire threaten extinction for salmonids? Responses of redband trout and bull trout following recent large fires on the Boise National Forest. Pages 47–57 in J. Greenlee, ed., Proceedings: Fire effects on rare and endangered species and habitats conference. International Association of Wildland Fire, Fairfield, Wash.

Robinson, C. T., and G. W. Minshall. 1996. Physical and chemical responses of streams in Yellowstone National Park following the 1988 wildfires. Pages 217–221 in J. Greenlee, ed., Proceedings of the Second Biennial Conference on the Greater Yellowstone Ecosystem: The ecological implications of fire in Greater Yellowstone. International Association of Wildland Fire, Fairfield, Wash.

Ruzycki, J. R., and D. A. Beauchamp. 1997. A bioenergetics modeling assessment of the lake trout impact in Yellowstone Lake. Pages 127–133 in R. E. Gresswell, P. Dwyer, and R. H. Hamre, eds., Wild Trout VI: Putting the native back in wild trout. Trout Unlimited and Federation of Fly Fishers, Vienna, Va.

Simon, J. R. 1962. Yellowstone fishes. Yellowstone Library and Museum Association, Yellowstone Interpretive Series 3, Yellowstone National Park, Wyo.

St-Onge, I., and P. Magnan. 2000. Impact of logging and natural fires on fish communities of Laurentian Shield lakes. Canadian Journal of Fisheries and Aquatic Sciences 57:165–174.

Stottlemyer, R. 1987. Ecosystem nutrient release from a large fire, Yellowstone National Park. Ninth Conference on Fire and Forest Meteorology. American Meteorological Society, San Diego.

Tabachnick, B. G., and L. S. Fidell. 1989. Using multivariate statistics. HarperCollins, New York.

Tarapchak, S. J., and H. E. Wright, Jr. 1986. Effects of forest fire and other disturbances on wilderness lakes in northeastern Minnesota, part 1. Limnology. Archiv fur Hydrobiologie 106:177–202.

Theriot, E. C., S. C. Fritz, and R. E. Gresswell. 1997. Long-term limnological data from the larger lakes of Yellowstone National Park. Alpine and Arctic Research 29:304–314.

Tiedemann, A. R., C. E. Conrad, J. H. Dieterich, J. W. Hornbeck, W. F. Megahan, L. A. Viereck, and D. D. Wade. 1979. Effects of fire on water: A state-of-knowledge review. U.S. Forest Service General Technical Report WO-10.

Warren, C. E. 1971. Biology and water pollution control. Saunders, Philadelphia.

Wright, R. F. 1976. The impact of forest fire on the nutrient influxes to small lakes in northeastern Minnesota. Ecology 57:649–666.

Zar, J. H. 1984. Biostatistical analysis, 2d ed. Prentice-Hall, Englewood Cliffs, N.J.

Chapter 8 Stream Ecosystem Responses to Fire: The First Ten Years

G. Wayne Minshall, Todd V. Royer, and Christopher T. Robinson

The fundamental importance of disturbance in stream ecosystem dynamics and, in particular, of maintaining a heterogeneous or patchy habitat templet in both space and time is widely accepted (Resh and others 1988, Pringle and others 1988, Townsend 1989). Wildfires represent one of the primary forms of large-scale disturbance to natural streams in the Pacific Northwest (Swanson and others 1994). In the inland Northwest, prior to active fire suppression and modern anthropogenic changes in land use patterns, historical (1540–1940) fire episodes at widespread landscape scales have occurred on average every twelve years (Barrett and others 1997). During this 400-y period an annual average of 243×10^4 ha burned, including 74.5×10^4 ha of forests, compared with the $81–122 \times 10^4$ ha that burned during the four largest known fires in the area in the twentieth century (1910, 1919, 1988, 1994).

In 1988, wildfires in the Greater Yellowstone Area (GYA) burned approximately 25×10^4 ha and provided an important opportunity to assess the effects of large-scale disturbance on stream ecosystems over time. Twenty separate river basins or major subbasins, constituting a third of the stream systems in Yellowstone National Park, were directly affected by the wildfires (Minshall and Brock 1991). We annually examined environmental and biological responses of twenty-one streams in Yellowstone National Park (Fig. 8.1) in the first five years following the extensive 1988 wildfires and again in 1998. Further, we evaluated a subset of these streams (mostly located on Cache Creek) in each of the intervening years (1994–1997) except 1996.

In this chapter we review the published results of the first five years of our research (1988–1992) (Minshall and others 1997 and references cited therein) and expand on those results using data from the next five years extracted from manuscripts in preparation or recently submitted for publication. Our findings demonstrate an integral relationship over time between a stream and its catchment (drainage basin) following such large-scale disturbances as wildfire. However, individual streams varied considerably in the magnitude and timing of response depending on such factors as stream size, proportion of the catchment burned, and localized differences in precipitation, geology, and topography. Temporally, major physical changes in streams that occurred from 1995 through 1997 (postfire years 7–9) were especially noteworthy. The physical environment of streams in some burned watersheds, such as parts of Cache Creek, changed more in those three years than in the first six postfire years.

Landscape changes over time and among streams were readily apparent in photographs taken from the same location and position on each visit—a form of documentation termed rephotography (Fig. 8.2). Habitat conditions then were documented using measurements of channel morphology, substratum particle-size distribution, and accumulations of large woody debris. We expected the changes in instream habitat would be reflected in differences in the abundance and kinds of organisms found in the streams. Documenting these changes was important because surprisingly little is known of the extended effects of wildfire, or other large-scale disturbances, on stream ecosystems and because of a strong interest in the integrity of the park's natural resources, including status of the trout sport fishery.

CHRONOLOGY OF CHANGES IN YELLOWSTONE STREAMS FOLLOWING WILDFIRES

It is instructive to separate the temporal responses of streams to wildfire into four periods: (1) immediate changes (the time of active burning to a few days after); (2) short-term changes (a few days after active burning to the end of the first year); (3) midterm changes (the second year to sometime beyond the tenth year); and (4) long-term changes (tens to hundreds of years). The precise length of each

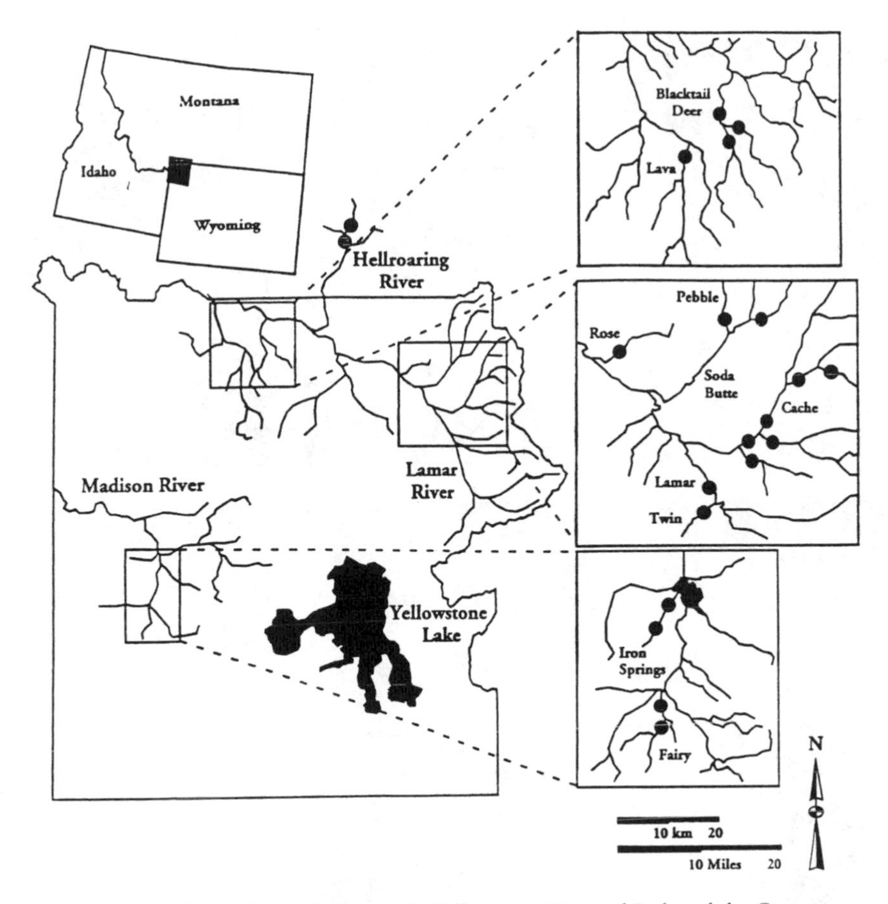

Figure 8.1. Location of the study streams in Yellowstone National Park and the Greater Yellowstone Area. Rose, Soda, Pebble, and Amphitheater (to the right of Pebble) Creeks are reference streams; the remainder was burned (39–92 percent) by wildfires in 1988.

period depends on the degree of disturbance by fire and the local environmental conditions of burned catchments such as weather and climate, topography, geology, soil conditions, and forest type. Basically, we expected a strong association between stream ecosystem components, the ecological integrity (sensu King 1993) of the watershed, and the forest and riparian conditions upstream of each focal point (stream reach, study site, and so on) (Vannote and others 1980). Biotic parameters (such as richness, density, biomass, and dominance) were expected to decline dramatically in Yellowstone streams within a year of disturbance by wildfire and then gradually recover to prefire conditions over an extended period (tens to hundreds of years) (Minshall and others 1989, Minshall and Brock 1991). We also predicted that the physical and biological impacts

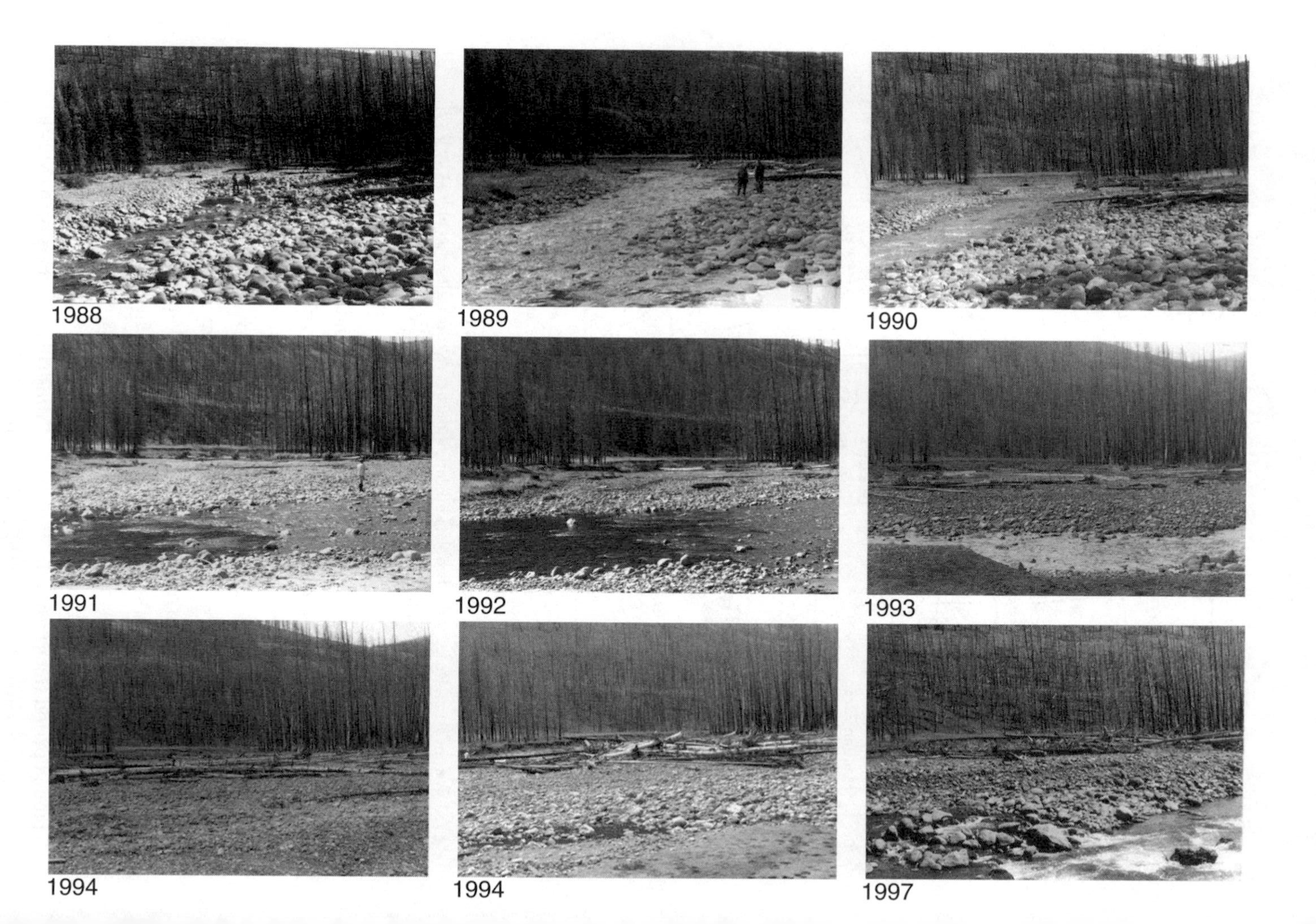
1988
1989
1990
1991
1992
1993
1994
1994
1997

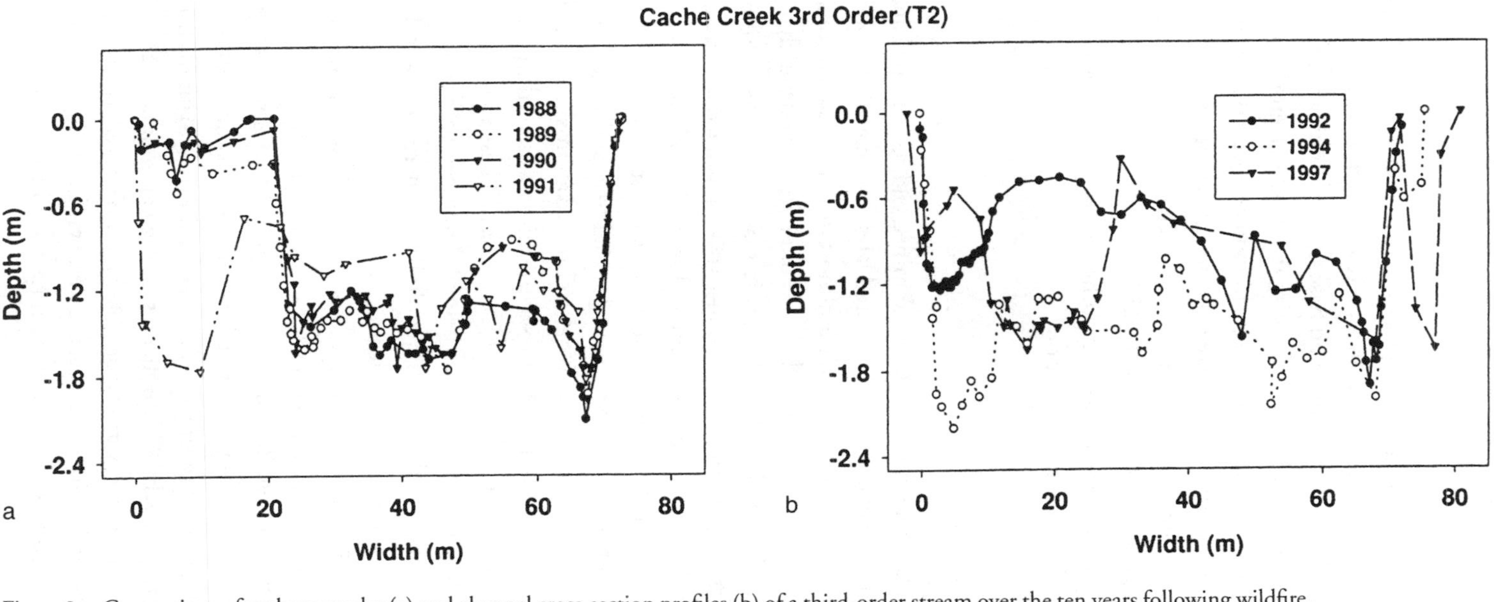

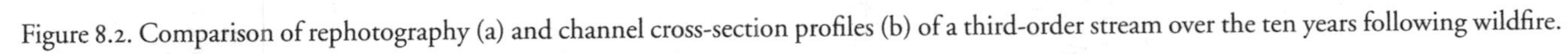
Figure 8.2. Comparison of rephotography (a) and channel cross-section profiles (b) of a third-order stream over the ten years following wildfire.

would vary in a downstream direction due to dilution by an increasing number of unburned catchments, changes in food resources from a dominance of an externally derived (allochthonous) food base to a predominance of an internally derived (autochthonous) one, and shifts in consumer group (macroinvertebrates) composition paralleling changes in food resources (Chapter 9). The immediate and short-term effects were expected to be the most dramatic and to profoundly alter stream habitats and biota, relative to those before the fires (Minshall and others 1989, Minshall and Brock 1991). The mid- and long-term changes in stream ecosystems were hypothesized to parallel the successional replacement of the terrestrial vegetation.

IMMEDIATE EFFECTS

Beginning in late September 1988, we examined eighteen burned and four reference streams (one of each was eliminated later from consideration). Losses in upland and riparian vegetation and the almost instantaneous conversion of terrestrial vegetation to charcoal and ash resulted in immediate changes in the amount of sunlight and quality of organic matter—in other words, food resources—entering the streams. The most striking immediate changes within stream channels were the incineration and scorching of emergent mosses and heat fracturing of rocks in and adjacent to smaller streams. Although most burned trees remained standing, many downed trees and large limbs were observed within or bridging streams. We also observed approximately ten dead cutthroat trout in our 250-m long study sections in midsized (third-order) Cache Creek and the West Fork of Blacktail Deer Creek. These were believed to have died as a direct result of the fire (see below). However, we also know of an instance on a tributary to the Little Firehole River where an errant drop of fire-retardant caused a number of fish deaths.

Most dissolved chemical concentrations increased in streams of burned catchments the first year following the fires. Based on studies by other researchers in 1988 on the effects of wildfire on Glacier National Park streams (Spencer and Hauer 1991), we believe dramatic and rapid increases in stream phosphorus and nitrogen levels occurred during the Yellowstone fires due to inputs from ash and smoke gases, respectively. We speculate that high ammonia levels that entered the water from the smoke were responsible for the observed fish mortalities. Few or no immediate deleterious effects of fire were evident in algae growing attached to rocks (periphyton) or macroinvertebrate assemblages, even in the smallest streams examined. These impacts are more difficult to discern due to the small size and rapid decay rates of the organisms involved, although lotic macroinvertebrates are adversely affected by exposure to ammonia (Gammeter and Frutiger 1990).

Within the fire perimeter, there was an extensive mosaic of burned, partially burned, and unburned areas (Turner and others 1994, Chapter 13). Our investigation revealed a comparable patchiness among and within watersheds and distinct differences in the effects of wildfire on streams of different size (Minshall and Brock 1991, Minshall and others 1997). Following fire, small (first- and second-order) headwater streams (for example, Fairy Creek and upper tributaries of Blacktail Deer Creek) were more physically and chemically variable than intermediate-size (third- and fourth-order) streams (for example, Cache and Hellroaring Creeks) or reference streams. In general, catchments of smaller streams that burned generally had a greater proportion of their area affected than did burned catchments of larger streams. For our study streams, the mean catchment burned was 75 percent for first- and second-order streams and 50 percent for third- and fourth-order streams. However, we noticed during aerial and ground reconnaissance that the catchments of many fire-affected third- and fourth-order streams throughout Yellowstone Park and along its northern boundary were less than 50 percent burned, and that this amount was much less for even larger streams. No streams larger than sixth order are found in the park. Consequently, the initial impact on biological properties also appeared more pronounced in smaller streams, although intermediate-size burned streams located in steep terrain with confined floodplains (for example, third-order Cache and Hellroaring Creeks) experienced greater amounts of overland flow and associated effects on the biota than other large study streams.

The most consistent exceptions from the general patterns found in this study were Fairy and Iron Springs Creeks, and their differences were attributable to one or more relatively distinctive features. These two streams were located along the west side of Yellowstone in an interior-type climate, characterized by a spring peak in precipitation and underlain by different base rock (rhyolite) than the other study streams (located in the north-central and northeastern portions of the park on andesite rock with a montane-type climate). In addition, Fairy Creek had the lowest gradient of any study stream. The second-order site was not forested and was strongly influenced by geothermal springs. A large proportion of flow in Iron Springs Creek is groundwater; thus the third-order site displayed little variation in flow and usually did not freeze over in winter.

SHORT-TERM EFFECTS

Between October 1988 and March 1989, macroinvertebrate density, biomass (not shown in figure), and richness tended to decrease in burned streams, whereas values increased or remained constant in reference streams (Fig. 8.3). No physical disturbance from runoff occurred during this period because rainfall was minor, the ground was frozen and snow-covered, and the streams were ice-covered for most of the time. Therefore we attribute these biotic changes in burn streams to high amounts of charcoal (greater than 40 percent) in the benthos resulting from the fires (Minshall and others 1997) and the absence of unburned leaf litter and algae (food resources). We had expected that burned material would be the principal source of allochthonous organic matter at this time; however, we had not anticipated that ice and snow cover would reduce the amount of light reaching the streambed and severely limit the growth of attached algae which normally is an alternative food resource (Fig. 8.3).

We believe that the input of charcoal decreased the palatability and quality (for example, increased carbon:nitrogen [C:N] ratio) of organic matter resources as food. For example, in a food utilization study of some selected stream invertebrates, only one of eleven taxa utilized burned organic matter as a food source (Mihuc and Minshall 1995, Chapter 9). Periphyton biomass also decreased in burned streams (except Iron Springs Creek) during this period, although comparable changes were observed in reference streams.

Melting of the 1989 snowpack was protracted (P. Farnes, Snowcap Hydrology, Bozeman, Mont., personal communication, Chapter 10). Consequently, although several periods of "blackwater," associated with overland flow from heavy rains, occurred between spring runoff and our August 1989 sampling, streambed erosion and channel alterations generally were much less than expected or than occurred in later years. However, several first- through third-order streams, particularly in the Cache Creek and Hellroaring Creek catchments, had substantial channel alteration and rearrangement of woody debris. In addition, reductions in flow and substrate heterogeneity occurred in burned streams, as indicated by changes in annual coefficients of variation for these measures between 1988 and 1990 (Minshall and others 1997). A number of studies in other areas of the western United States have documented similar changes in burned streams resulting from increased sediment loads and peaks in runoff (Helvey 1973, 1980, Beaty 1994).

Most dissolved constituents, nitrates in particular, were higher in August 1989 than in October 1988, apparently in response to rainstorms during or immedi-

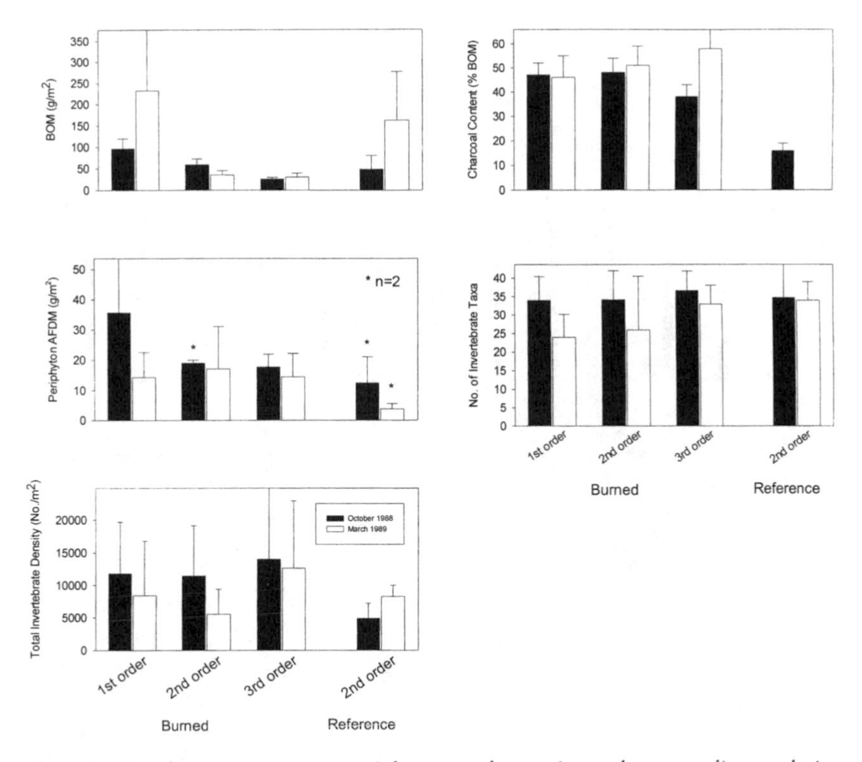

Figure 8.3. Benthic organic matter, periphyton, and macroinvertebrate standing stocks just after fire (October 1988) and prior to spring snowmelt (March 1989).

ately prior to the summer 1989 sample collections. We found a direct correlation between nitrate loss and percent catchment burned in the Yellowstone study streams (Minshall and others 1997). Other short-term changes observed for burn streams were the downstream movement of charcoal and fine sediment, and increased water temperatures in first- and second-order streams.

MIDTERM CHANGES IN POSTFIRE STREAM SYSTEMS

The midterm responses (1990 to present) of Yellowstone stream ecosystems to wildfire were driven primarily by increases in runoff from snowmelt and localized rainstorms, and by regrowth of terrestrial upland and riparian plant cover. Precipitation records for the northeast part of Yellowstone and nearby Parker Peak (the nearest meteorological station with continuous records) were chosen to represent annual conditions (Fig. 8.4a). The 1988 fires occurred in a dry year

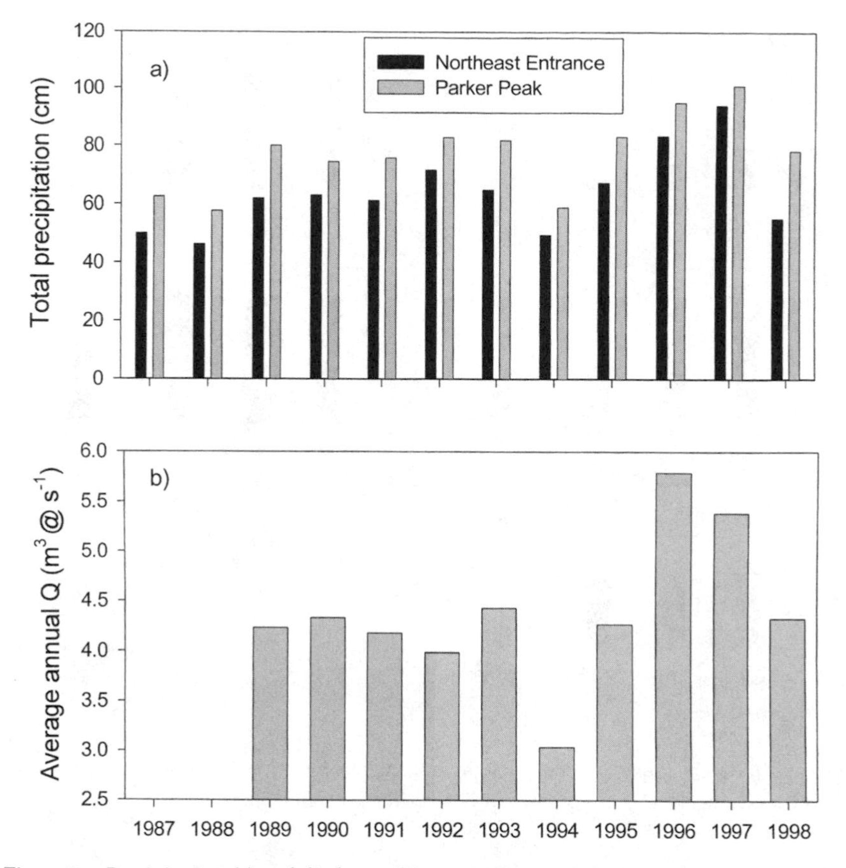

Figure 8.4. Precipitation (a) and discharge (b) at recording stations in northeastern Yellowstone NP.

following an extended drought. Precipitation in 1990–1993 was near normal; 1994 was similar to 1988, and 1996–1998 was the wettest of the period. Annual patterns in stream discharge for Soda Butte Creek, which lies in the same unburned watershed as the meteorological station and is immediately adjacent to Cache Creek, closely followed that of precipitation (Fig. 8.4b). The lowest discharge occurred in 1994 (and probably 1988) and the highest in 1996–1998 (see Chapter 10).

These precipitation and discharge records, however, are not totally representative of factors affecting conditions in such nearby burn streams as Cache Creek. For instance, precipitation in 1995 was one of the four lowest and discharge was average for the period, yet 1995 had the third-highest peak discharge resulting from a rain-on-snow event in June (A. Marcus, University of Oregon,

Eugene, personal communication). In addition, neither the precipitation nor discharge records agree precisely with observed changes in stream channel conditions in adjacent Cache Creek (Minshall and others 2001b). For example, Soda Butte Creek had similar, moderate discharges in 1989–1991, but our Cache Creek sites showed moderate downcutting and a loss of large woody debris from the channel in 1989, little change in 1990, and the first major episode of stream channel alteration in 1991. This difference, between discharge conditions in Soda Butte Creek and channel dynamics in Cache Creek, is probably due to a combination of variations in local meteorological conditions (see Chapter 3) and the effects of wildfire (for example, the loss of terrestrial plant cover, decreased evapotranspiration, and increased rate and amount of surface runoff).

Although some major changes in stream channel conditions were evident in the first three postfire years, the biota in the burn streams appeared to be on a "fast recovery track" (*sensu* Minshall and Brock 1991) associated with relatively little change in channel morphology and progressive regrowth of the riparian vegetation. However, postfire year 3 (1991) was marked by at least two large runoff events that caused major physical changes in all burn streams having moderate to steep gradients. Ewing (1996) also noted elevated suspended sediment loads in 1991 in the Lamar River downstream of its confluence with Cache Creek. In Cache Creek, channel morphology changed only moderately during the rest of the period (1993–1997) in headwater streams (Fig. 8.5). In many places along these streams, however, over the years the flow tended to move back and forth across the valley floor, in a temporally braided fashion, as deposition and erosion created new flow paths. In third-order Cache, in all years during this period except 1994, dramatic changes in channel conditions were recorded at most or all transects (see Fig. 8.2). In fourth-order Cache, year-to-year changes in channel form and substratum conditions were relatively minor until 1997, when a wave of cobble-sized stones entered the study section and the thalweg shifted from the left to the right side of the channel (Fig. 8.6). To summarize, major alterations in the stream channels over time appeared to move progressively downstream, from the headwater tributaries in 1989, 1991, and 1992, to third-order streams in 1991–1997, and, finally, to the fourth-order stream in 1997. Channel cross-section profiles (Figs. 8.2, 8.5, 8.6) showed a general narrowing and deepening of the channels of headwater streams (first- and second-order) during the ten-year period. In contrast, the larger streams tended to maintain a relatively constant active channel width through 1991, after which it expanded markedly and was accompanied by considerable bed movement that indicated substantial bank erosion and in-channel sediment deposition.

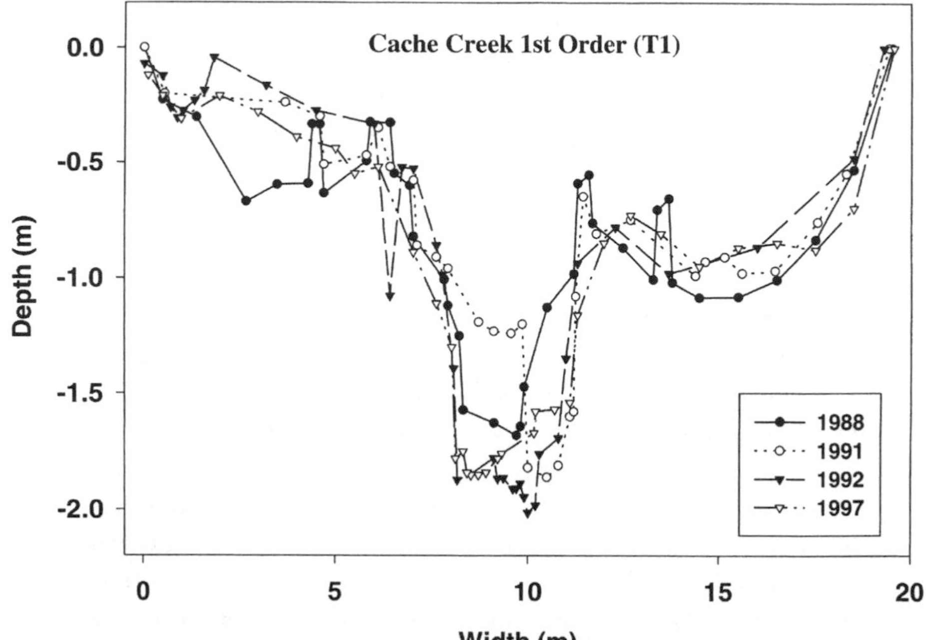

Figure 8.5. Rephotography and channel cross-sectional profiles of the Cache Creek first-order site at one (T-1) of five permanent transects established in 1988. Photos by G. Wayne Minshall.

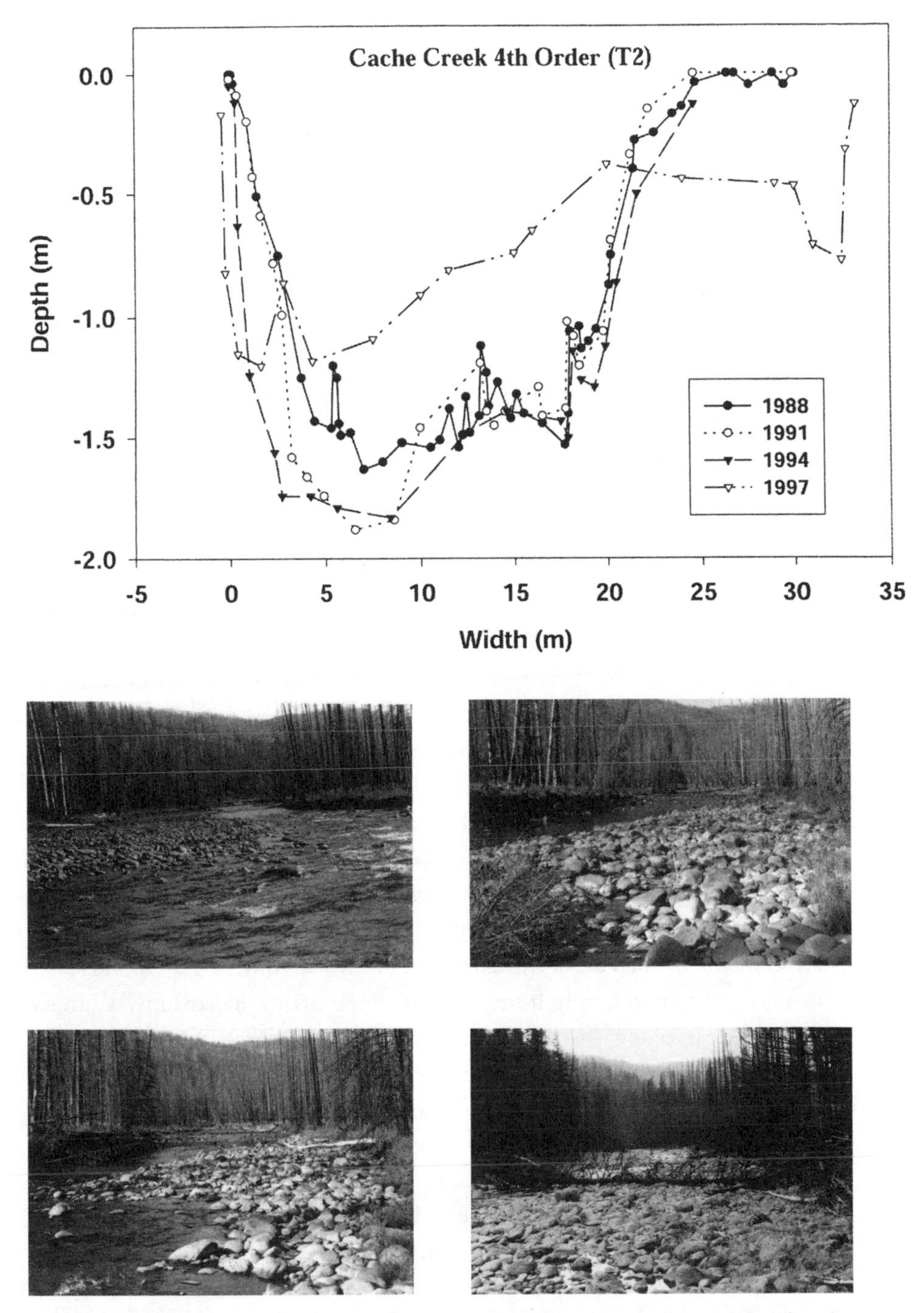

Figure 8.6. Rephotography and channel cross-sectional profiles of the Cache Creek fourth-order site at one (T-2) of five permanent transects established in 1988. Photos by G. Wayne Minshall.

Woody debris in streams retains smaller organic matter and sediment and provides valuable habitat for fish and macroinvertebrates (see also Chapter 12 for similar roles of woody debris in terrestrial systems). Within the burned catchments, substantial movement of woody debris occurred in all streams throughout the midterm period (Figs. 8.2, 8.5, 8.6). Initially, the first- through third-order burn streams had more large wood pieces than did fourth-order burn streams (Minshall and others 1997). This reflected the lower competency of high flows to move large pieces of wood and the close proximity of trees to the main channel in small streams. However, high flows in 1995–1997 undercut banks and felled many snags into the third- and fourth-order streams that were transported and left remaining during base flow (Figs. 8.2, 8.6). This large woody debris accumulated on point bars, at the heads of islands, and in the shallows of braided sections where it remained longer than a year or two. Small streams had lower woody debris volumes because a large portion of fallen trees remained outside the channel margin.

The amount of benthic organic matter (BOM) in Cache Creek was highest immediately after the fire and greater in the headwaters (first and second order) than in the mainstream (third and fourth order) resulting from a supplementation of existing (prefire) BOM by charcoal (for example, Fig. 8.7). Values at first- and second-order Cache declined from 1988 through 1990 and then leveled off at values near those of the reference stream. The decrease in BOM in third- and fourth-order Cache was a year later than at the two headwater sites, probably because of temporary retention and distance-related delays as the material was being transferred through the system. Just after the fires, charcoal constituted more than 60 percent of the benthic organic matter in the first- through third-order burn sites and about 50 percent at the fourth-order site (Fig. 8.7). The relative amount of charcoal decreased progressively with time, except for increases in 1995 and 1997 resulting from overland transport to the stream via surface runoff. Not until 1993–1995 did values of charcoal approach levels found in reference streams. However, even after 10 y, they were higher in the burn than in the reference streams (Minshall and others 2003).

The quantity of benthic organic matter in burn streams was initially higher than in the reference streams due to the addition of charcoal from the burned trees and shrubs. We expected a rapid decline and then a gradual recovery of BOM in the form of leaf litter as the riparian vegetation grew back relatively quickly from surviving roots (confirmed by observation of profuse sprouting the year following fire) (Chapters 4, 5). Instead, there has been an extended period of decline in BOM due to the subsequent inhibition and loss of riparian vege-

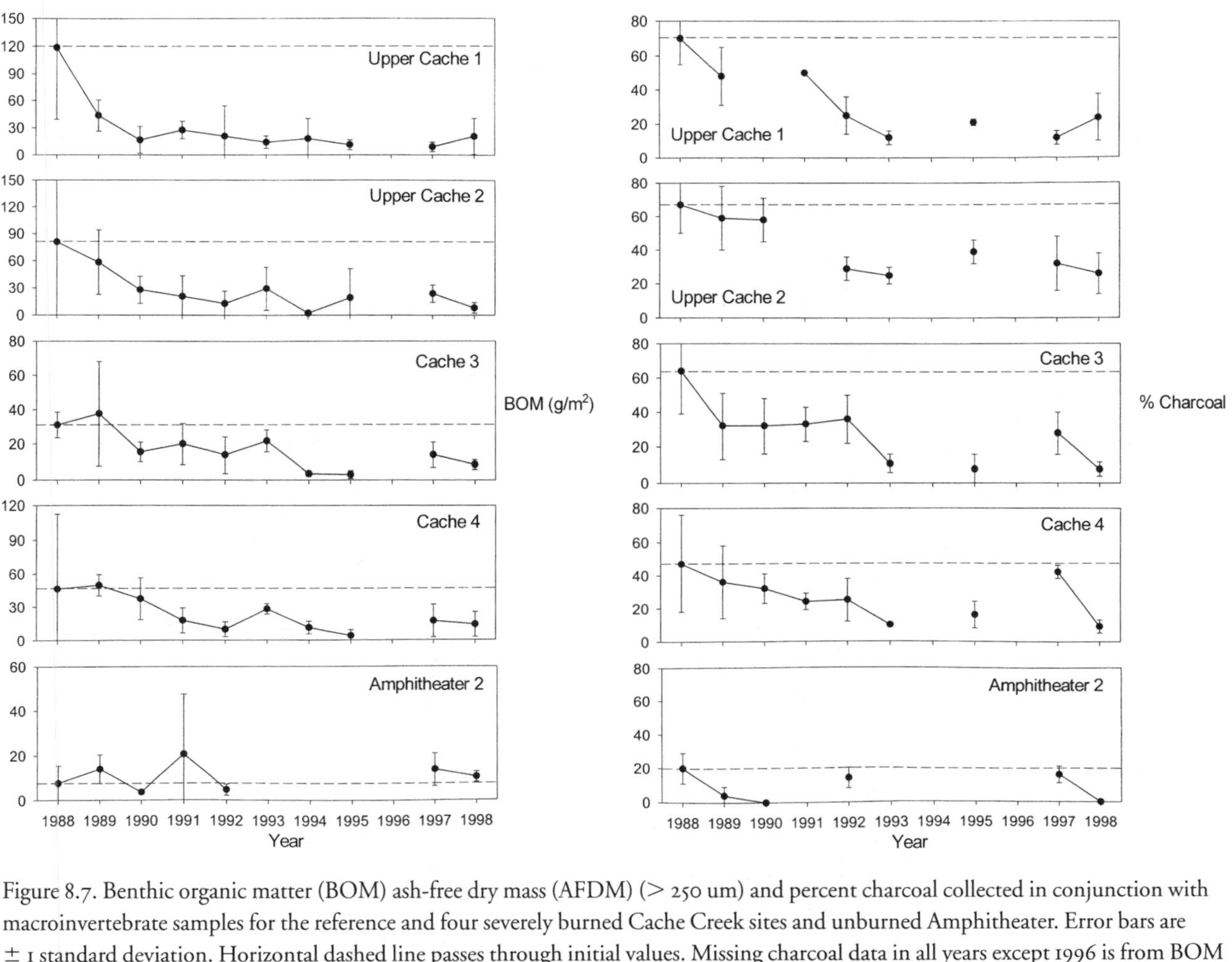

Figure 8.7. Benthic organic matter (BOM) ash-free dry mass (AFDM) (> 250 um) and percent charcoal collected in conjunction with macroinvertebrate samples for the reference and four severely burned Cache Creek sites and unburned Amphitheater. Error bars are ± 1 standard deviation. Horizontal dashed line passes through initial values. Missing charcoal data in all years except 1996 is from BOM samples, which were processed inadvertently for AFDM prior to compositional analysis.

tation by channel downcutting (resulting in reduced wetting of the root zone) and bank erosion.

Periphyton biomass (ash-free dry mass; AFDM) at the intensively burned Cache Creek sites declined the first two years following the fire, then increased to the highest recorded levels in 1991 and 1993 (Fig 8.8). In 1992 and the remainder of the 10-y period, AFDM values were near or below 1988 levels. The apparent decrease in 1992 is anomalous since it occurred in a year of low flow and little channel change in the second-, third-, and fourth-order sites but substantial channel change in the first-order site. A comparable decline also was registered at the reference site, suggesting the possibility of a systematic analytical error. Chlorophyll *a*, which is a better measure of the living component of periphyton, was higher in 1988 at all sites than at any other time during the study. However, the high variance at the burn sites in 1988 indicated that periphyton distribution was patchy. Periphyton abundance was much lower and more homogeneous during the remainder of the study. Our quantitative sampling of periphyton did not include macrophytic forms of filamentous green algae, which were abundant in some years. In particular, we observed that the two largest Cache Creek sites contained large amounts of a filamentous green algae (probably *Cladophora*) in 1992 and 1994, with filaments more than 20 cm in length extending from most cobble-sized rocks. We also found compositional changes in the diatom assemblage of burn streams that were associated with the amount of catchment burned (Robinson and others 1994). In all third- and fourth-order burn study sites, there was a progressive decline in periphyton biomass during the first five years that paralleled values in reference streams (Minshall and others 1997). There was a distinct decrease in biomass from 1988 to 1989 in the first- and second-order streams (and all burn streams), but values in the remaining years were highly variable and showed no general pattern. With one exception (Twin Creek), biomass values at burn sites were comparable to the low levels recorded in 1992 and were similar to reference streams throughout the remaining study period (Minshall and others 1997, and in press).

We expected periphyton biomass to increase substantially soon after the fires (in response to increases in light and nutrients) and to remain high throughout the 10-y period (Minshall and others 1989, Minshall and Brock 1991). We had not anticipated the overriding importance of turbid water and scour associated with years of active bed movement that impeded growth or removed the periphyton. Even more striking, based on the comparison of burn streams with the reference stream, is that the temporal pattern in periphyton biomass was independent of disturbances resulting from wildfire. This probably reflects the fact that

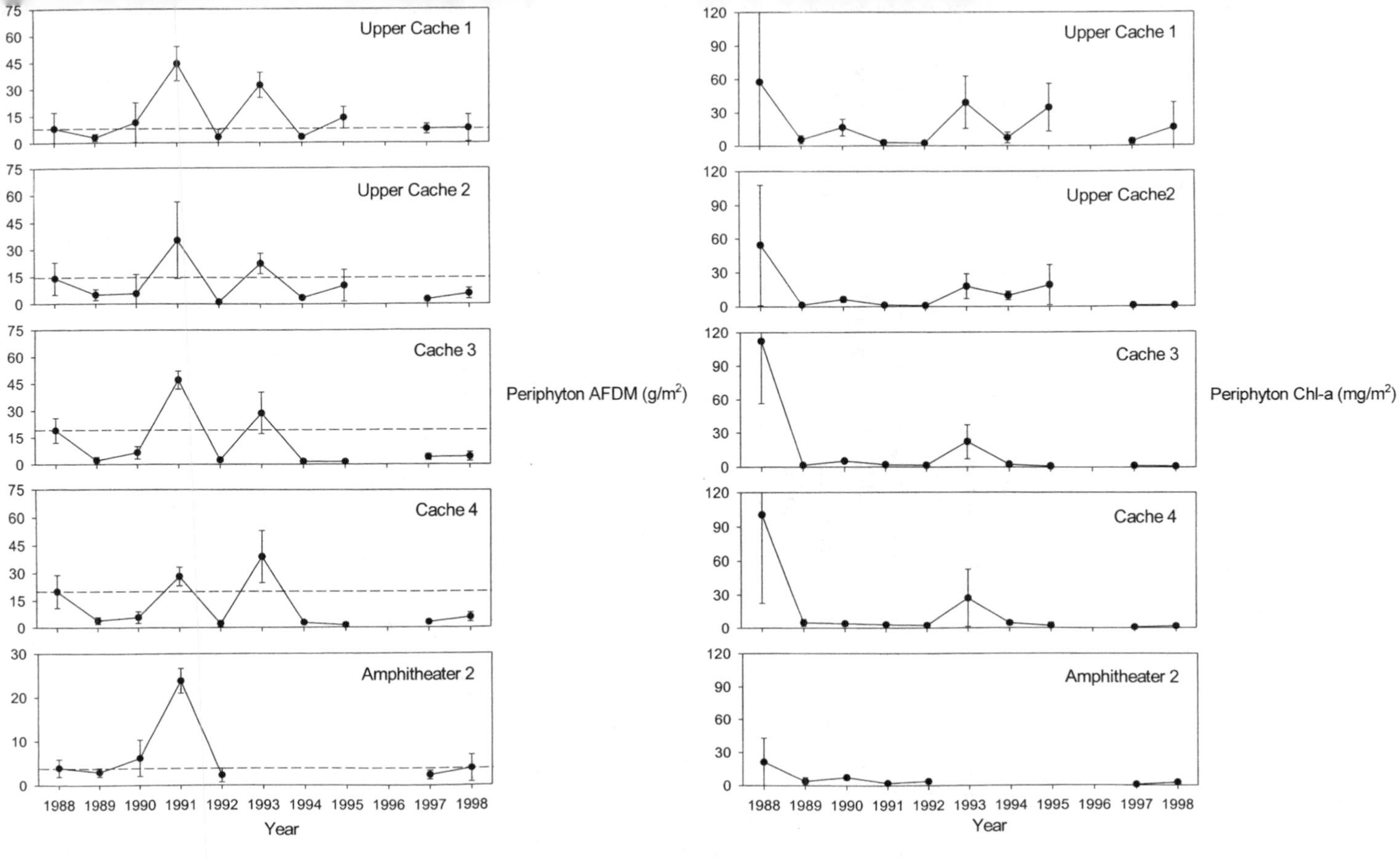

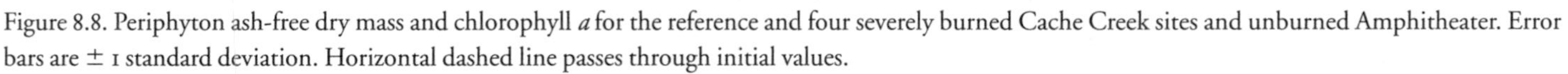

Figure 8.8. Periphyton ash-free dry mass and chlorophyll *a* for the reference and four severely burned Cache Creek sites and unburned Amphitheater. Error bars are ± 1 standard deviation. Horizontal dashed line passes through initial values.

periphyton operate on a short temporal scale (days to weeks) relative to the interannual changes in discharge and stream channel conditions following the fires. The sporadic occurrence of filamentous algae observed during the study also is consistent with this explanation. In contrast, benthic macroinvertebrates, with life cycles of months to years, better reflected the annual changes caused by fire (see below).

In general, macroinvertebrate densities in Cache Creek declined markedly in the first postfire year, returned to levels approaching or exceeding prefire values in 1990–1992 and decreased again in 1994 (Fig. 8.9). Densities remained low in Cache 3 and 4 for the remainder of the study period but increased or surpassed 1992 levels at Cache 1 and 2. Interannual values were variable at Cache 1 and 2, but notable decreases occurred in 1991 and 1994. Only Cache 2, which already had low values in 1988, showed no decrease in density in 1989. In contrast to the burned streams, macroinvertebrate density changed little during the study period at the reference site. Overall, total densities in 1998 were comparable to those measured in previous years (mean approximately 10,000/m^2) except for four burn streams (West Fork Blacktail Deer, Iron Springs 2 and 3, Cache 4) which had mean densities $\geq$ 20,000/m^2. Also, the densities in the first- and second-order burned streams were greater than those in their reference streams, but those in the third- and fourth-order burn streams were similar to those of their reference streams (Minshall and others 2003).

Except for the first-order site, total biomass in Cache Creek was much more variable (as SD) among years in the burn streams than that in the reference site (Fig. 8.9). Changes in biomass generally followed annual patterns for density at all sites but was the less variable measure of the two. Temporal patterns of biomass at the two largest Cache Creek sites were similar; mean values in 1993 greatly exceeded the initial (1988) levels and all other years. At the end of the ten years, total biomass values at the Cache Creek burn sites were higher than those found in 1988. This was generally true for the other burn sites too; however, the values often were comparable to those for the reference streams. The main exceptions were the first-order sites, which varied widely from below to nine times the reference stream values, and Hellroaring 4, which also was about nine times the Soda Butte reference stream value.

Mean macroinvertebrate richness at Cache 1 was consistently lower and that at Cache 2 generally higher (except in 1993) than the initial 1988 value, except for much higher values relative to 1988 in 1997 and 1998 at both sites (Fig. 8.10). In contrast, the two downstream burn streams varied between being similar to the initial value, during the early part of the study, and markedly lower in the

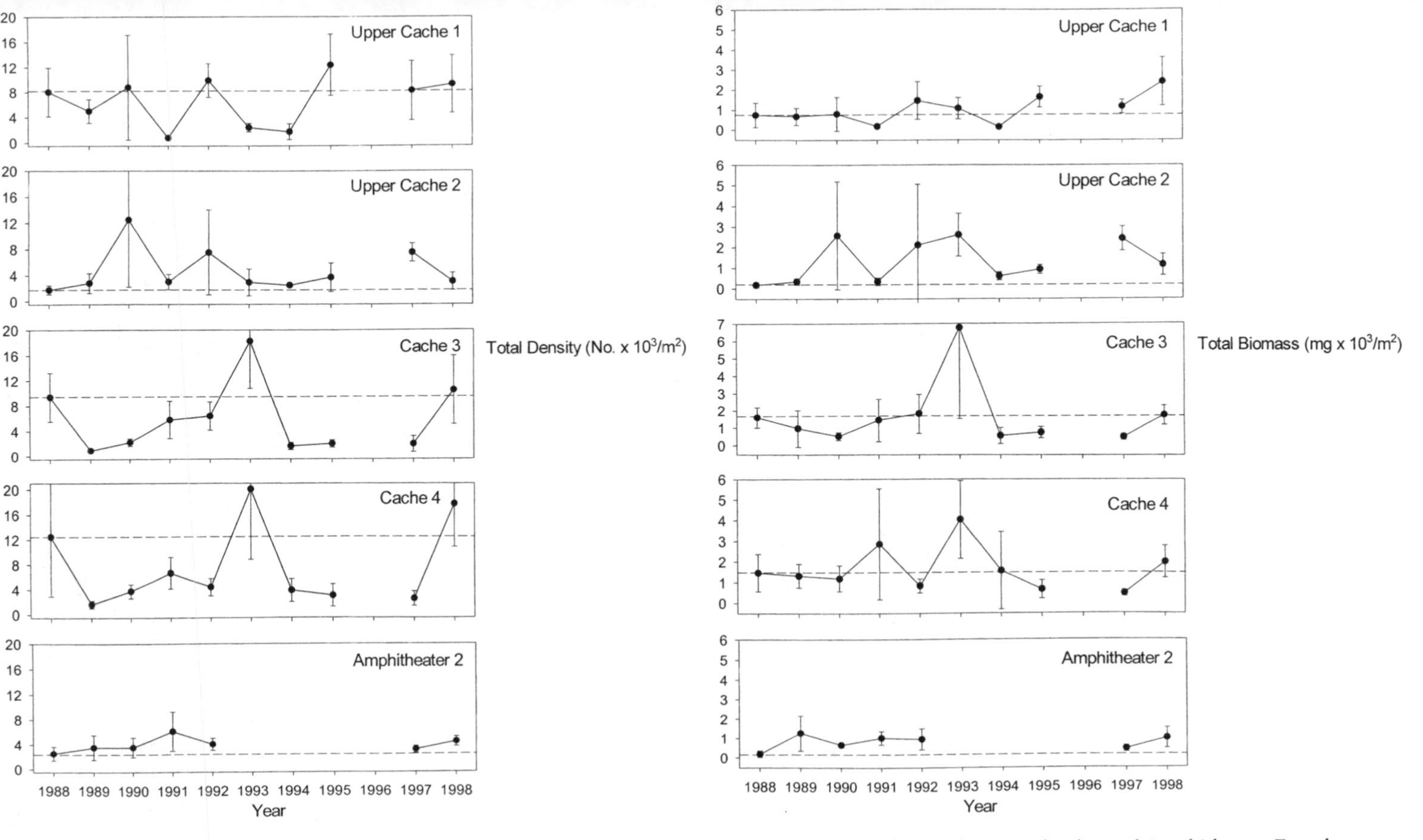

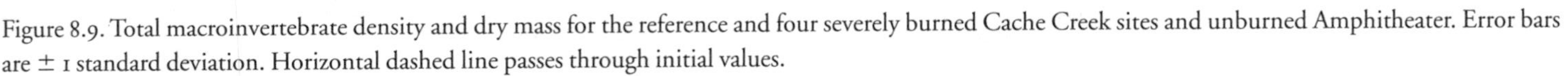

Figure 8.9. Total macroinvertebrate density and dry mass for the reference and four severely burned Cache Creek sites and unburned Amphitheater. Error bars are ± 1 standard deviation. Horizontal dashed line passes through initial values.

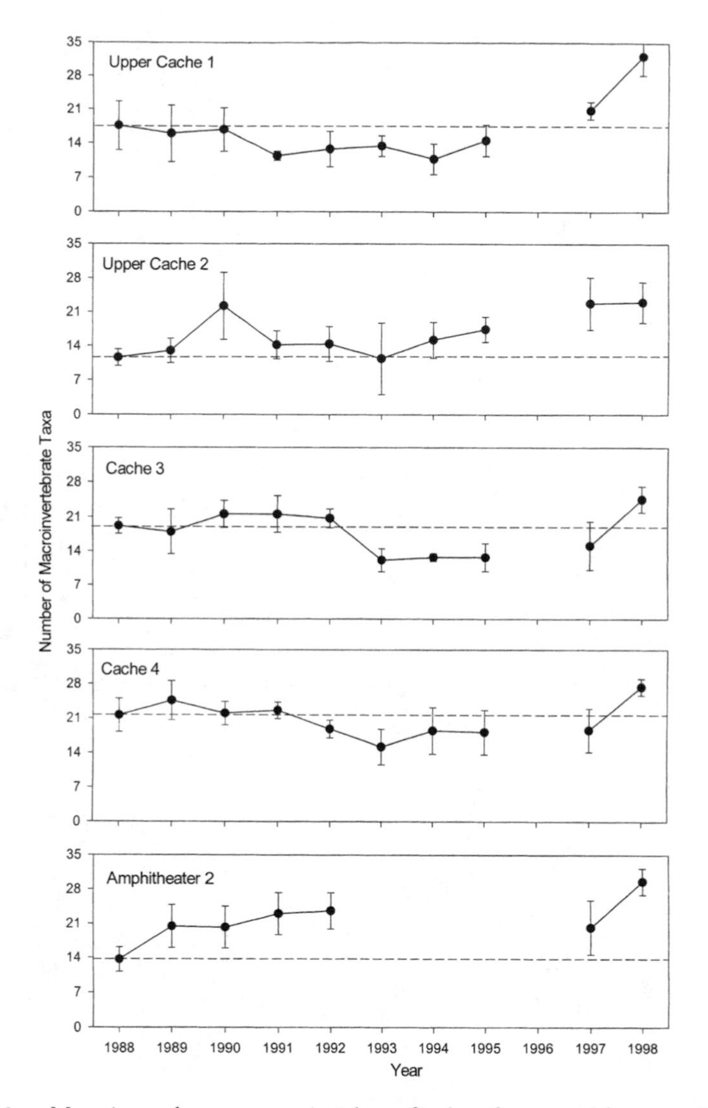

Figure 8.10. Macroinvertebrate taxonomic richness for the reference and four severely burned Cache Creek sites and unburned Amphitheater. Error bars are ± 1 standard deviation. Horizontal dashed line passes through initial values.

period 1993–1995. But by 1997 values at both sites had returned to values comparable to those of 1988. In contrast, richness for the reference stream was higher than the initial assessment for all subsequent years, including the last. The apparent increases in richness in three of the four burn streams in 1998 also occurred in the reference stream and thus cannot be attributed solely to wildfire.

Although this increase was unexpected, we attribute it to a shift from drought conditions at the time of the fire to average or above precipitation near the end of the study.

Macroinvertebrate communities in burn streams displayed major changes in species composition in response to the observed changes in stream habitats (Mihuc and others 1996, Minshall and others 1997, Minshall and others 2001b). For example, burned streams exhibited differences in trophic group composition from that found in reference streams, suggesting alterations in food resources and a shift to more trophic generalists. However, macroinvertebrate response appeared to be more individualistic rather than associated with community properties.

Changes caused by fire can affect macroinvertebrates in ways other than through alterations in food resources, such as via higher water temperatures. Species life histories and behaviors respond in different ways and degrees to these various changes in habitat characteristics. Opportunistic species, particularly those well-suited for dispersal through drift and with relatively short generation times (such as chironomids and *Baetis*), seem to be especially adapted to conditions following fire, regardless of their trophic niche (Minshall and others 2001b). In contrast, other species decreased in abundance soon after the fire and showed little or no recovery during the ten-year study. This was especially noticeable among the Ephemeroptera, especially the dorso-ventrally compressed taxa (for example, *Cinygmula, Epeorus,* and *Rhithrogena*) (Minshall and others 1997).

EXPECTED LONG-TERM CHANGES

The changes that have taken place over the past ten years, and that are likely to occur over the next decade, are expected to be the most dramatic to occur over the postulated 100- to 300-year recovery sequence. In the first ten years following wildfire, physical changes associated with runoff have overridden biotic changes associated with forest regeneration. However, as noted earlier, stream ecosystems are profoundly influenced by the condition of their watershed. For example, in 1998, many of the conifer seedlings that germinated in the year following the fires (1989) were already two meters or more in height, and many of the charred tree trunks of entire forests killed by fire were still standing. In another ten years, it is expected that these saplings will be 6 to 7 m tall and that almost all of the dead trees will have fallen (Lyon 1984). These changes, occurring over a relatively short time, will dramatically alter the kinds and amounts of food resources in the streams and change the availability of large woody debris. Nearly all headwater burn streams currently are accruing pieces of wood in their channels. Large woody debris is important in the development of pool habitat in steep-gradient streams. As the wood stabilizes, longer-lasting pools are expected to form that should increase habitat for fish in these smaller streams. However, because less wood was found in the active channels of the larger (third–fourth-order) burn streams toward the end of the first ten years, we expect fewer pools to develop in fire-affected larger streams than in comparably sized reference streams. In turn, a decrease in adult fish density may accompany this deterioration in habitat over the next decade.

Based on our short- and midterm results to date, long-term predictions for stream habitat development can be made for streams in burned catchments. Forest trees and riparian shrubs will progressively increase in height and diameter, thereby increasing shading, terrestrial-plant litter, and size of woody debris. Dead burned trees will continue to be added to the streams for another ten or so years, but their mean size will increase as the larger standing-dead trees progressively become more prone to falling and the number of dead smaller-diameter standing trees declines. These deadfalls are expected to enhance long-term complexity/stability and may even accelerate recovery to prefire conditions. We expect this effect to be especially pronounced in flow-stabilized intermediate-sized streams (for example, Hellroaring 3). Large trees should again become incorporated into stream channels about 150 years following the fires and increase the number of deep pools and between-habitat diversity. However, within-habitat substratum diversity should decrease and median particle size shift (increase in riffles, decrease in pools) due to the stabilizing action of the wood and more controlled runoff. The structure and function of stream flora and fauna will parallel these changes in landscape characteristics by influencing the long-term dynamics in habitat diversity and changes in the kinds and amounts of food resources.

Our research emphasizes the importance of studying stream ecosystems for many years following large-scale disturbance. Conclusions based on only one or a few years of data can be misleading in terms of overall trends, as evidenced by the apparent "devastation" of stream ecosystems immediately after the 1988 fires, their rapid progress toward "recovery" in postfire years 1–2, their equally abrupt downturn in postfire years 3–4, and their massive reorganization in years 7–9. Far too little data exist on conditions over extended periods after wildfire to know for certain whether our predictions for Yellowstone will prove correct. In fact, the initial recovery trajectory seen for Yellowstone streams is much different—faster initially, with longer time delays before major storm impacts were seen—than expected, based on research we have conducted in central Idaho (Minshall and others 2001a). The absence of comparable data on long-term dynamics, high interannual variability in postfire disturbance impacts among streams of different size, and differences in recovery trajectories from those found in other Rocky Mountain streams provide strong arguments for the need to obtain an extended temporal perspective for Yellowstone lotic ecosystems in the aftermath of the 1988 fires.

References

Barrett, S. W., S. F. Arno, and J. P. Menakis. 1997. Fire episodes in the inland Northwest (1540–1940) based on fire history data. U.S. Forest Service General Technical Report INT-GTR-370.

Beaty, K. G. 1994. Sediment transport in a small stream following two successive forest fires. Can. J. Fish. Aquat. Sci. 51:2723–2733.

Ewing, R. 1996. Postfire suspended sediment from Yellowstone National Park, Wyoming. Wat. Resour. Bull. 32:605–627.

Gammeter, S., and A. Frutiger. 1990. Short-term toxicity of NH_3 and low oxygen to benthic macroinvertebrates or running waters and conclusions for wet weather water pollution control measures. Water Sci. Techn. 22:291–296.

Helvey, J. D. 1973. Watershed behavior after forest fire in Washington. Pages 403–422 in Proceedings: Irrigation and Drainage Division, American Association Civil Engineers, Fort Collins, Colo.

———. 1980. Effects of a north central Washington wildfire on runoff and sediment production. Water Resources Bulletin 14:627–634.

King, A. W. 1993. Considerations of scale and hierarchy. Pages 19–45 in S. Woodley, G. Francis, and J. Kay, eds., Ecological integrity and the management of ecosystems. St. Lucie Press, Delray Beach, Fla.

Lyon, L. J. 1984. The Sleeping Child burn: 21 years of postfire change. U.S. Forest Service Research Paper INT-330.

Mihuc, T. B., and G. W. Minshall. 1995. Trophic generalists vs. trophic specialists: Implications for food web dynamics in post-fire streams. Ecology 76:2361–2372.

Mihuc, T. B., G. W. Minshall, and C. T. Robinson. 1996. Response of benthic macroinvertebrate populations in Cache Creek, Yellowstone National Park to the 1988 wildfires. Pages 83–94 in J. M. Greenlee, ed., The ecological implications of fire in Greater Yellowstone, International Association of Wildland Fire, Fairfield, Wash.

Minshall, G. W., K. E. Bowman, and C. D. Myler. 2003. Effects of wildfire on Yellowstone stream ecosystems: A retrospective view after a decade. Proceedings: First National Congress on Fire Ecology, Prevention, and Management, San Diego. Tall Timbers Research Station.

Minshall, G. W., and J. T. Brock. 1991. Anticipated effects of forest fire on Yellowstone stream ecosystems. Pages 123–135 in B. Keiter and M. S. Boyce, eds., Greater Yellowstone's future: Man and nature in conflict? Yale Univ. Press, New Haven.

Minshall, G. W., J. T. Brock, and J. D. Varley. 1989. Wildfires and Yellowstone's stream ecosystems: A temporal perspective shows that aquatic recovery parallels forest succession. BioScience 39:707–715.

Minshall, G. W., C. T. Robinson, and D. E. Lawrence. 1997. Immediate and mid-term responses of lotic ecosystems in Yellowstone National Park, U.S.A., to wildfire. Can. J. Fish. Aquat. Sci. 54:2509–2525.

Minshall, G. W., C. T. Robinson, D. E. Lawrence, D. A. Andrews, and J. T. Brock. 2001a. Benthic macroinvertebrate assemblages in five central Idaho (USA) streams over a 10-year period following disturbance by wildfire. Int. J. Wildl. Fire 10:185–199.

Minshall, G. W., T. V. Royer, and C. T. Robinson. 2001b. Response of the Cache Creek macroinvertebrates during the first ten years following disturbance by the 1988 Yellowstone wildfires. Can. J. Fish. Aquat. Sci. 58:1077–1088.

Pringle, C. M., R. J. Naiman, G. Bretschko, J. R. Karr, M. M. Oswood, J. R. Webster, R. L. Welcomme, and M. J. Winterbourn. 1988. Patch dynamics in lotic systems: The stream as a mosaic. J. North Am. Benthol. Soc. 7:503–524.

Resh, V. H., A. V. Brown, A. P. Covich, M. E. Gurtz, H. W. Li, G. W. Minshall, S. R. Reice, A. L. Sheldon, J. B. Wallace, and R. Wissmar. 1988. The role of disturbance theory in stream ecology. J. North Am. Benthol. Soc. 7:433–455.

Robinson, C. T., S. R. Rushforth, and G. W. Minshall. 1994. Diatom assemblages of streams influenced by wildfire. J. Phycol. 30:209–216.

Spencer, C. N., and F. R. Hauer. 1991. Phosphorus and nitrogen dynamics in streams during wildfire. J. North Am. Benthol. Soc. 10:24–30.

Swanson, F. J., J. A. Jones, D. O. Wallin, and J. H. Cissel. 1994. Natural variability: Implications for ecosystem management. Pages 80–94 in M. E. Jensen and P. S. Bourgeron, eds., Eastside Forest Ecosystem Health Assessment. Vol. 2: Ecosystem management: Principles and applications. U.S. Forest Service, Pacific Northwest Research Station, Portland, Ore.

Townsend, C. R. 1989. The patch dynamics concept of stream community ecology. J. North Am. Benthological Soc. 8:36–50.

Turner, M. G., W. Hargrove, R. H. Gardner, and W. H. Romme. 1994. Effects of fire on landscape heterogeneity in Yellowstone National Park, Wyoming. J. Vegetation Sci. 5:731–742.

Vannote, R .L., G. W. Minshall, K. W. Cummins, J. R. Sedell, and C. E. Cushing. 1980. The river continuum concept. Can. J. Fish. Aquat. Sci. 37:130–137.

Chapter 9 Food Web Dynamics in Yellowstone Streams: Shifts in the Trophic Basis of a Stream Food Web After Wildfire Disturbance

Timothy B. Mihuc

Food webs in lotic systems are poorly understood. The compilation of freshwater food webs by Briand (1985) included only nine streams, while that by Schoenly and others (1991) included eleven streams. Since those compilations, several studies have added to our knowledge of stream food webs primarily with respect to link relationships and connectance (Hildrew 1992, Closs and Lake 1994, Mihuc and Minshall 1995, Findlay and others 1996, Tavares-Cromar and Williams 1996, Benke and Wallace 1997). Recently several investigators have attempted to determine energy flow in aquatic systems (Junger and Planas 1994, Benke and Wallace 1997, Whitledge and Rabeni 1997), but we still have little empirical information on which to develop a synthesis of lotic food web energy patterns. An understanding of food web link relationships alone (who eats whom) does not address the magnitude and efficiency of resource use and energy flow through a trophic web. Watershed-scale disturbance events, such as wildfire, are more likely to alter energy flow pathways in stream food webs rather than link relationships (Mihuc and Minshall 1995, Minshall and others 1989).

WILDFIRE AND LOTIC SYSTEMS

Wildfire can affect lotic systems at temporal scales ranging from days to decades and spatial scales ranging from microhabitats to entire watersheds. General impacts of wildfire on lotic systems are discussed in Minshall and others (1989, 1997), Mihuc and others (1996), and Robinson and Minshall (1996). Predicted impacts include alteration of woody debris dynamics, nutrient dynamics, sediment suspension, leaf litter input, and changes in the composition of aquatic biota in postfire streams (Minshall and others 1989). Observed postfire impacts in Yellowstone streams included higher nitrate levels, higher summer water temperatures, increased channel restructuring, lower habitat heterogeneity, and an increase in burned organic material (Minshall and others 1997). Biotic responses included initial increases in Chironomidae abundance followed by increases after 3–5 yrs in density of the mayfly *Baetis bicaudatus* in postfire streams (Mihuc and others 1996, Minshall and others 1997). Predicted responses of macroinvertebrate functional feeding groups to wildfire (Minshall and others 1989), such as a decrease in shredder-detritivore taxa, were generally not observed in Yellowstone streams (Minshall and others 1997). Instead, invertebrate responses in postfire streams were individualistic, relating primarily to habitat preferences (flow/substrate conditions) and trophic niche strategies (generalist, specialist) for individual taxa (Mihuc and others 1996, Mihuc and Minshall 1995).

IMPACTS OF FIRE ON LOTIC FOOD WEBS

Impacts on aquatic food webs include short- and long-term alteration of resource pathways in postfire streams. A shift in trophic energy from detrital to grazing pathways in streams following wildfire has been predicted (Minshall and others 1989) but not tested until studies on food web dynamics in Cache Creek, YNP, after the 1988 wildfires. This prediction is based on a decrease in allochthonous (externally derived, detrital) inputs in postfire streams, coupled with an opening of the canopy resulting in an increase in autochthonous (internally derived, primary producer) inputs (Minshall and others 1989). This shift in energy source may also be coupled with or lead to changes in density, biomass, or resource use in the community. Other impacts could include a shift in dominant species along with altered link and/or connectance relationships.

Several research approaches exist to identify trophic pathways in lotic food webs, including (1) inference of pathways based on food material ingested; (2) estimation of pathways using isotopic tracers (for example, Junger and Planas

1994, Whitledge and Rabeni 1997); and (3) determination of consumer growth or assimilation rates (for example, Mihuc and Minshall 1995), and/or ingestion values (for example, Smock and Roeding 1986, Benke and Wallace 1980, 1997). Benke and Wallace (1997) took the last approach further by estimating the relative importance of energy pathways to consumers by coupling assimilation/ingestion values with production information. Approaches (2) and (3) were applied in studies of the Cache Creek food web for direct measurements of trophic energy pathways in a postfire stream and to make comparisons with a reference (unburned) stream (Amphitheater Creek, YNP). Dual isotope tracers (C^{12}/.C^{13}, N^{14}/N^{15}) coupled with invertebrate biomass (Mihuc and others 1996, Minshall and others 1997) were used to trace energy pathways from basal resources to primary consumers for each stream food web.

DETERMINATION OF POSTFIRE ENERGY PATHWAYS

Consumer-resource pathways were determined in two second-order Yellowstone National Park streams in the summer of 1990 and 1991 to characterize trophic pathways after the 1988 wildfires. The reference site (Amphitheater Creek) was unaffected by the 1988 wildfires (2 percent of watershed area burned), whereas the burned site, Cache Creek, was severely affected (39–71 percent of watershed area burned; Mihuc and others 1996). The two sites are located in adjacent valleys in the Lamar River basin and are similar in geologic parent material, slope, aspect, channel morphology, riparian vegetation, and forest cover. Resources collected included (1) coarse particulate organic matter (CPOM: $>$ 1 mm detritus including leaf litter, pine needles, wood fragments); (2) fine particulate organic matter (FPOM: $<$ 1 mm detritus); and (3) primary producers (periphyton, algae). Resources were collected by removing detrital leaf packs for CPOM, periphyton by rock scrapings, and sediment grabs for FPOM. Benthic macroinvertebrates were obtained using surber sampling and kick netting. Benthic macroinvertebrates (insects) used for isotope analysis included seven Ephemeroptera taxa (mayflies; *Ameletus cooki, Baetis bicaudatus, Cinygmula* sp., *Drunella coloradensis, D. doddsi, Epeorus albertae, Rhithrogenia* sp.), one Trichopteran (caddisfly; *Dicosmoecus atripes*), two Plecoptera (stoneflies; *Capnia* sp., *Zapada columbiana*), and the Dipteran family Chironomidae. Gut tracts were removed from all animals prior to stable isotope analysis. These eleven taxa represented 74 percent of the total community biomass in Cache Creek and 65 percent of the total biomass in Amphitheater Creek in 1990–1991. Taxa included representatives from three functional feeding groups (FFG): (1) shredders, which are adapted to consume large particles, presumably detritus *(Capnia* sp., *Dicosmoecus atripes, Zapada columbiana);* (2) gatherers, which are adapted for gathering smaller fine particulate material (Chironomidae, *Ameletus cooki, Baetis bicaudatus Rhithrogenia* sp.); and (3) scrapers, which are adapted to remove attached material, presumably algae and diatoms *(Cinygmula* sp., *Drunella coloradensis, D. doddsi, Epeorus albertae).* All samples were kept frozen until preparation for analysis by drying at 60°C for 24 h. To reduce the number of samples required for analysis, isotope analysis was performed on composite samples of 3–5 specimens for each taxon using a continuous flow dry combustion procedure for $\delta C^{13}$and N^{15}. Isotope data represent the composite "mean" signature for each taxon or resource type.

The relative contribution of autochthonous and allochthonous resources to primary consumer biomass was determined in the reference and burned stream using the generalist trophic dynamic model (Mihuc 1997). A linear stable isotope mixing model (Junger and Planas 1994) combined with previous trophic information and invertebrate biomass (Mihuc and Minshall 1995, Mihuc and others 1996) were used to determine the percentage of biomass for each taxon derived from allochthonous and autochthonous resources, respectively. End members for the linear mixing model were periphyton and CPOM δC^{13} values in each stream. Distance between the end members was used to reflect the relative amount of periphyton (autochthonous) and CPOM (allochthonous) derived biomass for each taxon. Data were summarized as the estimated amount of primary consumer biomass derived from the allochthonous and autochthonous resource pool for each taxon.

SHIFTS IN RESOURCE PATHWAYS IN CACHE CREEK

There were distinct shifts in isotope signatures among macroinvertebrates between the reference and burned streams (Fig. 9.1). This is reflected by a shift in resource use among several taxa *(Baetis bicaudatus,* Chironomidae, *Cinygmula, Dicosmoecus atripes, Zapada columbiana)* from allochthonous detrital sources in the reference stream to autochthonous periphyton in the burned stream. Resource isotopic signatures, however, remained similar between streams.

Based on the mixing model results, the relative contribution of allochthonous and autochthonous resources to primary consumers indicates a shift in resource base between the two streams (Fig. 9.2). In the postfire stream, a higher con-

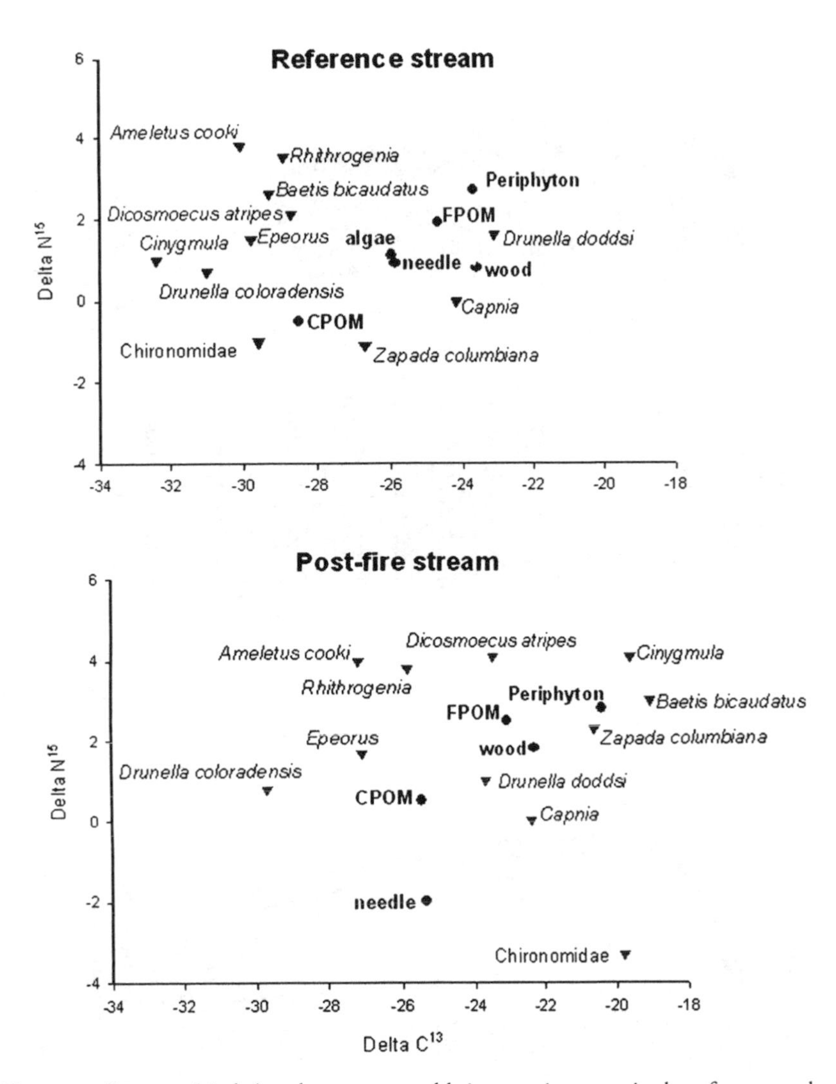

Figure 9.1. Resource (circles) and consumer stable isotope signatures in the reference and postfire stream. Signatures represent composite samples for each taxon or resource type.

sumer biomass is derived from autochthonous resources, while biomass derived from allochthonous resources declined (Fig. 9.2). The result was a "shift" in trophic energy pathways from detrital to periphyton resources in the postfire stream versus the reference stream. This occurred through trophic shifts among individual taxa; in particular, Chironomidae and *Baetis bicaudatus* were linked to detrital resources and had lower biomass (combined 11 percent of total) in the reference stream and shifted to periphyton use and increased biomass (com-

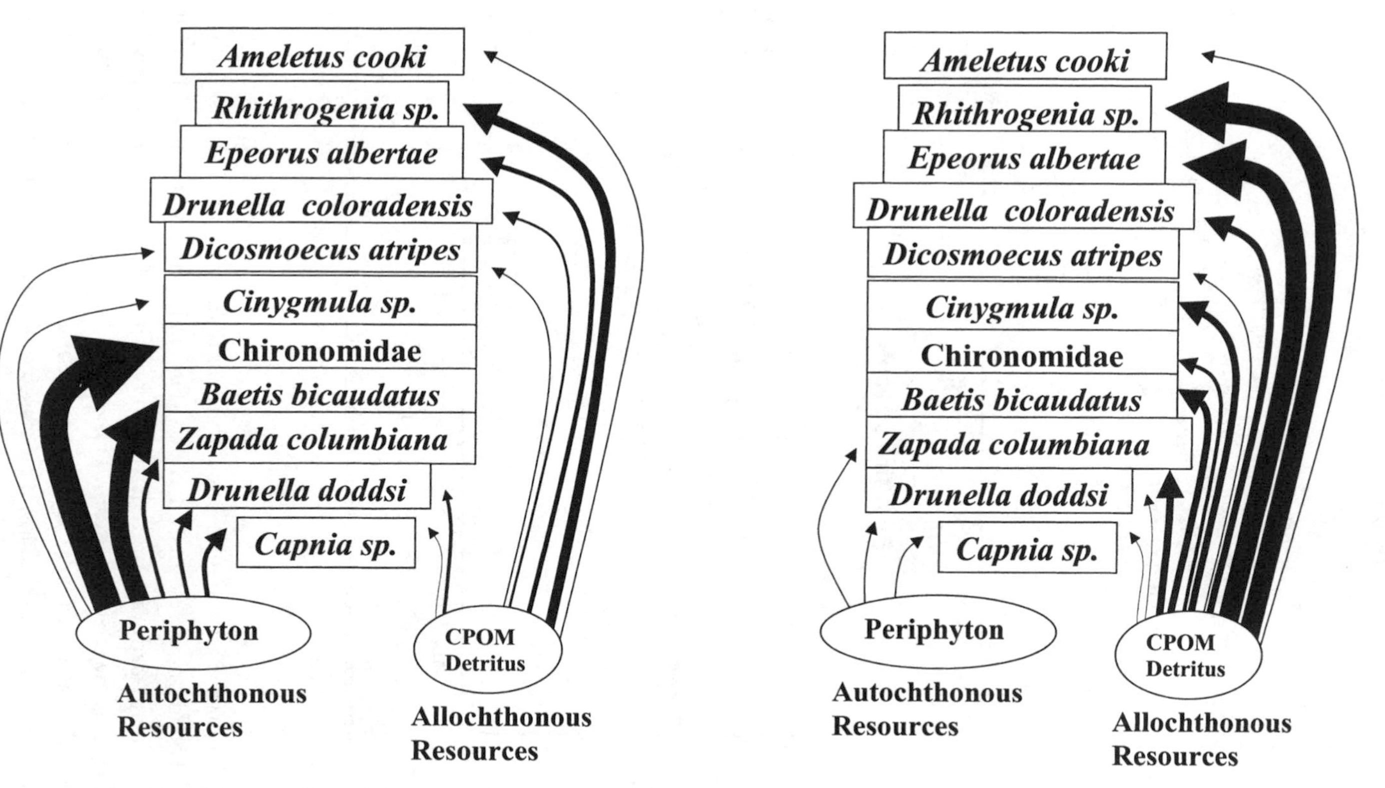

Figure 9.2. Relative contribution of autochthonous and allochthonous resources to macroinvertebrate biomass in the postfire stream (a), Cache Creek, and reference stream (b), Amphitheater Creek. Arrow thickness and direction indicates the percent of total community biomass derived from each resource. Energy pathways are shown for 74 percent and 65 percent of the total benthic invertebrate community biomass in the postfire and reference stream, respectively.

bined 47 percent of total) in the postfire stream (Fig. 9.2). *Cinygmula, Dicosmoecus atripes,* and *Zapada columbiana* also exhibited similar patterns in resource shifting between streams, although *Cinygmula* and *Dicosmoecus atripes* declined in biomass in postfire streams while *Zapada columbiana* biomass remained similar in both streams (Mihuc and others 1996).

IMPLICATIONS FOR RESOURCE PARTITIONING IN STREAMS

The relative importance of allochthonous and autochthonous energy resources in stream food webs is poorly understood (Mihuc 1997). Research in Yellowstone streams has demonstrated a shift from an allochthonous to an autochthonous energy base for a second-order Rocky Mountain stream following wildfire disturbance (Fig. 9.3), supporting the prediction by Minshall and others (1989) that "low-order streams, which were formerly dependent on exogenous sources for their organic energy, can be expected to shift to autotrophy until riparian communities develop sufficiently . . . for a return to dependence on allochthonous organic material."

Resource switching occurred among individual taxa, not through community shifts in functional feeding group (FFG) composition as predicted by Minshall and others (1989). In the reference stream most taxa relied on detrital resources for energy regardless of functional mouthpart morphology (for example, scraper, shredder, gatherer). Individual taxa from all three FFGs switched to autochthonous resources in the postfire stream, resulting in the observed food web shifts. All three shredder taxa exhibited shifts, with two of these taxa *(Capnia* sp., *Z. columbiana)* also demonstrating a generalist resource use strategy in the reference stream. This suggests that, in low-order Rocky Mountain streams, shredders may not function as obligate CPOM detritivores.

Consumer-resource growth experiments conducted on macroinvertebrate taxa from Cache Creek also indicate that generalist partitioning is a predominant strategy in Yellowstone streams (Mihuc and Minshall 1995). A few taxa that exhibited generalist resource switching based on isotope tracer data *(Cinygmula, Dicosmoecus atripes)* were identified as trophic specialists in the experiments conducted by Mihuc and Minshall (1995). This suggests that multiple approaches (for example, isotope tracers, growth-assimilation experiments, food habits) are required to document lotic primary consumer trophic strategies, including consideration of intraspecific spatial and temporal patterns in resource use. This supports the view that generalist resource use is predominant among lotic pri-

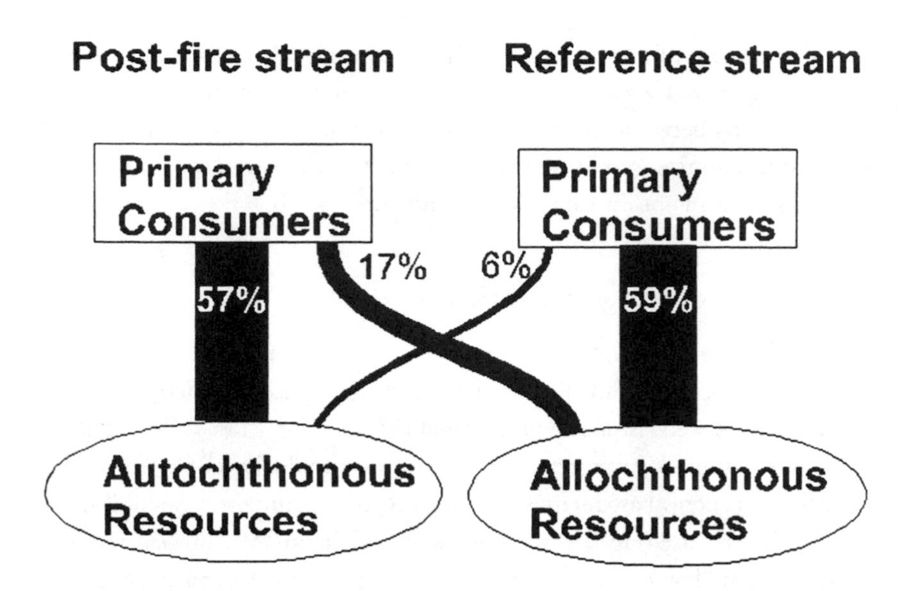

Figure 9.3. Relative contribution of autochthonous and allochthonous resources to macroinvertebrate biomass for selected food web members in Amphitheater Creek reference stream. Arrow thickness and direction indicate the percent of total community biomass derived from each resource and its pathway through primary consumer species. Energy pathways are shown for 65 percent of the total benthic invertebrate community biomass.

mary consumers and that resource switching is likely a common strategy (Mihuc 1997).

Resource switching may also be prevalent with ontogenetic shifts among individual taxa (Chapman and Demory 1963, Rader and Ward 1987). Based on partitioning theory (Schoener 1974), if most interspecific interactions are eliminated through partitioning of time and space, then generalist resource use should occur among taxa. Lotic macroinvertebrates demonstrate strong interspecific partitioning of space (flow and substrate conditions) and time through synchronous life history strategies whereby taxa "offset" life history stages seasonally from one another (Vannote and Sweeney 1980). Therefore under strong space and time partitioning one would expect generalist resource use strategies to predominate among lotic macroinvertebrates.

FUTURE RESEARCH NEEDS

Catchment scale wildfire acts to remove the primary allochthonous organic matter source for the aquatic food web, riparian vegetation, and produces an open canopy stream resulting in higher primary production. This probably persists during the first postfire decade with a return to prefire, allochthonous (detritus)-based pathways following establishment of riparian vegetation. This facilitated the wildfire-caused shift in resource base observed in Cache Creek, YNP. Future research directions include the need for further study of the importance of resource switching among primary consumers in stream systems, both disturbed and natural, and long-term research on trophic energy pathways in postfire streams beyond the initial postfire decade.

Most current consumer-resource paradigms in stream ecology are based on the use of functional feeding groups as trophic guilds (for example, McIntyre and Colby 1978, Cummins and Klug 1979, Vannote and others 1980, DeAngelo and others 1997). The prevalence of generalist resource use among lotic macroinvertebrates is poorly addressed in these paradigms which utilize FFGs as specialist-obligate trophic groups. Incorporation of generalist niche strategies into determining the trophic basis for lotic food webs is critical for development of more accurate theoretical and empirical consumer-resource models. A "shift" in that direction will necessitate that future paradigms avoid the use of assignment of taxa into obligate trophic guilds and address the importance of generalist trophic dynamics in lotic systems. In a sense this will require a return to the trophic dynamic ideas of Lindeman (1942) emphasizing resource quality along with magnitude and efficiency of use. Species-specific trophic shifts among generalist consumers may, in fact, be the most prevalent resource use strategy employed among lotic macroinvertebrates, particularly in response to watershed-scale disturbance events such as wildfire. This highlights the individualistic nature of stream benthic communities.

Acknowledgments

Isotope samples were processed by the University of Utah Stable Isotope Analysis Facility. Thanks to G. Wayne Minshall and the Idaho State University Stream Ecology Center for logistic support during field sampling. Janet R. Mihuc provided assistance in the collection of field samples and laboratory; thanks again, Janet!

References

Benke, A. C., and J. B. Wallace. 1980. Trophic basis of production among net-spinning caddisflies in a southern Appalachian stream. Ecology 61:108–118.

———. 1997. Trophic basis of production among riverine caddisflies: Implications for food web analysis. Ecology 78:1132–1145.

Briand, F. 1985. Structural singularities of freshwater food webs. Verh. Internat. Verein. Limnol. 22:3356–3364.

Chapman, D. W., and R. L. Demory. 1963. Seasonal changes in the food ingested by aquatic insect larvae and nymphs in two Oregon streams. Ecology 44:140–146.

Closs, G. P., and P. S. Lake. 1994. Spatial and temporal variation in the structure of an intermittent stream food web. Ecological Monographs 64:1–21.

Cummins, K. W., and M. J. Klug. 1979. Feeding ecology of stream invertebrates. Annual Review of Ecology and Systematics 10:147–172.

DeAngelo, D. J., S. V. Gregory, L. R. Ashkenas, and J. L. Meyer. 1997. Physical and biological linkages within a stream geomorphic hierarchy: A modeling approach. Journal of the North American Benthological Society 16:480–502.

Findlay, S., M. Pace, and D. Fischer. 1996. Spatial and temporal variability in the lower food web of the tidal freshwater Hudson River. Estuaries 19:866–873.

Hildrew, A. G. 1992. Food web and species interactions. Pages 309–330 in P. Calow and G. Petts, eds., The rivers handbook. Blackwell Science, Oxford.

Junger, M., and D. Planas. 1994. Quantitative use of stable carbon isotope analysis to determine the trophic basis of invertebrate communities in a boreal forest lotic system. Can. J. Fish. Aquat. Sci. 51:52–61.

Lindeman, R. L. 1942. The trophic-dynamic aspect of ecology. Ecology 23:399–416.

McIntyre, C. D., and J. A. Colby. 1978. A hierarchical model of lotic ecosystems. Ecological Monographs 48:167–190.

Mihuc, T. B. 1997. The functional trophic role of lotic primary consumers: Generalist versus specialist strategies. Freshwater Biology 37:455–462.

Mihuc, T. B., and G. W. Minshall. 1995. Trophic generalists vs. trophic specialists: Implications for food web dynamics in post-fire streams. Ecology 76:2361–2372.

Mihuc, T. B., G. W. Minshall, and C. T. Robinson. 1996. Response of benthic macroinvertebrate populations in Cache Creek, Yellowstone National Park, to the 1988 wildfires. Pages 83–94 in J. Greenlee, ed., The ecological implications of fire in Greater Yellowstone International Association of Wildland Fire: Proceedings Second Biennial Conference on the Greater Yellowstone Ecosystem. Fairfield, Wash.

Minshall, G. W., J. T. Brock, and J. D. Varley. 1989. Wildfires and Yellowstone's stream ecosystems. BioScience 39:707–715.

Minshall, G. W., C. T. Robinson, and D. E. Lawrence. 1997. Postfire responses of lotic ecosystems in Yellowstone National Park, USA. Canadian Journal of Fisheries and Aquatic Sciences 54:2509–2525.

Rader, R. B., and J. V. Ward. 1987. Resource utilization, overlap, and temporal dynamics in a guild of mountain stream insects. Freshwater Biology 18:521–528.

Robinson, C. T., and G. W. Minshall. 1996. Physical and chemical responses of streams in Yellowstone National Park following the 1988 wildfires. Pages 217–222 in J. Greenlee, ed., The ecological implications of fire in Greater Yellowstone International Association of Wildland Fire; Proceedings Second Biennial Conference on the Greater Yellowstone Ecosystem. Fairfield, Wash.

Schoener, T. W. 1974. Resource partitioning in ecological communities. Science 185:27–38.

Schoenly, K., R. A. Beaver, and T. A. Heumier. 1991. On the trophic relations of insects: A food-web approach. Am. Nat. 137:597–638.

Smock, L. A., and C. E. Roeding. 1986. The trophic basis of production of the macroinvertebrate community of a southeastern U.S.A. blackwater stream. Holarctic Ecology 9:165–174.

Tavares-Cromar, A. F., and D. D. Williams. 1996. The importance of temporal resolution in food web analysis: Evidence from a detritus-based stream. Ecol. Monographs 66:91–113.

Vannote, R. L., G. W. Minshall, K. W. Cummins, J. R. Sedell, and C. E. Cushing. 1980. The river continuum concept. Canadian Journal of Fisheries and Aquatic Sciences 37:130–137.

Vannote, R. L., and B. W. Sweeney. 1980. Geographic analysis of thermal equilibria: A conceptual model for evaluating the effect of natural and modified thermal regimes on aquatic insect communities. American Naturalist 115:667–695.

Whitledge, G. W., and C. K. Rabeni. 1997. Energy sources and ecological role of crayfishes in an Ozark stream: Insights from stable isotopes and gut analysis. Canadian Journal of Fisheries and Aquatic Sciences 54:2555–2563.

Chapter 10 Role of Fire in Determining Annual Water Yield in Mountain Watersheds

Phillip E. Farnes, Ward W. McCaughey,
and Katherine J. Hansen

This chapter presents the computational procedures for estimating average annual water yields based on annual precipitation and vegetation cover types. This procedure allows for an estimation of water yields under current conditions, under various levels of vegetation management, or under historic water yield based on fire history.

Historic water yield information is needed to determine the natural range of water yield variability so that hydrologic response can be correctly determined. Historically, the U.S. Geological Survey (USGS) has maintained stream gaging stations, the Natural Resources and Conservation Service (NRCS; formerly the Soil Conservation Service, SCS) has coordinated the snow survey program and forecast potential streamflow for agricultural and other uses, and the National Weather Service (NWS) has collected climatological data and forecast flood levels. Many other state and federal agencies, as well as private companies, have assisted with the collection of basic climatic data. Commonly there are minimal or no long-term historic streamflow data on undisturbed or disturbed watersheds prior to fire suppression. Where measured streamflow data are available, they are usually for only the past twenty to eighty years, and they may not accurately indicate the natural range of variability of historic water yields from mountain watersheds due to active forest fire suppression during this century.

Two examples of evaluating fire's role in determining annual water yield in mountain watersheds are presented. Annual water yield was estimated using developed procedures for the Tenderfoot Creek Experimental Forest in central Montana for the past 400+ years based on fire history records. Additionally, the increase in runoff for eleven years after the 1988 fires in Yellowstone National Park was estimated using runoff-forecasting equations developed with prefire data, and this was compared with the analytical procedures similar to Tenderfoot Creek.

Snowmelt accounts for 50 to 70 percent of the total annual runoff in lodgepole pine *(Pinus contorta)* and spruce/fir *(Picea engelmannii/Abies lasiocarpa)* stands of the Northern Rockies (Farnes 1978). Water yield can be increased in subalpine forests through the use of natural or prescribed fire or other vegetation management techniques such as harvesting. Openings created by fires or through forest management increase snow accumulation, decrease sublimation losses, and increase water yields (Gary and Troendle 1982, Hardy and Hansen-Bristow 1990, Hoover and Leaf 1967, Meiman 1987, Skidmore and others 1994). Partial cutting also increases snow accumulation by reducing interception loss and summer soil water depletion due to shading effects of the remaining overstory (Troendle 1986).

TENDERFOOT CREEK EXPERIMENTAL FOREST

Precipitation inputs, water yield outputs, and forest canopy interception data have been collected on the Tenderfoot Creek Experimental Forest (TCEF), Montana, since 1992. A stream gage on Tenderfoot Creek below Stringer Creek measures the majority of runoff generated on the experimental forest. The experimental forest was used for this study because it is representative of forests composed primarily of lodgepole pine, it is essentially undisturbed, and detailed fire history patterns have been mapped for the entire experimental forest by Barrett (1993). This forest type covers 6 million ha of commercial forest lands in the Rocky Mountains (Koch 1996, Wheeler and Critchfield 1984), and therefore the results of this study have wide geographic applicability. Mountain watersheds are important catchments for precipitation that generates runoff used downstream for drinking water, irrigation, hydroelectric power, and recreation. Any change in watershed vegetation may change the hydrology, which affects these downstream uses.

STUDY LOCATION

The Tenderfoot Creek Experimental Forest covers 3,709 ha of the headwaters of the Tenderfoot Creek watershed located in the Little Belt Mountains in central Montana. The experimental forest, established in 1961 for watershed research, has an average annual precipitation of 890 mm, varying from 595 mm in the lower elevations to 1,050 mm in the higher elevations. Elevations range from 1838 to 2421 m, with an average of 2206 m. Fire suppression since the early 1900s has reduced the area burned on the experimental forest (Barrett 1993).

The experimental forest is currently about 9 percent nonforested, 42 percent single-aged, and 49 percent two-aged stands. The geology, soils, and stand conditions within the experimental forest are fairly consistent, providing a near-optimum site to isolate the influence of canopy coverage and cover type on runoff. The oldest stands on TCEF were last burned in 1580, and the most recent burn affecting a significant area occurred in 1902 (Table 10.1). The 1580 burn is currently a mature Engelmann spruce/subalpine fir *(Picea engelmannii/Abies lasiocarpa)* stand with a few isolated mature lodgepole pine found where mortality has created gaps in the overstory. From 1580 to 1902, ten fires were identified with an average fire size of 621 ha and an average of thirty-two years between fires. Three small burns (average size of 6 ha) in 1921, 1947, and 1996 (thirty-two years between fires on the average) have affected less than 20 ha, or less than 1 percent of the experimental forest, in the past ninety-seven years. In contrast, 1660 and 2415 ha burned in each of the two preceding centuries (Table 10.1).

Methods

We used a seven-step approach to estimate average annual runoff on the Tenderfoot Creek Experimental Forest (McCaughey and others 1997). This process involved the development of mathematical relationships between a number of variables. In step one, it was necessary to estimate average annual runoff that was generated from precipitation. Farnes (1978) developed a relationship between average runoff and average precipitation for approximately one hundred gaged drainages in Montana and Yellowstone National Park. However, most gaged watersheds are mixed forest and nonforest.

To determine the effects of forest on runoff, it was necessary to determine how much runoff would be produced from these watersheds if they were nonforested. Using data on canopy interception of snow and precipitation and reduced runoff, it was possible to develop a relationship between runoff from a nonforested or a 100 percent forested watershed for various amounts of annual

Table 10.1. Year of the fire, area burned, and percent of the total area burned on the Tenderfoot Creek Experimental Forest in central Montana

Year	Size (ha)	Percent of total area
1580	1900	51
1676	32	1
1726	1108	30
1765	552	15
1831	41	1
1845	1008	27
1873	1317	36
1882	30	1
1889	19	<1
1902	206	6
1921	8	<1
1947	11	<1
1996	0.4	<1

Source: Barrett 1993, unpublished report.

precipitation for Montana drainages (Fig. 10.1). Runoff is minimal when average annual precipitation is less than about 320 mm because vegetation uses nearly all available soil moisture. Most of the gaged streams at this low level of precipitation have a very small percentage of their drainage forested. It usually requires more precipitation to support forest stands. Using Figure 10.1, the runoff measured from a partially forested watershed can be adjusted to estimate what runoff would be produced if all of the forest was removed. Using this relationship for gaged streams analyzed in Farnes (1978) and the percentage of forest cover for each gaged drainage, an estimate of the percentage of annual precipitation that would be runoff with zero forest stands was obtained (Fig. 10.2). If there are gaged streams in a specific area of interest, they can be used to adjust the standard curve up or down to represent more localized hydrologic conditions.

In step two, average annual runoff measured on Tenderfoot Creek over the past five years was correlated with the snow and precipitation data from nearby weather stations and *SNO*w survey *TEL*emetry (SNOTEL) sites. These correlations were used to estimate the thirty-year averages (1961–1990) of annual runoff, April 1 snow water equivalent (SWE, the amount of liquid water that would be present if snow were melted), April through June precipitation (rain), and average annual precipitation on TCEF (Fig. 10.3). July through September

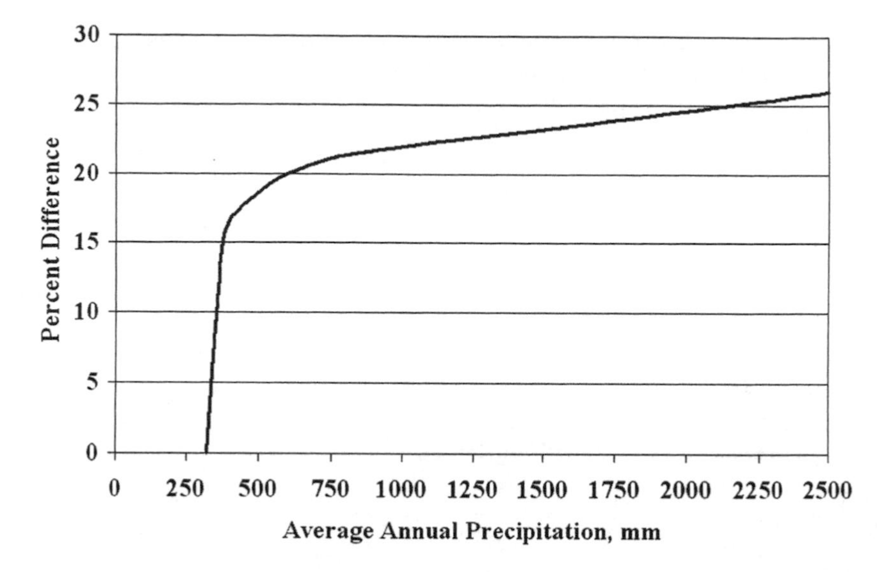

Figure 10.1. Estimated percent difference in runoff between nonforested and 100 percent forest drainages for various amounts of average annual precipitation for Montana drainages.

precipitation was not considered in this study because high evaporation rates, increased water use by vegetation, and high soil moisture deficits during this time add little to stream runoff.

In step three, we developed a relationship between forest canopy and the reduction in accumulation of snow (SWE) due to canopy interception (Fig. 10.4). Canopy was measured using a photocanopyometer (Codd 1959). Snow interception is expressed as the reduction of snow water equivalent (SWE) in lodgepole pine and spruce/fir stands compared with snow accumulation in openings. Snow water equivalent has been measured at six canopy-covered and eight open sites on the Tenderfoot Creek Experimental Forest since 1993. Since 1994, daily SWE has been recorded from two snow pillows on the experimental forest, one in the open and one under a forest canopy. Forest canopy has been measured and comparisons made between canopy cover and SWE at Natural Resource Conservation Service snow courses in Montana since the early 1960s (Farnes 1971). The effect of a logging-altered forest canopy on SWE in Montana was similar to fire or natural mortality as shown by Hardy and Hansen-Bristow (1990).

After the 1988 fires in Yellowstone National Park, additional forest canopy and SWE measurements were made in burned and unburned forest stands (Farnes 1989, Farnes and Hartman 1989). Canopy cover and SWE data for stands with varying densities of canopy cover on the experimental forest were combined

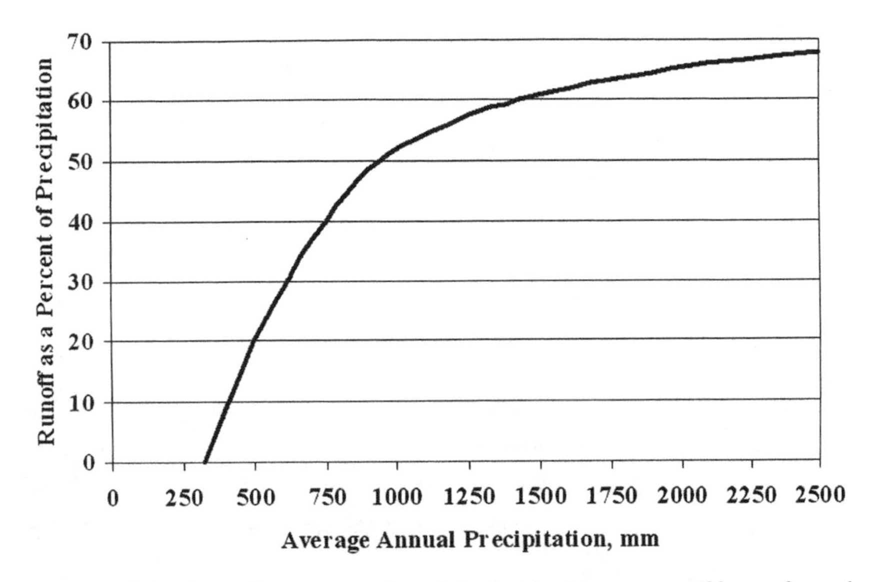

Figure 10.2. Percentage of average annual precipitation that becomes runoff for nonforested drainages as estimated from gaged drainages, percent of drainage that is forested, and adjustment for forest cover from Figure 10.1 for Montana watersheds.

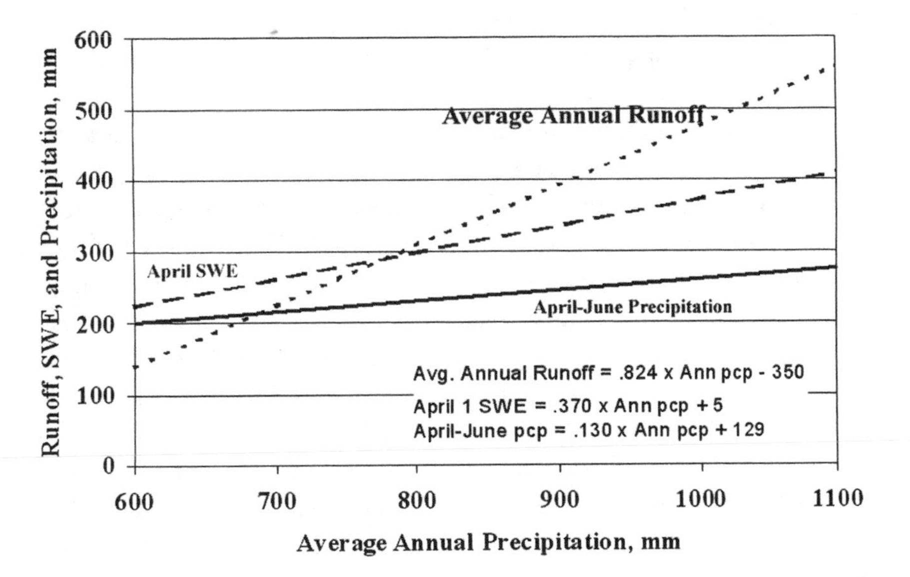

Figure 10.3. Relationship of average annual runoff, April 1 snow water equivalent (SWE), and April–June (AMJ) precipitation in open canopy areas to average annual precipitation (Pcp).

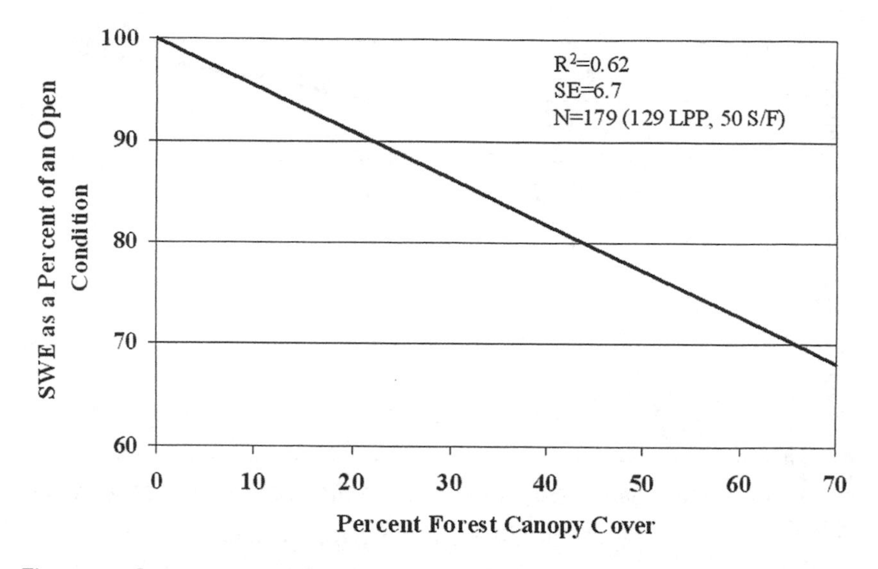

Figure 10.4. Snow water equivalent (SWE) calculated as a percent of an open condition (no forest canopy) compared with percent forest canopy cover for lodgepole pine and spruce/fir stands in Montana and Tenderfoot Creek Experimental Forest. Canopy cover is measured with a photo canopy meter and a 30-degree cone.

with data from these other studies (Hardy and Hansen-Bristow 1990, Skidmore and others 1994, Moore 1997, Moore and McCaughey 1997). Data combined from these studies yielded a reasonable correlation ($R^2 = 0.62$) between SWE and percent forest canopy cover.

In step four, data from TCEF and other studies (Farnes 1989, Farnes and Hartman 1989, Hardy and Hansen-Bristow 1990, Moore 1997, Moore and McCaughey 1997, Skidmore and others 1994) were used to develop a relationship between percent forest canopy and cover type (CT) index based on Despain's (1990) forest cover types (Fig. 10.5). Cover type classification is a method of classifying plant communities that result from a continuous successional process. The lower the CT number designation, the younger the stand. For example, an LP0 CT index represents a recently burned forest where lodgepole pine is expected to colonize the site or has already done so. An LP3 CT is a mature stand containing extensive lodgepole pine mortality and high amounts of large and small seedlings/saplings of Engelmann spruce and subalpine fir. The species designation for a particular cover type depends on the species that dominates a successional stage.

The cover type for a fire-generated stand was combined with stand age, as determined by fire history, to obtain a numerical cover type index value that more

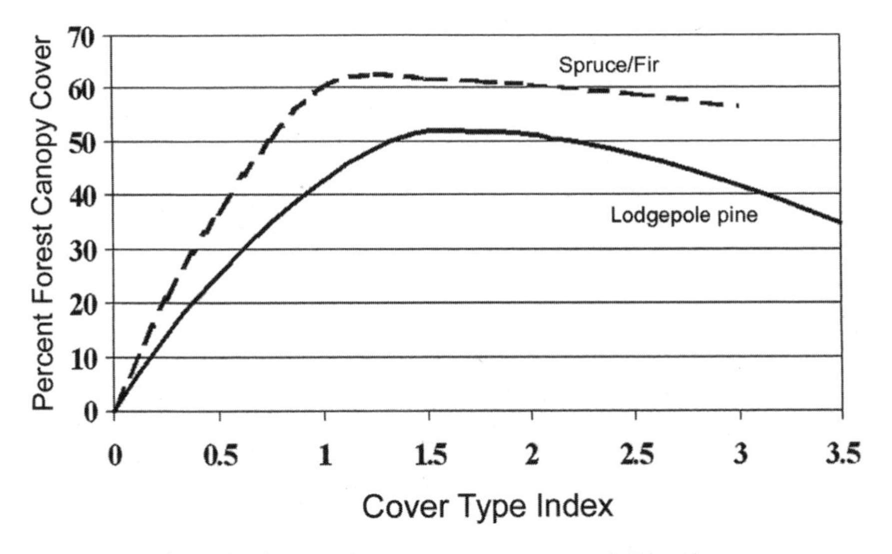

Figure 10.5. Relationship between forest canopy cover measured with a photo canopy meter and the cover type index developed by Despain (1990) for lodgepole pine and spruce/fir stands.

accurately reflects the successional stage of the stand. An LP1 is a mature lodgepole pine forest and may be from 50 to 150 years old, depending on site growing conditions. An LP3 is an overmature, climax stand, generally older than 300 years, with many dead and dying trees. Likewise, an SF1 is a mature spruce/fir stand approximately 150 to 200 years old, and an SF2.5 stand is near climax and approximately 400 to 500 years old. The age of mature stands in the TCEF study area was assumed to be about 100 years for lodgepole pine and about 200 years for spruce/fir stands on the experimental forest. The CT index for a lodgepole pine stand on TCEF that is 150 years old would be an LP1.5, and a 200-year-old spruce/fir stand would have an SF1.0 CT index. Although not all agencies and companies use the Despain (1990) vegetative classification system, it is possible to develop canopy and runoff relationships by comparing stand descriptions developed by Despain with those used by each entity.

Step five incorporates data from steps two, three, and four to quantify the reduction in SWE due to canopy interception and CT index for lodgepole pine and spruce/fir stands (Fig. 10.6). A relationship was developed between the April–June precipitation (rain) throughfall data collected at TCEF and other study areas (Arthur and Fahey 1993, Fahey and others 1988, Wilm and Niederhof 1941) and the cover type index (Fig. 10.7). This relationship accounts for spring precipitation loss due to canopy coverage when soils are typically saturated and rain has a direct effect on streamflow.

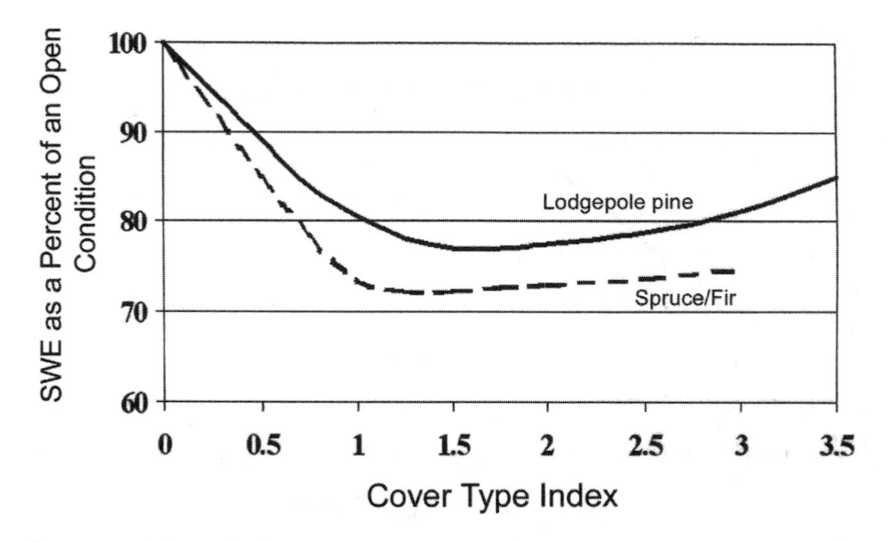

Figure 10.6. Relationship between snow water equivalent (SWE) calculated as a percent of an open condition (no forest canopy) from Figure 10.4 and percent forest canopy compared to cover type index developed by Despain (1990) from Figure 10.5 for lodgepole pine and spruce/fir stands.

Average annual runoff was calculated in step six for each 5-cm precipitation increment that had been measured and delineated for the precipitation gradient of the experimental forest. Runoff from watersheds adjacent to the Tenderfoot Creek Experimental Forest (Farnes 1971, 1978, Farnes and Hartman 1989) was used to estimate past runoff for the experimental forest using Figures 10.1 and 10.2. The average annual runoff was computed for each precipitation zone assuming no forest canopy cover. The canopy interception reduction in SWE and rain throughfall for each CT was then used to calculate runoff for each cover type in all precipitation zones. The reduction in SWE and April–June precipitation, used to compute the runoff for the 900-mm precipitation zone within a lodgepole pine CT, is shown in Table 10.2. Average annual runoff for each precipitation zone and for lodgepole pine CT was then computed (Table 10.3), using procedures like those shown in Table 10.2. We acknowledge that changes in snowfall and precipitation may have occurred over time, but for this study, the reconstruction of runoff for the past 400-plus years was based on precipitation amounts received in the current thirty-year average base period (1961 to 1990).

In the final step, we estimated average annual runoff from each fire-generated stand using cover type before and after each fire. A CT index value was assigned for each stand immediately prior to a fire based on its age. The stand was then

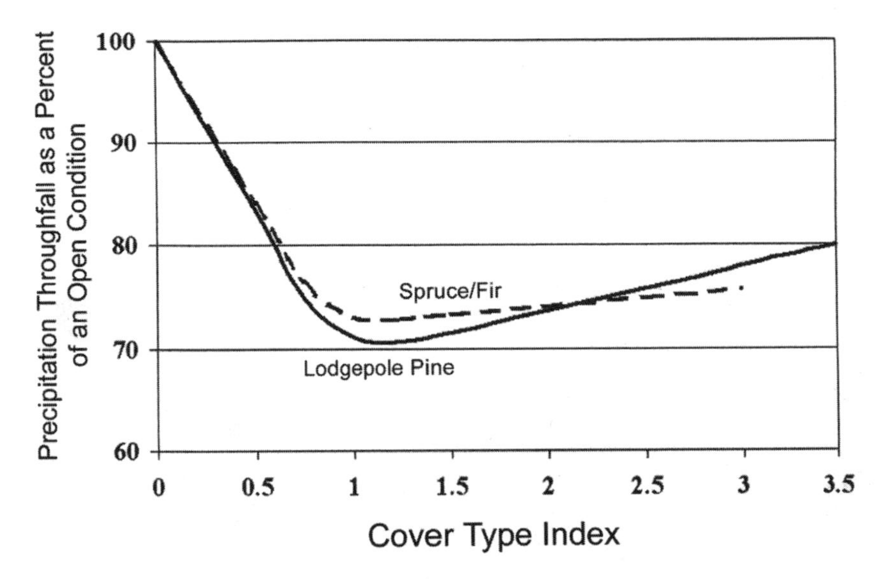

Figure 10.7. Relationship between April–June precipitation (rain) throughfall calculated as a percent of an open condition (no forest canopy) and cover type index developed by Despain (1990) for lodgepole pine and spruce/fir stands.

reassigned a new CT index value for the year after the fire, reflecting the stand's new open canopy condition. Net precipitation accumulation for each CT and precipitation zone was estimated by taking the total precipitation input and estimating the runoff before and after the fire due to canopy interception of snow (SWE) and April–June rain. Average annual runoff was computed for all CT stands and open areas based on precipitation zone and accumulated to obtain average annual runoff for the entire watershed for the years before and after each fire. Table 10.4 shows how the difference between the prefire and postfire runoff for the 1726 fire is computed.

Results and Discussion

Average annual runoff, the amount of precipitation falling as snow (April 1 SWE) and the amount falling as rain (April–June) increased at different rates as the average annual precipitation increased (Fig. 10.3). Runoff is a function of precipitation received and the amount remaining after vegetation use, soil and groundwater recharge, and evaporation and sublimation losses. Vegetation significantly affects soil and groundwater recharge and subsequent runoff. Clearcuts or openings retain the highest percentage of soil moisture in the top 1.2 m of soil and mature stands the lowest (Newman and Schmidt 1980). Evaporation

Table 10.2. Calculations of runoff for a lodgepole pine (LP) cover type (CT) for a 900 millimeter (mm) precipitation zone on the Tenderfoot Creek Experimental Forest in central Montana

CT LP	SWE reduction[a]	Precipitation reduction[b]	Total reduction[c]	Net precipitation[d]	Runoff[e] (mm)
.0	0	0	0	900	382
.1	7	7	14	886	377
.2	16	17	33	867	368
.3	24	25	49	851	362
.4	30	33	63	837	356
.5	37	41	78	822	349
.6	44	49	93	807	343
.7	49	58	107	793	337
.8	55	65	120	780	332
.9	61	69	130	770	327
1.0	65	71	136	764	325
1.1	69	71	140	760	323
1.2	72	71	143	757	322
1.3	75	71	146	754	320
1.4	77	70	147	753	320
1.5	78	69	147	753	320
1.6	78	69	147	753	320
1.7	78	68	146	754	320
1.8	78	66	144	756	321
1.9	77	65	142	758	322
2.0	77	64	141	759	323
2.1	76	64	140	760	323
2.2	75	63	138	762	324
2.3	74	62	136	764	325
2.4	73	60	133	767	326
2.5	72	59	131	769	327
2.6	71	59	130	770	327
2.7	69	58	127	773	328
2.8	67	57	124	776	330
2.9	65	55	120	780	332
3.0	63	54	117	783	333

Note: Annual precipitation zone: 900 mm; annual runoff: 382 mm; April 1 SWE: 338 mm; runoff, precipitation 42.5%; April–June precipitation 246 mm.

[a]338 (% open × 338)

[b]246 (% open × 246)

[c]Sum SWE and precipitation reduction in mm

[d]Precipitation zone minus total reduction

[e]Net precipitation times 0.425

Table 10.3. Average annual runoff for lodgepole pine (LP) cover types (CT) and average annual precipitation zones on Tenderfoot Creek Experimental Forest in central Montana

	Average annual precipitation zone (mm)									
CT LP	600	650	700	750	800	850	900	950	1000	1050
	Runoff (mm)									
.0	144	180	217	256	298	340	382	428	470	511
.1	141	177	213	253	293	334	377	420	462	503
.2	138	173	209	247	287	328	368	412	454	493
.3	135	170	205	242	282	322	362	405	446	485
.4	132	166	200	237	277	316	356	398	438	477
.5	130	163	197	233	272	310	349	391	431	469
.6	127	159	193	228	266	304	343	384	423	461
.7	124	156	189	224	261	299	337	378	415	452
.8	122	153	186	220	257	294	332	371	409	446
.9	120	151	183	218	254	290	327	367	404	441
1.0	119	150	181	215	251	288	325	364	401	437
1.1	119	149	180	214	250	286	323	362	399	435
1.2	118	148	180	213	249	285	322	360	397	433
1.3	117	148	179	212	248	284	320	359	395	431
1.4	117	148	179	213	248	284	320	359	395	431
1.5	118	148	179	213	248	284	320	359	395	431
1.6	118	148	179	213	248	284	320	359	395	431
1.7	118	148	179	213	249	284	320	360	396	431
1.8	118	148	180	213	249	284	321	360	397	432
1.9	118	149	180	214	250	285	322	360	397	433
2.0	119	149	181	215	250	286	323	361	399	434
2.1	119	149	181	215	250	286	323	362	399	434
2.2	119	150	181	215	251	287	324	363	400	436
2.3	120	150	182	216	252	288	325	364	400	437
2.4	120	151	183	217	253	289	326	365	402	438
2.5	120	151	183	218	254	290	327	366	403	439
2.6	120	152	184	218	254	290	327	367	404	440
2.7	121	152	184	219	255	292	329	369	406	442
2.8	122	153	185	220	256	293	330	370	407	444
2.9	122	154	186	221	258	294	332	371	409	446
3.0	123	154	187	221	258	295	333	373	410	447

Table 10.4. Estimated average annual runoff for each cover type (CT) and for the years before and after the 1726 fire in Tenderfoot Creek Experimental Forest in central Montana

Water Year	Area (ha)	CT	Average Annual Precipitation (mm)	Runoff (mm)	Runoff $m^3 \times 10^6$
1726	343	NF	850	340	1.166
	39	SF0.7	900	330	0.129
	732	LP2.0	900	323	2.364
	1862	LP1.5	850	284	5.288
	32	LP0.5	1000	431	0.138
	701	LP3.0	900	333	2.334
Total	3709				11.419
1727	343	NF	850	340	1.166
	39	SF0.7	900	330	0.129
	325	LP2.0	900	323	1.050
	1862	LP1.5	850	284	5.288
	32	LP0.5	1000	431	0.138
	1108	LP0	900	382	4.233
Total	3709				12.004

Notes: Increase in average annual runoff is 12.004 minus 11.419 or 0.585 $m^3 \times 10^6$ from the 1726 burn of 1108 ha.

NF = Nonforested. LP0 in 1727 is area burned in 1726. Ha is hectares and m^3 is cubic meters.

losses can be considerable in warm, dry years. Sublimation is a function of the area in the forest, the age and species structure of the forest canopy, and the season in which precipitation occurs. Water stored in the winter's snowpack gives the greatest contribution to annual runoff because the soils are saturated and vegetation water use is minimal during snowmelt.

Regression coefficients of percent canopy cover and SWE calculated as a percentage of an open condition were similar for lodgepole pine and spruce/fir and were therefore combined (Fig. 10.4). The SWE showed a linear relationship with canopy.

Canopy cover increased quickly as lodgepole pine and spruce/fir stands approached maturity (CT index of 1 to 1.5) and decreased slowly as stands advanced into late successional stages (Fig. 10.5). Mortality due to age, disease, wind throw,

Table 10.5. Estimated runoff from Tenderfoot Creek Experimental Forest in central Montana, 1581–1998

Period		Runoff ($m^3 \times 10^6$)
1581–1998	Average annual runoff for period of record	11.752
1581–1998	Maximum average annual runoff (1581)	12.547
1581–1998	Minimum average annual runoff (1872)	11.321
1961–1990	Average annual runoff	11.569
Current	Average annual runoff with all forest removed	13.278
Current	Maximum annual runoff, no forest, wet year	21.245
Current[a]	Average annual runoff	11.462
Current[a]	Maximum annual runoff, wet year	18.339
Current[a]	Minimum annual runoff, dry year	5.731

[a]Currently 91 percent of Tenderfoot Creek Experimental Forest is timbered and 9 percent is meadows and rock outcrops.

and insects creates gaps in late successional stands allowing increased amounts of snow and rain to reach the forest floor, enabling multistructured stands to develop.

The amount of snow (SWE) and rain (April–June precipitation) that reached the forest floor decreased sharply as lodgepole pine and spruce/fir stands approached maturity (Figs. 10.6, 10.7). Precipitation throughfall and SWE remained constant for a short period after stands matured and then increased slowly as stands approached late successional stages and opened up due to tree mortality. A higher percentage of snow reached the forest floor under lodgepole pine than under spruce/fir stands because of the dense foliage characteristics of spruce and fir (Figs. 10.6, 10.7). Precipitation throughfall of rain was nearly equal for lodgepole pine and spruce/fir stands.

The average annual runoff from 1581 to 1998 was estimated to be 11.752 $m^3 \times 10^6$. The largest average annual runoff during the past 400-plus years was estimated at 12.547 $m^3 \times 10^6$ after the fires burned 51 percent of TCEF in 1581 (Table 10.5, Fig. 10.8). The minimum average annual runoff (11.321 $m^3 \times 10^6$) was calculated to occur just prior to the 1873 fires. The runoff for a wet year would be about 160 percent of average based on data from gaged streams near TCEF. The annual runoff for a very dry year would be about 50 percent of average. The estimated average annual runoff for the thirty-year period ending in 1998 is approaching the minimum yield estimated to have occurred in the past 400-plus years (Table 10.5, Fig. 10.8).

Fire and its effect on canopy type and cover have played a significant role in determining the average annual runoff patterns for Tenderfoot Creek Experi-

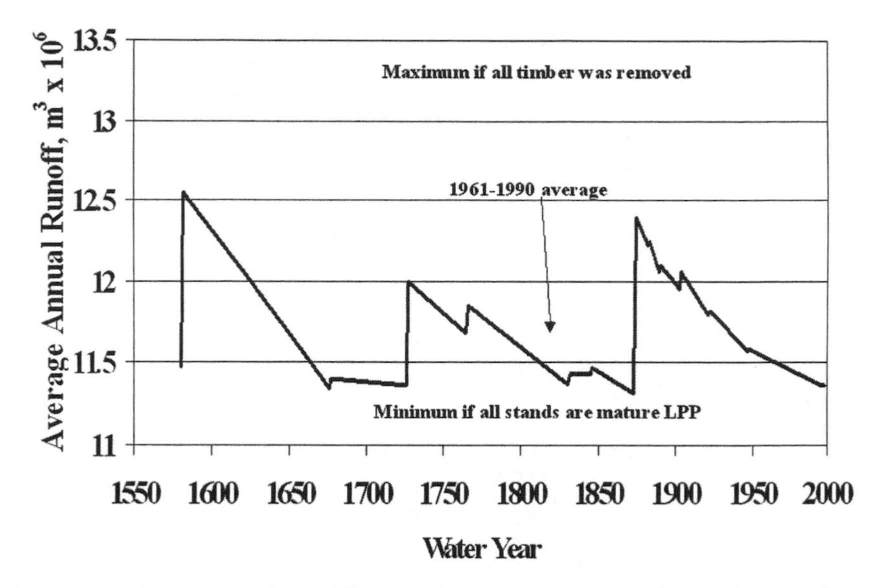

Figure 10.8. Average annual runoff for Tenderfoot Creek Experimental Forest based on fire history data and canopy cover/precipitation/runoff relationships under the 1961–1990 average precipitation regime.

mental Forest and throughout the North American West over the past 400-plus years. Prior to fire suppression, when average annual runoff approached a minimum level in a mature forest, fires removed large portions of the forest canopy, resulting in increased runoff (Fig. 10.8). Fire suppression and succession over the past ninety years have allowed the forested areas to reach a mature stage, resulting in water yields that are now approaching the estimated minimum level. Continued fire suppression over large forested areas will create a fuel buildup that could result in catastrophic fires during drought years. The effects of such major fires could severely affect stream channels, fisheries, wildlife, and recreation.

We feel that this approach to estimate historic runoff patterns can be used in other forest types if precipitation throughfall and other input variables are known. Management strategies could then be developed and applied that create a mixture of successional stages which reduce fire potentials and maintain adequate water yields.

YELLOWSTONE FIRES OF 1988

The extensive wildfires of 1988, centered around Yellowstone National Park, Wyoming, allowed studies of the potential changes in streamflow volumes and the timing and amount of peak flows. Downstream water users and agencies responsible for disaster and emergency services were concerned with these potential changes in runoff beginning in the 1989 runoff year and have had a continued interest in any changes in runoff patterns. High streamflow in 1996 and 1997 renewed interest in the effects that the 1988 fires may have had on streamflow.

Study Area

The largest impact of the 1988 fires in and near Yellowstone National Park was in the Yellowstone, Madison, and Snake River Drainages (Despain and others 1989). Evaluations of burned areas suggested the water yield changes in these drainages would be more extensive than in other drainages in and near the park. Only the Yellowstone and Madison River Drainages were analyzed for this study because of the availability of long-term stream gaging data near the burned watersheds. These gages are maintained by the U.S. Geological Survey, and they are identified as Yellowstone River at Corwin Springs, Montana, located about 11 km downstream from Gardiner and Yellowstone National Park, and Madison River near Grayling, Montana, located just downstream from Hebgen Lake and about 27 km northwest of West Yellowstone and Yellowstone National Park. Burned areas north of YNP, but within the Yellowstone River Drainage, that flow into the park and above the Corwin Springs gage, were included in these studies. In the Madison River Drainage, approximately 1460 ha of burn west of YNP were included in the drainage above the Grayling gage. (Approximately 75 percent of the drainage above the Grayling gage is within the boundaries of Yellowstone National Park.)

Methods

Two methods were used to evaluate the extent of volume runoff modification due to the 1988 fires: an analytical procedure, which was described in detail in the previous section on Tenderfoot Creek Experimental Forest, and a forecast procedure.

The forecast procedure involved developing a snowmelt runoff forecast equation for the prefire period. To develop the equation, fall soil moisture, snow water equivalent (SWE) of the winter snowpack, and spring precipitation were regressed against snowmelt runoff volume. The forecast equation was then used to estimate the postfire snowmelt runoff, assuming no changes in the hydrologic relationship because of the fire. Since the measured runoff included effects of fire, any difference was then attributed to changes in runoff resulting from the

1988 fires. This section will evaluate these changes based on eleven runoff years after the fire.

The analytical procedure related the extent and type of forest modification to changes in effective snow and precipitation interception, which in turn was related to changes in the annual runoff. In some situations, it might be possible to use a double-mass curve analysis (Searcy and Hardison 1960) between adjacent drainages to assess runoff changes, but this method has limited usefulness in evaluating the effects of the 1988 Yellowstone fires because nearly all adjacent drainages were affected to some extent by the fires. Also, adjacent gaged drainages are somewhat different, hydrologically, than burned drainages within Yellowstone National Park.

Forecast Procedure

The NRCS Snow Survey Office in Bozeman, Montana, has used a spring runoff (April–July) forecast equation composed of three variables: a soil moisture surrogate (November–January runoff), a spring precipitation variable (April–June), and a winter snowpack variable (April 1 SWE). The weight given to each variable and to each station was determined by multiple regression analysis (Schermerhorn and Barton 1968). Forecasting equations were developed for the Yellowstone River at Corwin Springs and the Madison River near Grayling stream gaging stations.

An April–July runoff forecast equation was generated for the Yellowstone River at Corwin Springs for the period 1961 through 1988 with the assistance of the NRCS and YNP (Roy Ewing, formerly with the Research Division, YNP). The procedure used the November–January runoff at Corwin Springs as a surrogate index of the soil moisture under the snowpack, April 1 SWE weighted for seven snow pillow locations, and April–June precipitation weighted for six locations in or near the drainage. None of the snow pillows or climatological stations was significantly altered by the fires. This equation had an R^2 of 0.952 and standard error of 110 $m^3 \times 10^6$ that was 5.6 percent of the average 1961–1990 April–July runoff.

With the assistance of NRCS, an April–July runoff forecast equation was generated for the Madison River near Grayling using data from 1968 through 1988. This procedure used the November–January runoff near Grayling as the surrogate for the soil moisture variable, April 1 SWE weighted for five snow courses and snow pillow stations, and April–June precipitation weighted for six stations. This equation had an R^2 of 0.948 and standard error of 28 $m^3 \times 10^5$ that was 6.0 percent of the 1961–1990 average April–July runoff.

Analytical Procedure

The analytical procedure required an accurate assessment of precipitation patterns, locations, and size and type of forest canopy alteration. The procedure was similar to that used in the Tenderfoot Creek Experimental Forest.

Acreage of areas burned that were completed immediately after the 1988 fires was calculated using the perimeter of the burn areas. This resulted in larger areas (used in Farnes and Hartman 1989) than were obtained after more detailed studies (Despain and others 1989) in the mid-1990s. More recently, burned areas were reevaluated using aerial photos taken a few years after the 1988 fires, and forested areas were reclassified as either no burn or canopy burn. The newer canopy burn data were used in this study.

Since the CT before and after the fires and the 1961–1990 average annual precipitation map by Farnes were already entered into the Yellowstone National Park GIS for drainages in and adjacent to Yellowstone National Park, it was possible to identify canopy burn areas by average annual precipitation zone, by CT, and by drainage. (This detailed information was provided by the Spatial Analysis Center, Yellowstone Center for Resources, Yellowstone National Park, Wyoming.) Drainages analyzed are shown in Figure 10.9.

Average annual April 1 SWE was compared to average annual precipitation for eighteen stations in the Yellowstone and Madison River Drainages (Fig. 10.10). There was no significant difference between stations in the Yellowstone or Madison Drainages. The April 1 SWE in mm was then estimated for each precipitation zone as 0.72 times average annual precipitation minus 173 for areas with annual precipitation greater than 500 mm. For this equation (greater than 500 mm) the square of the correlation coefficient (R^2) was 0.970 and the standard error (SE) was 42. For less than 500 mm precipitation, the relationship was curved due to midwinter snowmelt at lower elevations.

Average April–June precipitation was compared to average annual precipitation. There were some differences between the Yellowstone and Madison River Drainages, and the data were analyzed separately. For the Yellowstone River Basin, the April–June precipitation was estimated to be 0.21 times annual precipitation plus 41 (Fig. 10.11). For this equation, R^2 was 0.956 with a SE of 17. There were eleven stations used to develop this relationship. In the Madison River Drainage, the April–June precipitation was estimated to be 0.18 times the average annual precipitation plus 53. There were five stations used in this equation and the R^2 was 0.994 with a SE of 6 (Fig. 10.12). By using the previously

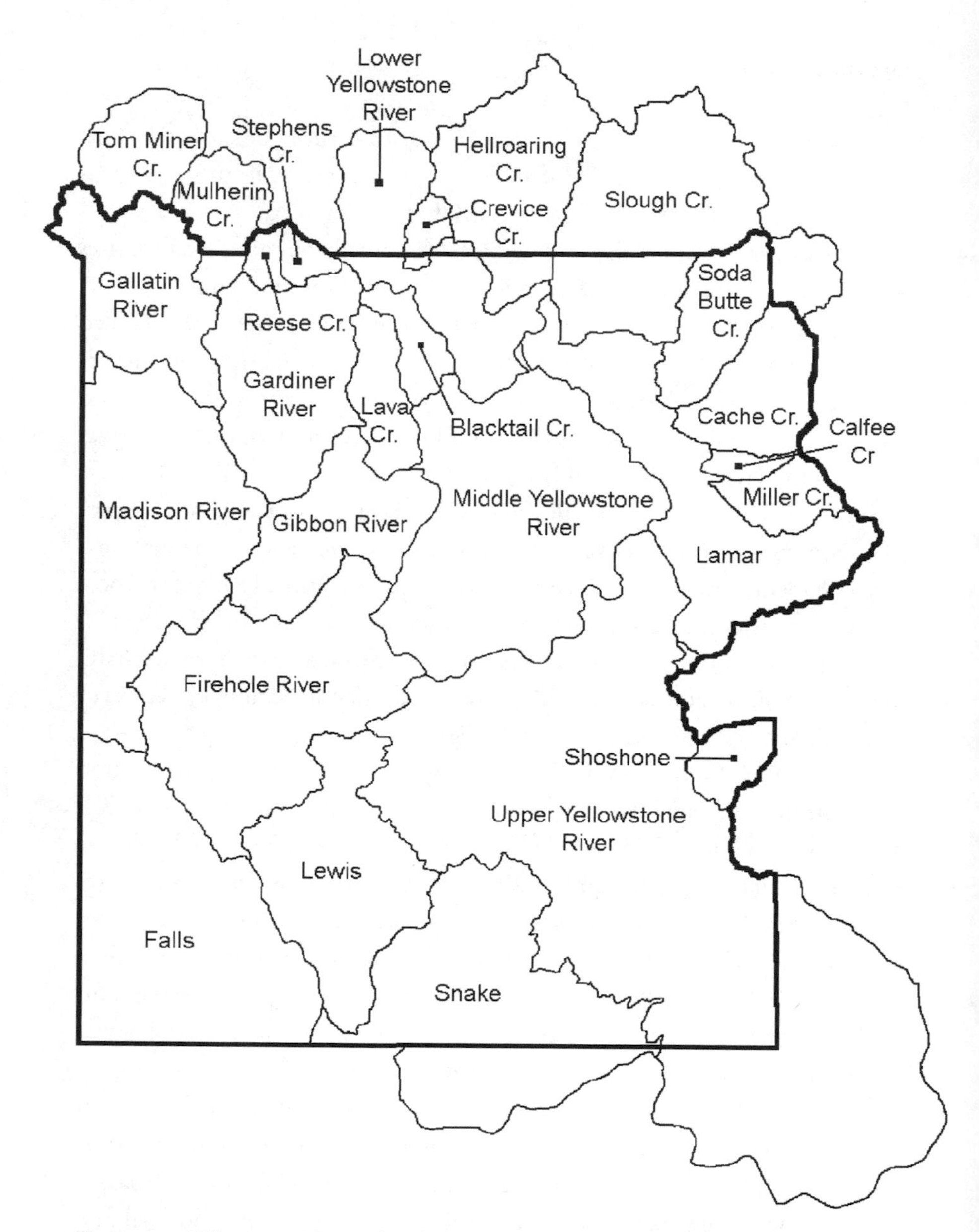

Figure 10.9. Yellowstone National Park drainages that were analyzed for increased runoff created by the 1988 fires.

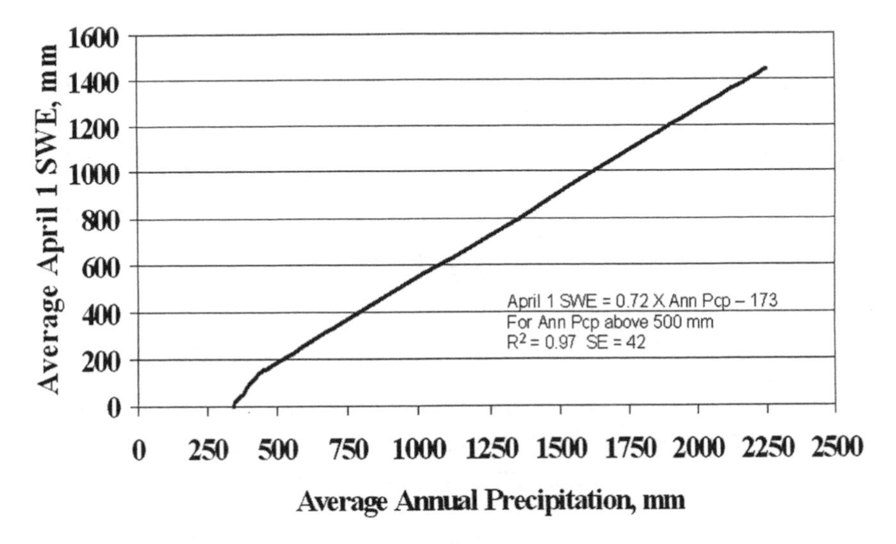

Figure 10.10. Relationship between the 1961–1990 average April 1 snow water equivalent (SWE) and the 1961–1990 average annual precipitation (Pcp) for eighteen stations in the Yellowstone and Madison River Drainages in and adjacent to Yellowstone National Park.

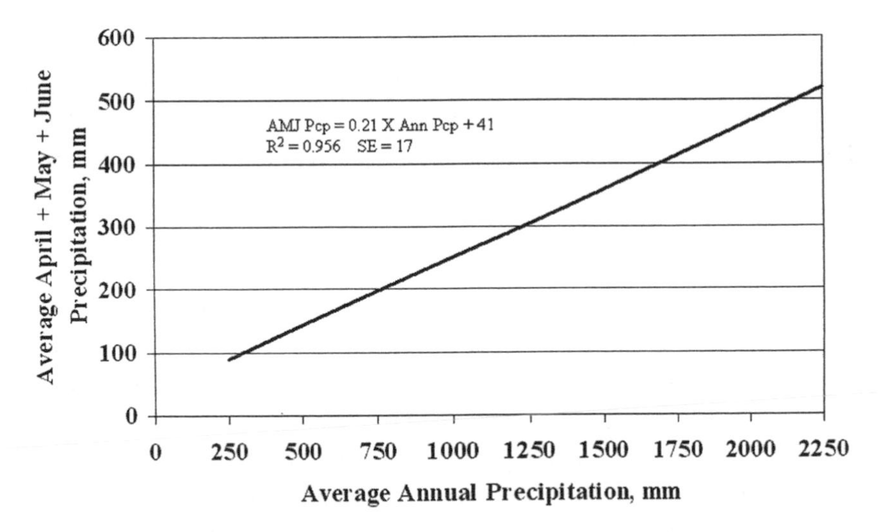

Figure 10.11. Relationship between the 1961–1990 average April–June (AMJ) precipitation and the 1961–1990 average annual precipitation (Pcp) for eleven stations in the Yellowstone River Drainage in and adjacent to Yellowstone National Park.

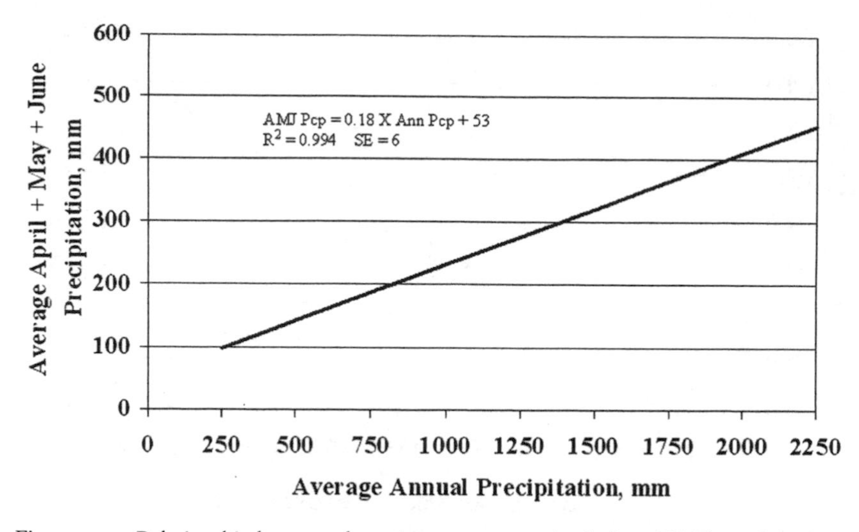

Figure 10.12. Relationship between the 1961–1990 average April–June (AMJ) precipitation and the 1961–1990 average annual precipitation (Pcp) for five stations in the Madison River Drainage in and adjacent to Yellowstone National Park.

presented data, it was possible to develop annual runoff for each precipitation zone for each CT (Table 10.6).

Using areas that were burned in 1988 and their CT prior to burn, and runoff from Table 10.6, the increase in runoff due to the fires of 1988 was calculated for each precipitation zone and CT and then accumulated for the total drainage. An example of computation for the Lava Creek Watershed is shown in Table 10.7. The calculated prefire runoff and increased runoff created by the 1988 fires for the drainages shown in Figure 10.9 in and adjacent to Yellowstone National Park are shown in Table 10.8.

Results and Discussion

The 1997 runoff for both the Yellowstone and the Madison was the largest on record, with continuous records at both stations starting in 1911. Annual water-year precipitation over both drainages was generally below average for approximately half the years after the fires. The remaining years had above-average precipitation, with some high-elevation sites measuring record to near-record amounts of precipitation in 1997. Overall, the average precipitation for the eleven years following the fires was near the 1961–1990 average in both drainages.

Climatic data used as input to the forecast equations for the 1989–1999 period compared as follows with the 1961–1990 averages. The November–January flow for the Yellowstone River was 100 percent of the 1961–1990 average. The April 1 SWE was 102 percent of the average, and the spring precipitation was 113 percent of the average. In 1997, the snow variable was 154 percent of the average, the highest for the period of record. The April–July runoff for four of the eleven postfire years was below average. Comparing the measured runoff to the runoff forecast with this equation for the eleven years after the 1988 fires shows that the measured April–July runoff averaged 7.1 percent more than the forecast runoff. Comparison of forecasts and runoff for postfire years for Yellowstone River at Corwin Springs and the highest and lowest years for the period of record is shown in Table 10.9.

Regression analysis, between April–July runoff and August–September runoff at Corwin Springs for the period of record, indicates that when spring runoff increases, there is also an increase in the August–September runoff. The slope of this regression is 0.181 ($R^2 = .633$). A 7.1 percent increase in the April–July runoff would translate to a 1.3 percent ($7.1 \times .181$) increase in the August–September runoff. There is no correlation between the April–July runoff and the following October–March runoff ($R^2 = .069$). This suggests there is no increase in the October–March runoff following years of high spring runoff. The increase in annual runoff, as a result of the 1988 fires for the Yellowstone River at Corwin Springs, based on the forecast procedure was estimated to be 5.3 percent. This is calculated as 7.1 percent times 1985 (average April–July runoff) plus 1.3 percent times 404 (average August–September runoff) plus 477 (average October–March runoff), all divided by 2866 (average annual runoff). The average annual increase for the Yellowstone River at Corwin Springs using the analytical procedure was estimated to be 4.0 percent compared with 5.3 percent obtained from the forecast procedure.

For the Madison River near Grayling forecast, the November–January runoff was 99 percent of the 1961–1990 average for the postfire period. The April 1 SWE was 101 percent average and the April–June precipitation was 108 percent average. In 1997, the April 1 SWE was 158 percent average, which was the highest for the period of record. The April–July runoff for five of the eleven postfire years was below the 1961–1990 average. Comparing the measured runoff to the runoff forecasted with this equation for the eleven years following the 1988 fires shows that the average actual April–July runoff averaged 7.6 percent greater than the forecast runoff. The lowest and highest years for the period of record and

Table 10.6. Average annual runoff for different precipitation zones and cover types (CT) in Yellowstone and Madison River Drainages in and adjacent to Yellowstone National Park

	1961–1990 Average Annual Precipitation Zone (mm)											
CT	250–299	300–349	350–399	400–449	450–499	500–749	750–999	1000–1249	1250–1499	1500–1749	1750–1999	2000–2249
	Average Annual Runoff for Yellowstone River Drainage, mm											
LP0	0	1	26	55	90	206	433	652	859	1064	1266	1466
LP1	0	.9	23	47	76	172	358	536	703	868	1029	1191
LP2	0	.9	23	47	76	170	351	525	688	850	1006	1153
LP3	0	.9	23	49	78	176	366	546	717	886	1050	1214
SF0	0	1	26	55	90	206	433	652	859	1064	1266	1466
SF1	0	.9	23	47	76	166	342	509	666	821	972	1123
SF2	0	.9	23	47	76	167	343	511	668	823	975	1126
SF3	0	.9	23	48	78	170	350	522	682	841	997	1152
WB0	0	1	26	55	90	206	433	652	859	1064	1266	1466
WB1	0	.9	24	51	83	186	388	580	762	942	1119	1294
WB2	0	.9	24	51	83	187	388	582	764	943	1120	1296
WB3	0	.9	24	52	84	188	392	587	770	952	1132	1309

					Average Annual Runoff for Madison River Drainage (mm)							
LP0	0	1	22	42	69	169	372	585	777	967	1153	1339
L01	0	.9	20	36	58	141	309	484	640	795	946	1097
LP2	0	.9	20	36	58	140	303	475	627	777	924	1071
LP3	0	.9	20	37	60	144	315	493	652	809	962	1116
SF0	0	1	22	42	69	169	372	585	777	967	1153	1339
SF1	0	.9	20	36	58	136	294	460	606	751	892	1034
SF2	0	.9	20	36	58	137	294	461	608	753	895	1037
SF3	0	.9	20	37	59	139	300	471	621	769	915	1059
WB0	0	1	22	42	69	169	372	585	777	967	1153	1339
WB1	0	.9	21	39	64	152	333	522	692	859	1022	1136
WB2	0	.9	21	39	64	153	333	522	692	860	1024	1188
WB3	0	.9	21	40	64	154	336	528	699	868	1034	1199

Note: Whitebark Pine (WB) values estimated as one-half the decrease for spruce/fir (SF) cover type.

Table 10.7. Example of prefire and postfire runoff computations for Lava Creek Drainage in Yellowstone National Park

Pcp Zone	CT	Area (ha)	Runoff (mm)	Runoff ($m^3 \times 10^6$)
Prefire				
350–399	NF	173	26	.045
	LP2	4	23	.001
	SF3	112	23	.026
400–449	NF	664	55	.365
	LP2	67	47	.031
	SF3	334	48	.160
450–499	NF	238	90	.214
	LP1	2	76	.002
	LP2	366	76	.278
	SF3	206	78	.161
500–749	NF	493	206	1.016
	LP1	30	172	.052
	LP2	86	170	.146
	LP3	1902	176	3.348
	SF3	201	170	.342
750–999	NF	190	433	.828
	LP1	324	358	1.160
	LP2	2434	351	8.543
	LP3	3393	366	12.418
	SF3	107	350	.374
	WB3	610	392	2.391
1000–1249	NF	2	652	.013
	LP1	170	536	.911
	LP2	486	525	2.552
	LP3	248	546	1.354
Sum		12842		36.726
Postfire				
350–399	NF	173	26	.045
	LP2	4	23	.001
	SF0	9	26	.002
	SF3	103	23	.024
400–449	NF	664	55	.365
	LP0	52	55	.029
	LP2	15	47	.007

(continued)

Table 10.7 (*Continued*)

Pcp Zone	CT	Area (ha)	Runoff (mm)	Runoff ($m^3 \times 10^6$)
	SF0	174	55	.096
	SF3	160	48	.077
450–499	NF	238	90	.214
	LP0	353	90	.318
	LP1	2	76	.002
	LP2	13	76	.010
	SF0	173	90	.156
	SF3	33	78	.026
500–749	NF	493	206	1.016
	LP0	1900	206	3.914
	LP1	30	172	.052
	LP2	86	170	.146
	LP3	2	176	.004
	SF0	186	206	.383
	SF3	15	170	.026
750–999	NF	190	433	.823
	LP0	2543	433	11.011
	LP1	324	358	1.160
	LP2	2434	351	8.543
	LP3	850	366	3.111
	SF0	34	433	.147
	SF3	73	350	.256
	WB0	542	433	2.347
	WB3	68	392	.267
1000–1249	NF	2	652	.013
	LP0	192	652	1.252
	LP1	170	536	.911
	LP2	486	525	2.552
	LP3	56	546	.306
Sum		12842		39.612

Note: NF is nonforested. LP0, SF0, and WB0 are areas that burned in the 1988 fires.

comparison of forecast and measured runoff for postfire years for Madison River near Grayling is shown in Table 10.10.

The lower elevations of the Madison River Drainage have considerable areas of obsidian sand, which readily transmits water from the surface to subsurface

Table 10.8. Prefire yield, increase in yield from the 1988 fires, and drainage area of watersheds in Yellowstone National Park and adjacent drainages shown in Figure 10.9

Drainage	Pcp (mm)	Area (ha)	Prefire Runoff ($m^3 \times 10^6$)	Increase Runoff ($m^3 \times 10^6$)	Percent Increase	Burn Area (ha)	Percent Watershed Burned
Blacktail	622	8093	15.947	1.711	10.7	3259	40
Cache	942	21000	90.129	6.754	7.5	10969	52
Calfee	831	3043	10.679	.331	3.1	692	23
Crevice	688	3923	9.438	.001	—	7	1
Gardner	871	42079	162.566	9.931	6.1	14050	33
Hellroaring	837	38043	136.724	4.315	3.2	8682	23
Lamar	833	55749	195.925	12.625	6.4	20977	38
Lava	777	12842	36.726	2.886	7.9	5685	44
Lower Yellowstone	576	34590	53.866	.812	1.5	3126	9
Middle Yellowstone	770	91059	272.493	11.216	4.1	17212	19
Miller	894	9894	38.088	3.106	8.2	4802	48
Slough	925	54552	233.076	10.546	4.5	18482	34
Soda Butte	987	26046	123.005	1.321	1.1	1970	8
Reese	767	3123	9.485	.000	0	0	0
Stephens	424	3362	2.110	.003	0.1	10	0
Yellowstone Lake	971	256710	1191.993	38.202	3.2	53881	21
Yellowstone at Corwin Springs[a]	876	665108	2582.250	103.760	4.0	163804	25
Mulherin	1118	2481	13.916	.436	3.1	322	13
Firehole	1133	72612	362.457	31.533	8.7	30336	42
Gibbon	894	32579	111.312	10.953	9.8	16479	51
Madison in YNP	958	71507	276.028	34.012	12.3	45734	64
Total Madison	1019	176698	749.797	76.498	10.2	92549	52
Madison Outside YNP	852	57702	181.761	1.898	1.0	1460	3
Madison near Grayling[a]	978	234400	931.558	78.396	8.4	94009	40
Gallatin in YNP	1222	34887	219.883	10.729	4.9	7934	23
Shoshone in YNP	1102	8185	43.992	.013	0	18	0
Falls in YNP	1504	79150	585.391	20.931	3.6	12796	16
Lewis	1270	47834	283.438	18.811	6.6	15343	32
Snake	1439	76652	538.973	48.740	9.0	37553	49

[a]Sum of the drainages above USGS stream gaging stations.

Table 10.9. Yellowstone River at Corwin Springs, Montana, April–July streamflow, $m^3 \times 10^6$

Water Year	Forecast Runoff	Measured Runoff	Difference	Percent Difference[a]
1989	2021	1907	−114	−6.0
1990	1747	1876	+129	+6.9
1991	1847	2091	+117	+11.7
1992	1706	1727	−59	−3.4
1993	1764	2286	+522	+22.8
1994	1495	1504	+9	+0.6
1995	1940	2137	+197	+9.2
1996	2732	3026	+294	+9.7
1997	3050	3426	+376	+11.0
1998	1978	2203	+69	+3.1
1999	2375	2224	−151	−6.8
Average 1989–1999	2060	2219	+159	+7.1

Notes: Lowest measured April–July runoff was 1087 $m^3 \times 10^6$ in 1919; highest was 3426 in 1997. Average 1961–1990 April–July runoff is 1985 $m^3 \times 10^6$. Average 1961–1990 August–September runoff is 404 $m^3 \times 10^6$; average October–March runoff is 477 $m^3 \times 10^6$. Average annual runoff is 2866 $m^3 \times 10^6$.

[a]Difference divided by measured runoff.

zone. It appears from the inflow hydrographs to Hebgen Reservoir that there is significant subsurface storage that feeds water into the surface system year around. Comparing the April–July snowmelt runoff with the August–September flows shows a reasonable correlation ($R^2 = 0.663$), as does the snowmelt runoff with the following October–March runoff ($R^2 = 0.622$). The April–July runoff correlates reasonably ($R^2 = 0.687$) with the August–March runoff. Since both periods have similar correlations, the flows for August through March were analyzed as one period. The slope of this regression is 0.64. For the Madison River near Grayling, the increase for the eight months following snowmelt would be 64 percent of the increase from snowmelt runoff, or 4.9 percent. This indicates a 6.3 percent increase in annual runoff using the forecast procedure. This would be calculated as 7.6 percent times 4862 (average April–July runoff) plus 4.9 percent times 5087 (average August–March runoff), all divided by 9769 (average annual runoff). The average annual increase for the Madison River near Grayling using the analytical procedure was calculated to be 8.4 percent, compared with 6.3 percent for the forecast procedure.

In summary, the analytical and forecast procedures suggest the increase in an-

Table 10.10. Madison River near Grayling, Montana, April–July streamflow, $m^3 \times 10^5$

Water Year	Forecast Runoff	Measured Runoff	Difference	Percent Difference[a]
1989	4614	4367	−247	−5.7
1990	3577	3861	+284	+7.4
1991	4108	4305	+197	+4.6
1992	3442	3577	+135	+3.8
1993	4626	5884	+1258	+21.4
1994	3442	3627	+185	+5.1
1995	5613	5798	+185	+3.2
1996	5946	6748	+802	+11.4
1997	7180	7513	+333	+4.4
1998	5441	5798	+359	+6.2
1999	5218	6106	+888	+14.5
Average 1989–1999	4837	5235	+398	+7.6

Notes: Lowest measured April–July runoff was 2048 $m^3 \times 10^5$ in 1934; highest was 7513 in 1997. Average 1961–1990 April–July runoff is 4682. Average 1961–1990 August–September runoff is 1317 $m^3 \times 10^5$; average October–March runoff is 3770 $m^3 \times 10^5$. Average annual runoff is 9769 $m^3 \times 10^5$.

[a]Difference divided by actual runoff.

nual runoff resulting from the 1988 fires was between 4.0 and 5.3 percent for the Yellowstone River at Corwin Springs and between 6.3 and 8.4 percent for the Madison River near Grayling.

MANAGEMENT IMPLICATIONS

Additional runoff will be generated as a result of the 1988 fires until regenerated trees approach maturity and a more closed canopy. Changes in snow and rain interception will be most noticeable for the next 80 to 90 years, and then snow and rain interception will approach prefire conditions with a CT between 1 and 1.5 or in about 100 to 150 years (Figs. 10.5, 10.6). The magnitude of this increase will continue to be a function of the area having had a canopy burn, the precipitation and snowfall patterns over the entire drainage, and the age of trees regenerated after the 1988 fires. To place these changes in perspective, consider the natural variation in recorded streamflow. The lowest annual flow for the Yellowstone River near Corwin Springs since 1911 was 1701 $m^3 \times 10^6$ in 1934, which was 59 percent of the 1961–1990 average runoff (2866). The maximum runoff was 4607 $m^3 \times 10^6$ in 1997, which was 161 percent of the average. The 4–5 percent increase as a result of the 1988 fires is quite small compared to the 102 percent change associated with annual climatic variability in snow and precipitation patterns.

For the Madison River near Grayling, the lowest annual runoff since 1911 was 5233 $m^3 \times 10^5$ in 1934, which was 54 percent of the 1961–1990 average runoff (9769). In 1997, the annual runoff was 13931 $m^3 \times 10^5$, which was 143 percent of the 1961–1990 average. As with the Yellowstone, the increase in annual runoff of about 6–8 percent due to the 1988 fires is small when compared with the 89 percent change resulting from the annual climatic variation. It appears the increased runoff resulting from the fires may need to be accounted for in the yearly operations of Hebgen Lake on the Madison River. However, the magnitude of the fire-related runoff changes on streams not having reservoirs might not be noticeable in the context of variability related to annual climatic fluctuations.

References

Arthur, M. A., and T. J. Fahey. 1993. Throughfall chemistry in an Engelmann spruce-subalpine fir forest in north central Colorado. Canadian Journal of Forest Research 23:738–742.

Barrett, S. W. 1993. Fire history report for the Tenderfoot Creek Experimental Forest. Unpublished report on file at U.S. Department of Agriculture, Forest Service, Rocky Mountain Research Station (formerly known as Intermountain Research Station), Forestry Sciences Laboratory, Bozeman, Mont.

Codd, A. R. 1959. The photocanopyometer. Pages 17–22 in Proceedings: Western Snow Conference, April 21–23, Reno, Nev.

Despain, D. G. 1990. Yellowstone vegetation: Consequences of environment and history in a natural setting. Roberts-Rinehart, Boulder, Colo.

Despain, D. G., A. Rodman, P. Schullery, and H. Shovic. 1989. Burned area survey of Yellowstone National Park: The fires of 1988. Divisions of Research and Geographic Information Systems Laboratory. Yellowstone National Park.

Fahey, T. J., J. B. Yavitt, and G. Joyce. 1988. Precipitation and throughfall chemistry in *Pinus contorta* spp., *latifolia* ecosystems in southeastern Wyoming. Canadian Journal of Forest Research 18:337–345.

Farnes, P. E. 1971. Mountain precipitation and hydrology from snow survey. Pages 44–49 in Proceedings: Western Snow Conference, April 20–22, Billings, Mont.

———. 1978. Preliminary report: Hydrology of mountain watersheds. U.S. Department of Agriculture, Soil Conservation Service, Bozeman, Mont.

———. 1989. Internal report: Relationships between snow water equivalents in burned and unburned areas, habitat cover types, canopy cover, and basal area. Department of Agriculture, Natural Resources Conservation Service, Bozeman, Mont.

Farnes, P. E., and R. K. Hartman. 1989. Estimating effects of wildfire on water supplies in

Northern Rocky Mountains. Pages 90–99 in Proceedings: Western Snow Conference, April 18–20, Fort Collins, Colo.

Gary, H. L., and C. A. Troendle. 1982. Snow accumulation and melt under various stand densities in lodgepole pine in Wyoming and Colorado. Res. Note RM-417. U.S. Department of Agriculture, Forest Service, Rocky Mountain Forest and Range Experiment Station, Fort Collins, Colo.

Hardy, J. P., and K. J. Hansen-Bristow. 1990. Temporal accumulation and ablation patterns of the seasonal snowpack in forests representing varying stages of growth. Pages 23–24 in Proceedings: Western Snow Conference, April 17–19, Sacramento, Calif.

Hoover, M. D., and C. F. Leaf. 1967. Process and significance of interception in Colorado subalpine forest. Pages 213–224 in W. E. Sooper and H. W. Lull, eds., Forest hydrology. Pergamon, New York.

Koch, Peter. 1996. Lodgepole pine in North America, vol. 1. Forest Products Society Press. Madison, Wis.

McCaughey, W. W., P. E. Farnes, and K. J. Hansen. 1997. Historic role of fire in determining the natural variability of annual water yield in mountain watersheds. Pages 52–60 in Proceedings: Western Snow Conference, May 4–8, Banff, Alberta.

Meiman, J. R. 1987. Influence of forests on snowpack accumulation. Pages 61–67 in Management of subalpine forest: Building on 50 years of research. Gen. Tech. Rep. RM-149. U.S. Department of Agriculture, Forest Service, Rocky Mountain Forest and Range Experiment Station, Ogden, Utah.

Moore, C. A. 1997. Snow accumulation under various successional stages of lodgepole pine. M.S. thesis, Montana State University, Bozeman.

Moore, C. A., and W. W. McCaughey. 1997. Snow accumulation under various forest stand densities at Tenderfoot Creek Experimental Forest, Montana, USA. Pages 42–51 in Proceedings: Western Snow Conference, May 4–8, Banff, Alberta.

Newman, H. C., and W. C. Schmidt. 1980. Silviculture and residue treatments affect water used by a larch/fir forest. Pages 75–110 in Proceedings: Environmental consequences of timber harvesting in Rocky Mountain coniferous forests. Gen. Tech. Rep. INT-90. U.S. Department of Agriculture, Forest Service, Intermountain Forest and Range Experimental Station, Ogden, Utah.

Schermerhorn, V., and M. Barton. 1968. A method for integrating snow survey and precipitation data. Pages 27–32 in Proceedings: Western Snow Conference, Lake Tahoe, Nev.

Searcy, J. K., and C. H. Hardison. 1960. Double-mass curves. Manual of hydrology, part 1, General surface-water techniques. Geological Survey Water Supply Paper 1541-B. United States Government Printing Office, Washington, D.C.

Skidmore, P., K. Hansen, and W. Quimby. 1994. Snow accumulation and ablation under fire-altered lodgepole pine forest canopies. Pages 43–52 in Proceedings: Western Snow Conference, April 18–21, Santa Fe, N.M.

Troendle, C. A. 1986. The potential effect of partial cutting and thinning on streamflow from the subalpine forest. Res. Pap. RM-274. U.S. Department of Agriculture, Forest Service, Rocky Mountain Forest and Range Experiment Station, Fort Collins, Colo.

Wheeler, N. C., and W. B. Critchfield. 1984. The distribution and botanical characteristics of lodgepole pine: Biogeographical and management implications. Pages 1–13 in D. M.

Baumgartner, R. G. Krebill, J. T. Arnott, and G. R. Weetman, eds., Lodgepole pine: The species and its management. Symposium proceedings, May 8–10, Spokane, Wash., May 14–16, Vancouver, B.C., Pullman, Wash. Office of Conferences and Institutes, Cooperative Extension.

Wilm, H. G., and C. H. Niederhof. 1941. Interception of rainfall by mature lodgepole pine. Journal of Trans. Am. Geophys. Union 22:660–666.

Part IV **Terrestrial Ecosystem and Landscape Perspective**

A . . . dilemma involves the very character of a natural ecosystem.
It is, most of all, a *changing* ecosystem.
—*Don Despain, Doug Houston, Mary Meagher, and Paul Schullery,*
Wildlife in Transition

Chapter 11 Early Postfire Forest Succession in the Heterogeneous Teton Landscape

Kathleen M. Doyle

Standing on Wildcat Peak just south of Yellowstone National Park, one can look west into the glacially carved basin of Jackson Hole and beyond to the towering peaks of the majestic Teton Range. The rolling, high-elevation, forested terrain of the Pinyon Peak Highlands to the east in the Bridger Teton National Forest provides yet another contrast. In the heterogeneous landscapes of the Teton region, topography, disturbance history, geology, vegetation, and weather patterns vary over relatively short distances. The topographic variation creates a mosaic of moisture conditions for plants and soil development, and creates barriers to the spread of fire. A variety of rock types are represented in the Teton region, ranging from ancient Precambrian rocks to more recent glacial deposits. The different upland forest types in the Tetons are dominated by lodgepole pine *(Pinus contorta),* Douglas fir *(Pseudotsuga menziesii),* aspen *(Populus tremuloides),* subalpine fir *(Abies lasiocarpa),* Engelmann spruce *(Picea engelmannii),* and whitebark pine *(Pinus albicaulis).* Elevational and orographic effects also create markedly different weather conditions across the region. Superimposed upon such heterogeneous landscapes are patterns of vegetation development after fire that are quite variable, but few studies have investigated the role that interacting factors play in causing this heterogeneity.

Six fires that burned in the Tetons between 1974 and 1991 provide an ideal opportunity to examine the importance of various factors on early succession for a wide range of environmental conditions and forest types. It is recognized that numerous factors contribute to variation in succession after fire, and that frequently complex interactions exist between various factors (Stahelin 1943, Bradley and others 1992). Stahelin (1943) hypothesized that fire severity, aspect, ground cover, soil factors, and prefire species composition interact to influence pathways of succession in high-elevation coniferous forests of the Central Rocky Mountains. Other studies across a range of ecosystems have shown that multiple pathways of succession can occur after disturbance on sites with similar environmental conditions (Jackson 1968, Glenn-Lewin 1980, Abrams and others 1985, McCune and Allen 1985b, Borgegård 1990, Baker and Walford 1995, Fastie 1995, Chappell and Agee 1996, Turner and others 1997).

I investigated Stahelin's hypothesis and the potential for multiple early successional pathways across the Teton region by examining the relative contribution of various factors (elevation, topography, soil nutrients and texture, prefire tree composition, distance to seed source, fire severity, and percent serotiny) on the postfire tree composition. To accomplish these goals, I first identified the range of variability in tree species composition in early successional forest communities. Second, I examined the correspondence between pre- and postfire forest composition. Third, I investigated the factors that most influenced the composition of early postfire forest types and compared these factors across the different prefire forest types.

Numerous investigators working in the Rocky Mountains (Daubenmire 1943, Langenheim 1962, Peet 1981, Romme and Knight 1981, Wentworth 1981, McCune and Allen 1985a, Allen and Peet 1990, Barton 1993, Barton 1994) have studied the distribution of late successional forest vegetation along environmental gradients. Other studies have examined pioneer vegetation at high-elevation sites (Stahelin 1943, Tomback and others 1993) and in more homogeneous terrain, such as in Yellowstone National Park (Chapters 4, 14, Turner and others 1997). Recent fires in the Teton region provide a striking contrast to the large fires of 1988 that burned across the Central Plateau of Yellowstone. Although the fires that burned across the Central Plateau are characterized by their heterogeneity in fire conditions (Chapter 4), they burned across much more homogeneous terrain dominated primarily by lodgepole pine forests and underlain by Quaternary rhyolite. It is important to understand how varying conditions contribute to different patterns of early succession so that patterns are not incorrectly extrapolated from one area to another. I know of no other study from the Rocky Mountains that has analyzed the importance of various factors on early postfire succession for such a wide range of environmental conditions and forest types as I have tried to do.

STUDY AREA, SAMPLING DESIGN, AND METHODS

The study area is located between 43° 40′ and 44° 06′ N latitude and 110° 25′ and 110° 50′ W longitude. Data were collected from forested areas that burned during seven fires between 1974 and 1991 in Grand Teton National Park (GTNP), the Pinyon Peak and Mount Leidy highlands of the adjacent Bridger Teton National Forest (BTNF), and in the J. D. Rockefeller Parkway (JDR) north of GTNP. The fires are the 1974 Waterfalls Canyon fire (WFC), the 1981 Mystic Isle fire (MYS), the 1985 Beaver Creek fire (BC), the 1987 Dave Adams Hill fire (DAH), the 1988 Huckleberry Ridge fire (HUCK), the 1988 Hunter fire (HUNT), and the 1991 Lozier Hill fire (LOZ, Fig. 11.1). Eighty-one burned stands were stratified across the seven burned areas to represent the variability in environmental and landscape conditions that exist in the region (Table 11.1). Stand boundaries were determined so that each stand would be relatively homogeneous in topography, burn severity, prefire vegetation, and mapped geologic and soil types (for example, Draft Manuscript, BTNF and USDA Soil Conservation Service 1985, Young 1982, Love and others 1992, Rodman and others 1992). As with many studies that utilize unplanned natural disturbances over large scales to understand ecological phenomena (Knight 1997), it was not possible to have good replication of all specific combinations of variables across different fires; but where possible, I sampled similar conditions across the different fires.

Continental weather patterns characterize the region. Mean annual precipitation is 59 cm at Moran, Wyoming (43° 51′ N, 110° 35′ W), at 2067 m elevation, where 71 percent of the annual precipitation falls between November and May, primarily as snow. The summers are short and cool (mean July temperature 15.1°C), and the winters are long and cold (mean January temperature −10.6°C, Martner 1986). Distance from the Teton Mountain Range greatly influences annual precipitation as a result of orographic effects. On the west side of Jackson Lake at Moran Bay (adjacent to the WFC fire), the annual precipitation is 30 cm greater than the annual precipitation at the Moran weather station located

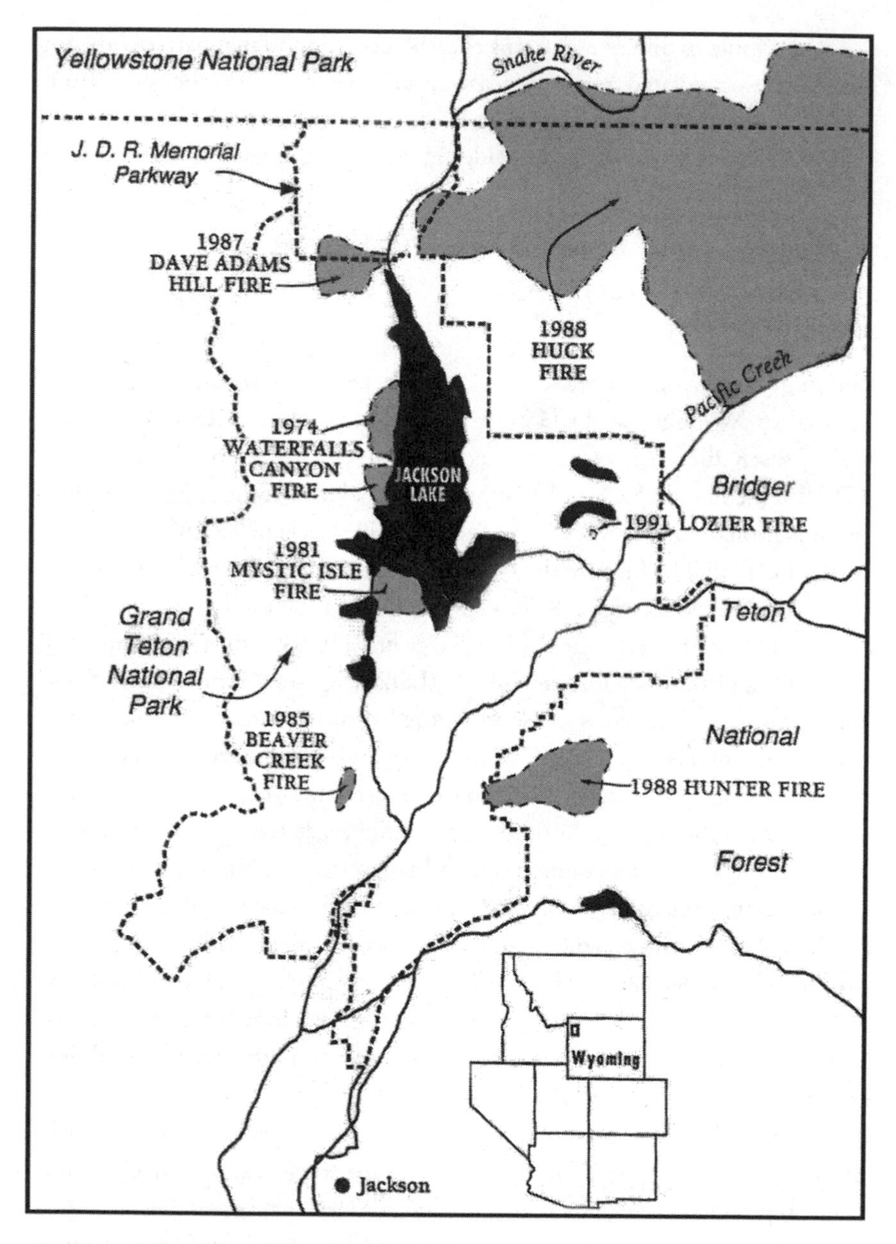

Figure 11.1. Map of Grand Teton National Park and adjacent areas in the Bridger Teton National Forest and the J. D. Rockefeller Parkway, showing the locations of the seven fire study areas (in gray). The fires occurred between 1974 and 1991. Eighty-one stands were stratified across the burned landscape to encompass the variability in prefire forests, geology, soils, topography, elevation, and burn severity.

at a similar elevation east of Jackson Lake. Annual precipitation at Moran Bay is comparable to that of Huckleberry Divide in the HUCK fire, a site approximately 150 m higher in elevation (2225 m, Farnes and others 1998).

Weather during the first growing season after the fires varied greatly (Fig. 11.2, data summarized from Western Regional Climate Center, maintained by K. Redmond, Reno, Nev.). In 1975, one year after the WFC fire, weather conditions were considerably cooler and wetter than average, and the winter of 1975 was a heavy snow year (Farnes and others 1998). In contrast, the weather was relatively warm and dry in 1988, one year after the DAH fire.

Geologic substrates within the study region include Precambrian rocks of the Teton Range, plus a variety of Paleozoic and Mesozoic sedimentary rocks, Tertiary rhyolitic tuff, Quaternary glacial debris, and other unsorted deposits such as landslide debris and colluvium (Table 11.1). Glacial deposits originated from two separate regions: the Teton Range to the west and the Yellowstone area to the north (Thornbury 1965). Glacial drift primarily from the Teton Mountains (such as is found in the WFC fire) includes more gneiss, schist, and amphibolite, and the resulting soils are higher in fertility than soils derived from glacial drift rich in quartzite and volcanic rocks that originated both from the Teton Range and the Yellowstone region (Love and others 1992).

Plot Sampling

Data were collected from 226 plots within the eighty-one stands during May through September 1991–1994. Typically, three 5 × 30–m plots were located per stand. However, the number of plots varied depending on the size of the stand. Data were summarized for each stand by averaging data from the replicate plots. The density and height classes by species for all live trees that established following fire were recorded for each plot. Depending on the total density of tree seedlings, the plot width was increased or decreased from the standard 5 m.

PREFIRE STAND COMPOSITION AND CONE SEROTINY

The prefire relative density and relative basal area of each tree species were estimated for each stand by counting and measuring the diameters of trees greater than 10 cm dbh (at 1.4 m) in and adjacent to the 226 plots. Trees that survived the fire, as well as standing and fallen dead trees killed by the fire, were measured. An importance value (IV) was calculated for each species by summing the relative basal area and relative density and dividing by two. When the sample size of prefire stems was low, or when a high proportion of the stems could not be identified (especially in older fires), prefire composition data were supplemented

Table 11.1. Description of fires and sampled stands

	1974 Waterfalls Canyon Fire	1981 Mystic Isle Fire	1985 Beaver Creek Fire	1987 Dave Adams Hill Fire	1988 Huck Fire	1988 Hunter Fire	1991 Lozier Fire
Attribute							
Total extent of burn (ha)	1487	810	416	952	49,171	2203	6
Duration of burn	July 19– Dec. 11	July 30– Oct. 14	Aug. 30– Nov. 4	Sept. 2– Nov. 14	Aug. 20– Nov. 12	Aug. 20– Nov. 4	Oct. 16– 25
Geographic description	Gently sloping to very steep terrain; dense forests to open slopes	Rolling glaciated terrain and island in Leigh Lake; dense forests to open slopes	Flat to rolling glaciated terrain; dense forests to open slopes	Gentle to steep slopes; dense forests with scattered parks	Gentle to steep terrain; dense forests with some parks at high elevations	Mostly steep terrain; conifer and aspen forests in a non-forested matrix	Open woodlands in matrix of grasslands and shrublands
Sample size	13 stands [37 plots]	12 stands [32 plots]	9 stands [26 plots]	10 stands [29 plots]	25 stands [71 plots]	10 stands [28 plots]	2 stands [3 plots]
Characteristics of stands							
Elevation range (m)	2064–2675	2073–2125	2012–2113	2061–2438	2134–2900	2140–2517	2158–2219
Percent serotiny	0%	27–42%	12–36%	4–21%	0–51%	27–38%	—

20 stands sampled	N = 1 stand	N = 4 stands	N = 3 stands	N = 5 stands	N = 5 stands	N = 2 stands	N = 0 stands
Major geologic substrates of sampled stands	Glacial drift—Precambrian origin; Schist; Gneiss; Sedimentary (calcareous members)	Glacial drift (quartzite and volcanic)	Glacial drift—Precambrian origin; Alluvial fan/gravel deposits	Glacial drift—Sedimentary (calcareous members); Rhyolite and tuff	Rhyolite and tuff; Sedimentary (lacking calcareous members)	Limestone; Colluvium; Sedimentary (calcareous members)	Sandstone/ conglomerate
Soil Analysis							
Percent sand	28–64 (45)	24–71 (52)	24–70 (51)	28–60 (40)	26–69 (46)	28–66 (39)	56–65 (60)
Percent clay	7–19 (11)	5–14 (9)	5–25 (9)	7–24 (13)	7–21 (12)	8–21 (11)	8 (8)
Ca (p.p.m)	478–3709 (2049)	536–2164 (1024)	502–3063 (1447)	610–5181 (1893)	330–3209 (1266)	1012–6311 (2535)	4506–5529 (5126)
Total Exchange Capacity	8–28 (17)	7–22 (13)	7–24 (14)	8–33 (17)	6–28 (15)	11–38 (20)	29–33 (31)
pH	4.9–7.0 (5.8)	4.6–5.9 (5.2)	5.0–6.6 (5.5)	5.0–7.1 (5.6)	4.5–6.3 (5.1)	5.2–7.3 (6.0)	6.6–7.0 (6.8)

Note: For soil parameters, numbers in parentheses are the mean values for all plots for each fire. Values separated by a dash are the minimum and maximum values for soil parameters for each fire.

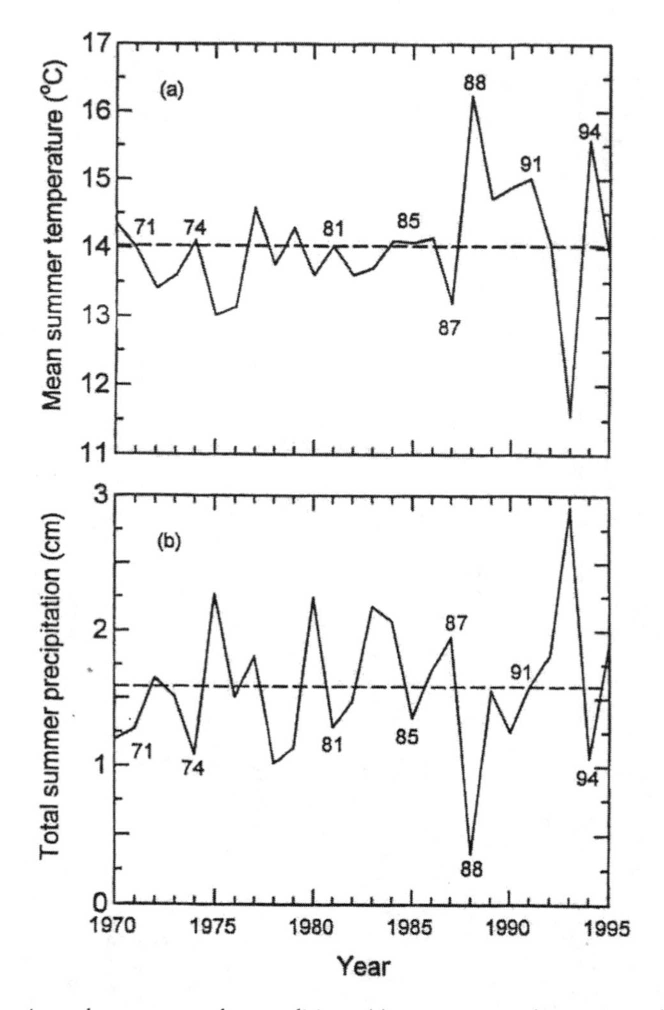

Figure 11.2. Annual summer weather conditions: (a) mean summer (June–August) temperature and (b) total summer (June–August) precipitation for a twenty-five-year period at Moran, Wyoming. A dashed line shows the mean value for the period (1970–1995). Major fires in the region are indicated by the year of the fire.

by vegetation maps (Dole and others 1936, Interagency Grizzly Bear Study team unpublished habitat type/cover type maps for GTNP and the JDR Parkway), field notes, and data from nearby stands believed to be similar in forest type before fire (Doyle 1997).

The IVs for each prefire tree species were used as variables in the ordination analyses (see "Data Analysis," below). Stands were also classified into ten prefire forest types based on the IV for each prefire tree species and using a K means algorithm and Euclidean distance metric (Systat 6.0 for Windows, SPSS Inc., 1996). The following ten prefire forest types were identified: (1) aspen, (2) lodgepole pine, (3) lodgepole pine/subalpine fir, (4) subalpine fir, (5) subalpine fir/Engelmann spruce, (6) Engelmann spruce/subalpine fir, (7) Douglas fir/subalpine fir, (8) Douglas fir/lodgepole pine, (9) Douglas fir, and (10) whitebark pine.

An important adaptation for regenerating after fire is the degree of cone serotiny. Serotinous cones remain closed until sufficiently heated (typically by fire), at which point resins melt and seeds are released (Chapters 4, 10). Lodgepole pine trees have the capability of producing both serotinous and nonserotinous cones on the same tree, but the degree of serotiny varies among individuals. The percent serotiny (percentage of lodgepole pine trees within a stand that have $>$ 50 percent serotinous cones) was estimated for twenty burned stands by evaluating live lodgepole pine in adjacent unburned stands that were comparable in site conditions, canopy species composition, and structure. Percent serotiny was determined by examining as many as fifty-five lodgepole pine trees objectively located along a transect (see Doyle 1997). In an effort to estimate prefire cone serotiny for the sixty-one stands for which there were no comparable unburned stands, correlation and multiple linear regression analyses were conducted to ascertain if percent serotiny could be predicted based on measured environmental variables as done by Tinker and others (1994).

Compared with any other combination of environmental variables, prefire importance of lodgepole pine was found to be the best estimator of percent cone serotiny (Spearman Rank Correlation = 0.62, $p < 0.005$). Thus, other than prefire lodgepole pine, no other estimator of serotiny is included in the ordinations. Although percent sand and phosphorus are positively correlated with serotiny, and percent clay and distance from ridge are negatively correlated with serotiny ($p < 0.025$), because of collinearity with prefire lodgepole pine, these variables did not significantly improve the ability to predict serotiny. Elevation, aspect, slope, and topographic moisture were not significantly correlated with percent serotiny in the twenty stands I sampled as opposed to significant correlations with these features, particularly elevation, in Yellowstone National Park (Chapter 4).

SOILS AND GEOLOGY

For each plot, the percent cover of bedrock, boulders, cobbles, gravel, bare mineral soil, and deadwood was estimated using the line intercept method (Mueller-Dombois and Ellenberg 1974). Rock specimens were collected and soil pits were excavated within each stand to ascertain whether the substrata were consistent with previously mapped geologic and soil types (for example, BTNF unpublished manuscript, Young 1982, Love and others 1992, Rodman and others 1992). Within each plot, three soil samples were collected from the surface horizon (top 10 cm) and pooled for subsequent analyses of texture, available nutrients, and chemical properties. Soil texture was estimated using the hydrometer method (Gee and Bauder 1986). Chemical analyses were performed by Brookside Laboratories, Inc. (New Knoxville, Ohio), and included cation exchange capacity (Brown and Warncke 1988), Mehlich 3-extractable P (Mehlich 1984), ammonia-N and nitrate-N (Gelderman and Fixen 1988), percent soil organic matter (Schulte 1988), pH (Eckert 1988), and calcium, magnesium, potassium, sodium, aluminum, boron, copper, iron, manganese, zinc, and soluble sodium (Mehlich 1984). Some soil parameters were transformed using a log or square root transformation to achieve a normal distribution prior to statistical analysis.

A principal component analysis (PCA) was conducted to reduce the number of soil parameters used in the ordination analyses. Four principal components were selected and account for 73 percent (78 percent stand data) of the total variation in the soils data: cations, texture, nitrogen, and phosphorus. Axes were rotated using a varimax rotation. The seven fires encompass much of the variability in soils and geologic substrates that exist in the region (Table 11.1, Doyle 1997).

TOPOGRAPHY

For each plot, measurements of slope and aspect were made in the field using a clinometer and compass. Plots were classified based on their slope position (1 = ridgetop, 2 = upper slope, 3 = middle slope, 4 = lower slope, 5 = valley bottom) and slope shape (1 = convex, 2 = convex-straight, 3 = straight, 4 = concave-straight, 5 = concave, Parker 1982). Distances from closest ridge or closest stream were measured in the field or planimetrically from topographic maps and aerial photos. Three separate measures of moisture based on topography were developed and used as potential explanatory variables in ordination analyses (Loucks 1962, Austin 1972, Greig-Smith 1982). The measures include a topographic position scalar, a moisture index, and a topographic moisture scalar.

The topographic position scalar was developed by averaging the values for slope shape and slope position for each plot to create a single variable that ranged from 1 to 5. Class 1 represents the driest topographic positions (convex slopes on ridges) and class 5 the most mesic (concave valley bottoms).

The moisture index was developed based on potential solar radiation. Adjustments were made to account for the fact that west slopes are subjected to more extreme drying than east slopes of the same inclination, because radiation is received later in the day, when temperatures are warmer and humidity is lower (see Doyle 1997). Potential solar radiation (PSR, total annual potential insolation received on a surface) was determined from the tables of Frank and Lee (1966), which provide calculations of PSR for increments of slope and aspect at different latitudes. PSR is greatest on south slopes, least on north slopes, and equivalent on aspects equidistant from south and north (east and west slopes). PSR has the advantage of combining slope and aspect into a single variable and frequently is used in studies of vegetation in heterogeneous regions (see Allen and Peet 1990, Busing and others 1993). High values of the moisture index (ranging from 0 to 1) indicate greater moisture.

The third scalar, the topographic moisture scalar, combines the moisture index and topographic position scalar into a single nomogram (Doyle 1997). The nomogram was partitioned into five segments with lines of equal slope, as the effect of each variable was assumed to be constant over the range of the other variable (see Doyle 1997). Plots were thereby classified into five topographic moisture classes ranging from 1 (driest) to 5 (most mesic).

FIRE SEVERITY AND CONIFER SEED AVAILABILITY

Because quantitative data on fire-line intensity (kW m^{-1}) for each fire were not available, qualitative indicators were used instead (De Ronde 1990, Bradley and others 1992, Agee 1993, Turner and others 1994). Criteria used to assign a stand to one of three fire severity classes (moderate, moderate-severe, and severe) include fire-caused tree mortality, the amount of fuel remaining and scorch height on trees, and the amount and degree of scorching of ground fuels (Table 11.2). A light surface burn (Turner and others 1994) or low-severity class (Agee 1993) was not sampled.

Seed availability is a function of seed survival on site in the canopy of live or dead trees, or in the unburned duff or soil, and of seed dispersal from adjacent areas. Thus the availability of conifer seed in a plot is influenced by fire severity, distance to live trees within a stand, and distance to larger unburned patches of conifer forest. I developed four measures of seed availability: (1) distance to closest unburned conifer tree, (2) weighted distance to closest unburned conifer patch, (3) a "seed survival scalar" which combines distance to closest unburned conifer tree and fire severity class, and (4) a "seed source scalar," which integrates weighted distance to unburned forest and the seed survival scalar (see Doyle 1997). For all measures, higher values correspond to lower seed availability.

Table 11.2. Classification of fire severity classes

Attribute	Fire Severity Class		
	Moderate	Moderate–Severe	Severe
Tree mortality	Mortality of canopy trees 40–90% within first two decades after fire. Live trees and saplings dispersed throughout stand.	Mortality of canopy trees >90% within first two decades after fire. Widely spaced live canopy trees or clumps of trees found in the stand.	Mortality of large and small trees 100%.
Fuel on dead trees	Dead conifer trees often have unscorched bark. Large and small branches and fine twigs, and often needles, remain from the base to the top of the tree.	Bark on dead trees scorched, but not deeply charred by fire. Large and many small branches, and some fine twigs, remain on dead canopy trees. Dead leaves or needles may be present on trees or on soil surface.	Bark on dead trees is deeply charred. Large and some small branches remain on dead trees. Some fine twigs remain, but most have been consumed by fire. Dead needles usually absent.
Scorch height on trees	Many or all trees scorched only near the base of the tree, but some trees may be scorched into the canopy.	Scorch height variable within a stand. Fire reaches the canopy of some but not all trees.	Most or all trees scorched into the canopy.
Ground layer fuel	Effect of fire is patchy on the understory. Fire may consume litter, upper duff, understory plants, and foliage of understory trees. Little to no mineral soil exposed.	Fire consumes understory plants and foliage of understory trees. Patches of exposed mineral soil exist, but often much of forest floor remains.	Fire consumes understory plants and foliage of understory trees. Organic matter on soil surface deeply charred or consumed. Much bare mineral soil exposed.
Woody debris on ground	Logs on ground that predate fire are partially blackened.	Logs on ground that predate fire are scorched to charred.	Logs on ground that predate the fire are deeply charred.

a. Moderate burn in the Huckleberry Ridge fire in 1994 (six years after the fire).

b. Moderate–severe burn in the Huckleberry Ridge fire in 1994 (six years after the fire).

c. Severe burn in the Huckleberry Ridge fire in 1993 (five years after the fire). Photos by Kathleen M. Doyle.

Distances to unburned trees or patches were measured in the field from the center of the plot. However, if the unburned trees or patches were located long distances from the plot, the closest unburned tree or patch was delineated on color infrared aerial photos (1:40,000 scale). To reduce distortion in the photos caused by heterogeneous terrain, a Bausch and Lomb stereo zoom transfer scope was used to overlay 7.5-minute USGS topographic maps with the aerial photos. Distances were then measured planimetrically from the maps.

I measured the distance to the closest conifer forest patch at least 3 ha in size. Often, however, there were closer patches which were smaller than 3 ha. To account for the potential seed provided by the small patches, I calculated a weighted average between the closest small (< 3 ha) and closest large (> 3 ha) patch. The relative weight given to the small patch was determined by its size (Doyle 1997).

TIME SINCE FIRE AND GEOGRAPHIC LOCATION

The year of the fire (1974–1991) is included as a variable in the ordinations. A low fire year (for example, 1974) indicates an older fire. Because all the fires occurred in different years—except for two in 1988 (HUNT and HUCK)—fire year encompasses any variation between the oldest and more recent fires. This might include variation linked with geographic location (for example, geologic substrates, topography, site history, weather) as long as it varies chronologically.

Data Analysis

Stands were classified into postfire vegetation types using the classification program TWINSPAN (Hill 1979), using PC-ORD software (McCune and Mefford 1995). The density (number m^{-2}) of seedlings, saplings, or sprouts after fire for each tree species and their combined density in a stand were used in the analysis. The program was run using the default options, except that three density levels were selected (0, 0.05, 0.5 trees m^{-2}), corresponding to sparse, low-moderate, and high density. Twelve postfire forest types were identified based on tree species composition and tree density (Table 11.3). Postfire forest types are named based on the species with the highest maximum density and frequency, plus qualifiers such as sparse, moderate, and dense, indicating overall tree density. As vegetation varies continuously, each type is not a homogeneous entity and can include a considerable amount of variation. One stand was reassigned to its own community type, subalpine fir, as it is dissimilar to the other stands in the original Douglas fir cluster. One stand had no trees within the plots, but was assigned to the lodgepole pine sparse community because lodgepole pine occurred

Table 11.3. Abundance of tree species by postfire forest type

Postfire Forest Type	Aspen	Lodgepole Pine/ Aspen	Lodgepole Pine (Sparse)	Lodgepole Pine (Moderate)	Lodgepole Pine (Dense)	Lodgepole Pine/ Douglas Fir	Lodgepole Pine/ Aspen/ Mixed Conifer	Whitebark Pine/ Subalpine Fir	Engelmann Spruce/ Subalpine Fir/ Lodgepole Pine	Engelmann Spruce/ Douglas Fir/ Subalpine Fir	Douglas Fir	Subalpine Fir
Number of stands	6	4	4	11	10	8	10	9	5	11	2	1
Mean elevation	2175	2290	2077	2146	2171	2280	2427	2646	2057	2325	2193	2594
Total tree density (range)	mod–dense	mod–dense	sparse	low–mod	high–dense	high–dense	sparse–mod	sparse–mod	mod–high	low–mod	low–mod	low
POTR *Populus tremuloides*	0.47–4.41	0.09–1.06	0.04	T–0.06	T–0.03	0.01–0.08	T–0.10	T–0.01	T–0.02	T–0.03		
Aspen	100%	100%	25%	54%	60%	62%	90%	22%	80%	45%		
PICO *Pinus contorta*		0.12–11.06	0.01–0.04	0.07–0.26	0.49–4.15	0.05–3.28	T–0.44	T	0.02–0.12	T–0.24		
Lodgepole pine		100%	75%	100%	100%	100%	100%	22%	100%	63%		
PIAL *Pinus albicaulis*		0.04				0.01	T–0.07	T–0.29	0.01	T–0.08		
Whitebark pine		25%				12%	70%	88%	20%	72%		
ABLA *Abies lasiocarpa*				T–0.02	T–0.02	T–0.03	T–0.03	T–0.02	0.01–0.17	0.01–0.11	0.04	0.08
Subalpine fir				54%	60%	50%	90%	88%	100%	100%	50%	100%
PIEN *Picea engelmannii*				T	T		T–0.01	T	0.09–0.48	T–0.34		
Engelmann spruce				18%	10%		50%	22%	100%	90%		
PSME *Pseudotsuga menziesii*				T–0.03		T–1.22		0.02	T–0.01	T–0.24	0.03–0.12	
Douglas fir				18%		100%		11%	40%	100%	100%	

Note: Classification was provided by TWINSPAN based on density per m^2 of trees (seedlings/saplings/sprouts) that established after fire. Minimum and maximum tree density (for stands where the species is present) is provided on the first line. The frequency of stands where the tree species was present is given on the second line. T = present but density ≤ 0.005 per m^2; sparse = 0–0.05 trees per m^2; low = 0.05–0.1 trees per m^2; mod. = 0.1–0.5 trees per m^2; high = 0.5–1.0 trees per m^2; dense = > 1.0 trees per m^2.

within the stand but outside the plot boundary. Each species is present in six or more of the postfire types, although all species have unique patterns of frequency and density (Table 11.3).

To explore the relationships between the measured explanatory variables and postfire tree composition, the ordination technique canonical correspondence analysis (CCA) was employed using the CANOCO software (version 3.12, ter Brak 1991). Detrended Correspondence Analysis was also used to evaluate how well the measured variables account for the tree composition after fire (see Doyle 1997).

CCA arranges plots in relation to one another based on their similarity in species composition (in this case postfire tree composition) and predictive variables. The following explanatory variables were included in the CCA: fire year, percent cover deadwood, percent cover rock, percent cover mineral soil, PCA soil axes (cations, soil texture, nitrogen, phosphorus), percent slope, moisture index, topographic position scalar, topographic moisture scalar, elevation, distance to ridge, distance to stream, fire severity, distance to unburned conifer tree, distance to unburned conifer forest, seed survival scalar, seed source scalar, and prefire importance of each canopy species (aspen, lodgepole pine, Engelmann spruce, subalpine fir, Douglas fir, and whitebark pine).

Successive canonical ordinations were performed on subsets of plots to compare the major prefire forest types in terms of total variance explained and the relative importance of different explanatory variables. The subsets included here are as follows: (1) prefire Douglas fir stands (IV of Douglas fir > 33 percent), (2) prefire Engelmann spruce/subalpine fir stands (IV of Engelmann spruce plus subalpine fir > 67 percent), and (3) prefire lodgepole pine stands (IV of lodgepole pine > 38 percent).

To evaluate the relative importance of factors, I grouped major variables into the following sets of related variables: (1) prefire vegetation, which includes the prefire importance values of all six tree species, (2) topographic variables, which include elevation, moisture index, topographic moisture scalar, topographic position scalar, percent slope, distance to ridge, and distance to stream, (3) soil variables, which include the four axes identified in the PCA (cations, texture, nitrogen, phosphorus), (4) fire effects, which include distance to live conifer tree, distance to unburned conifer forest, seed survival scalar, seed source scalar, and fire severity, and (5) fire year (single variable).

Partial canonical ordination (partial CCA) coupled with CCA for each subset of stands (Borcard and others 1992) was used to identify the total and unique contribution of the sets of factors in explaining the variation in species distribution. Because of interactions between factors such as prefire vegetation and topography, the variance explained by one set of variables often partially overlaps with the variation explained by a second set of factors. Thus variation can be partitioned into separate components, and the amount of variation explained by a set of variables independent of other factors can be identified.

For subsets of stands, I performed the following analyses. First, to obtain the total variance in species composition explained by all measured variables, I conducted a CCA constrained by all variables. Second, to identify the total variation explained by each subset of variables (for example, prefire vegetation), I performed a CCA constrained by only those variables. Finally, to ascertain how much variance was explained solely by the set of variables of interest independent of all other variables (for example, prefire vegetation), I performed a partial CCA. Thus the total variance accounted for by a set of variables and the portion of the variance that is unique to a set of variables (independent of all other variables) can be determined.

PATTERNS OF EARLY POSTFIRE FOREST SUCCESSION

Consistent with Stahelin's observations from high-elevation forests in the Central Rocky Mountains, a number of interacting factors lead to multiple pathways of early succession for a given prefire forest type. I found that the relative importance of factors is not the same in all prefire forest types, and factors differ somewhat from those Stahelin (1943) identified as important. However, as Stahelin observed, prefire tree species composition does influence postfire regeneration considerably, and the most successful reestablishment often occurs in areas where lodgepole pine and aspen were present before fire (Chapters 4, 5, 10, 14). Although physical factors, especially topographic variables and to a lesser extent soils, account for considerable variability in tree establishment in early succession, distance to seed source, weather patterns, and site history also appear to influence whether the prefire vegetation will reestablish soon after fire.

In this section, I describe the extent to which multiple pathways of succession occur for each major forest type, how the independent variables influence postfire tree species composition, and the projected successional trajectories. In the next section, I discuss the relative importance of factors in influencing early postfire succession in the Teton region in contrast with other areas in the Rocky Mountains.

Douglas Fir Stands

Douglas fir reestablished in all but one stand where it was dominant before fire (prefire IV 3 33 percent), but it was abundant in only four of fourteen stands (Fig. 11.3). In addition to Douglas fir, the most frequently encountered species after fire in Douglas fir stands is subalpine fir, followed by lodgepole pine (71 percent and 57 percent of stands, respectively). Engelmann spruce and whitebark pine are present in 43 percent, and aspen was observed in 36 percent of the postfire stands.

Three pathways of early succession are identified for burned Douglas fir stands: (1) stands dominated by Douglas fir with relatively high density (observed only for some stands where Douglas fir was the sole prefire dominant), (2) mixed conifer stands, sometimes with aspen, with sparse Douglas fir and lodgepole pine density, and (3) stands dominated by lodgepole pine (density $>$ 0.18 trees m^{-2}) with sparse Douglas fir (Fig. 11.3).

Plots with the highest densities of Douglas fir trees after fire (0.08–1.22 m^{-2}) are associated with high cations and pH (pH 6.4–6.9), high prefire Douglas fir (65–100 percent), and lower elevation ($<$ 2230 m). There is also an association with higher moisture index, lower fire severity, shorter distance to unburned forest, and fewer prefire lodgepole pine (Doyle 1997). Plots with abundant Douglas fir regeneration are not typically found on drier topographic positions; however, two plots with abundant Douglas fir regeneration occur on a steep west-facing slope, where the very coarse-textured soils may offset the dry topographic position (as water is typically more available in sandy soils in arid areas; Noy-Meir 1973). The only stand with no Douglas fir regeneration, classified as lodgepole pine/aspen/mixed conifer, has low moisture index (0.24), low prefire Douglas fir (0.41), and lower cations and pH (pH = 5.8). A mixed conifer stand with limited regeneration of Douglas fir, despite high prefire importance of Douglas fir (0.80), occurs on a steep, dry, southeast slope at high elevation (2562 m) and is a long distance from the closest unburned Douglas fir (130 m). A stand where the density of lodgepole pine greatly exceeds that of Douglas fir after fire is found on a relatively dry topographic position with a limited seed source of Douglas fir, but with a large potential seed source of lodgepole pine from serotinous trees in an adjacent burned stand (Doyle 1997).

The early postfire dominance of Douglas fir in some stands is consistent with the observation that the species is both a seral and climax species in the Rocky Mountains (Despain 1973, Steele and others 1983, Peet 1988, Bradley and others 1992). In two early successional stands within the boundaries of the fire classi-

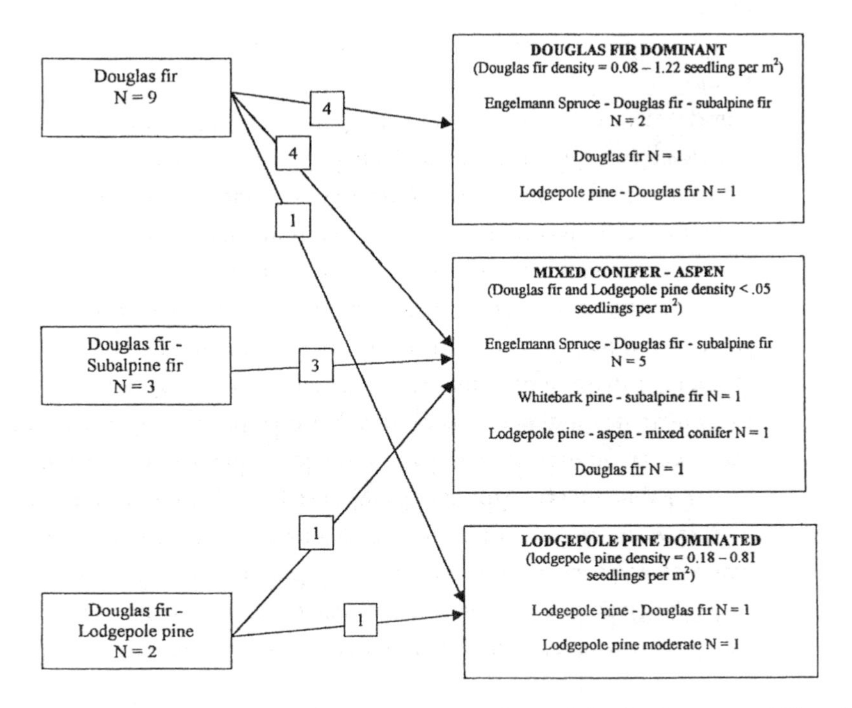

Figure 11.3. Pathways of early succession for Douglas fir stands (prefire Importance Value of Douglas fir >33 percent). The postfire forest types (boxes on the right) that emerged from each prefire forest type (boxes on the left) are shown. The groupings of postfire forest types are derived from the canonical correspondence analysis. N = number of stands.

fied as Douglas fir, Douglas fir is the sole prefire dominant before and after fire and is likely to remain the dominant species over time. In early successional lodgepole pine/Douglas fir stands, Douglas fir is likely to replace lodgepole pine given enough time and an adequate seed source of Douglas fir (Peet 1981, Steele and others 1983). In mixed conifer stands, the importance of Douglas fir over time is likely to be more variable. Tesch (1981) found Douglas fir to dominate both early and late succession in a mesic situation in Montana where it established with other conifers after fire. Likewise, I expect that Douglas fir will continue to dominate a north-facing stand on limestone in the HUNT fire where it is associated with lodgepole pine. In early successional stands on mesic sites classified as Engelmann spruce/Douglas fir/subalpine fir, Douglas fir may persist for centuries, although the more shade-tolerant spruce and fir are likely to increase in importance over time (Peet 1988, Bradley and others 1992). On some sites where Douglas fir is uncommon after fire, lack of adequate seed may result in the slow reestablishment of Douglas fir, and other conifers may dominate for

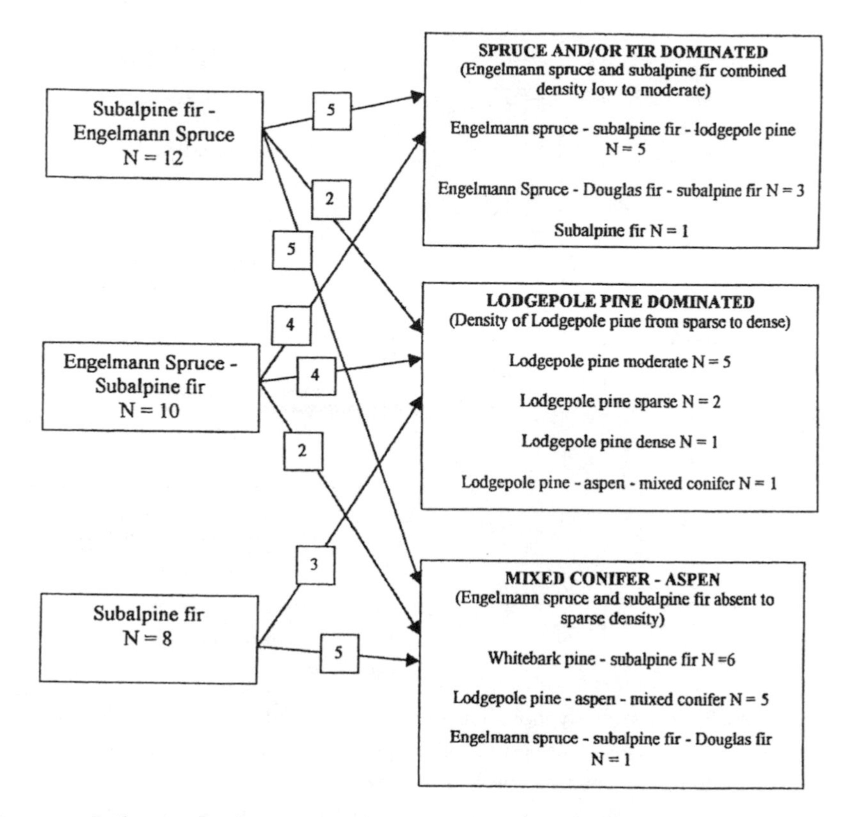

Figure 11.4. Pathways of early succession for Engelmann spruce/subalpine fir stands (prefire Importance Value of spruce and fir > 67 percent). The postfire forest types (boxes on the right) that emerged from each prefire forest type (boxes on the left) are shown. The groupings of postfire types are derived from the canonical correspondence analysis. N = number of stands.

many decades, especially if growing conditions for Douglas fir are not optimal (low cations and low moisture).

Engelmann Spruce/Subalpine Fir Stands

Using the predictive variables, three early successional pathways can be identified following fire in stands dominated or codominated by subalpine fir or Engelmann spruce (prefire IV of spruce and fir > 67 percent, Fig. 11.4). They are (1) stands with substantial regeneration of Engelmann spruce and/or subalpine fir (0.05–0.5 m^{-2}, Fig. 11.5), (2) stands dominated by lodgepole pine (Fig. 11.6), and (3) mixed conifer and aspen stands with relatively sparse tree regeneration (Fig. 11.7). The three pathways have emerged following fire in stands codomi-

Figure 11.5. Regeneration of Engelmann spruce (dominant species), lodgepole pine, and subalpine fir in a severely burned stand in 1994 (twenty years after the Waterfalls Canyon Fire). The faster-growing lodgepole pine is substantially taller; however, all three species emerged together soon after the fire. The prefire forest type was classified as subalpine fir/ Engelmann spruce. Elevation of stand is 2070 m. The combination of availability of seed, high annual precipitation (compared with other areas in the Tetons at similar elevations), and unusually cool and moist weather conditions in the first postfire year appears to have created ideal conditions for spruce regeneration in the Waterfalls Canyon fire. Photo by Kathleen M. Doyle.

nated by Engelmann spruce and subalpine fir regardless of whether spruce or fir was the most important prefire species. However, in stands where subalpine fir was the prefire dominant without spruce, substantial establishment of spruce or fir has not been observed, probably because sites without spruce tend to be less fertile and drier. Although lodgepole pine–dominated stands emerged after fire for all three prefire spruce/fir types, the highest densities of lodgepole pine occurred after fire in subalpine fir forests. The mixed conifer/aspen type is common following fire in all three prefire types. The whitebark pine/subalpine fir type is relatively common after fire in stands where subalpine fir was the domi-

Figure 11.6. Moderate density regeneration of lodgepole pine on Mystic Isle in 1994, thirteen years after a moderately severe fire. No regeneration of spruce or fir was evident in the stand or elsewhere on the island, despite the fact that the prefire forests were dominated by subalpine fir and Engelmann spruce. The prefire vegetation, topographic moisture, elevation (2091 m), and proximity to nearest live trees in burned stands on Mystic Isle were comparable to stands in the WFC, which exhibited substantial postfire regeneration of Engelmann spruce. Less available seed, drier postfire weather conditions, and lower annual precipitation may explain why regeneration of spruce has been absent on this site and others outside of the Waterfalls Canyon fire. Photo by Kathleen M. Doyle.

nant species before fire, but it did not develop in any stands classified as Engelmann spruce/subalpine fir before fire (Fig. 11.4).

Plots with the highest densities of spruce and fir after fire (the Engelmann spruce/Douglas fir/subalpine fir and Engelmann spruce/subalpine fir/lodgepole pine types) are correlated with lower elevations (< 2300 m), short distance to closest live tree, high cations, high prefire spruce (0.26–0.52 m^{-2}), and moist topographic situations (Doyle 1997). Seventy-eight percent are found where the topographic moisture class is 4 (more mesic). There are striking differences in early succession between the oldest fire, the 1974 WFC fire, and the more recent fires. All but two of the stands with spruce regeneration, more than 0.04 trees m^{-2}, are in the 1974 WFC fire (Fig. 11.5). Notably, there are other stands in the MYS (Fig. 11.6) and HUCK fires at lower elevations, on moist topographic positions with relatively high cations, similar prefire importance of Engelmann

Figure 11.7. Low regeneration of lodgepole pine and sparse regeneration of subalpine fir and aspen in 1994, seven years after a moderate burn in the Dave Adams Hill fire. The prefire forest was Engelmann spruce/subalpine fir; soils were rich in cations, and the site had high topographic moisture. Elevation was 2261 m. Although Engelmann spruce was the prefire dominant, no spruce regeneration was observed in the stand. Photo by Kathleen M. Doyle.

spruce, and relatively close proximity to the nearest live tree (30–185 m), but reestablishment of spruce and fir is absent or limited.

Plots dominated by varying densities of lodgepole pine since fire (sparse, moderate, and dense lodgepole pine types) are correlated with prefire lodgepole pine and are found in the BC, MYS, HUCK, and DAH fires (Figs. 11.6, 11.7). However, it is significant that prefire tree importance varies widely for this group (lodgepole pine [0–25 percent], Engelmann spruce [0–68 percent], and subalpine fir [24–82 percent]). Plots occur at low-mid-elevations (< 2320 m) and on relatively moist topographic positions. Seventy-five percent occur on topographic moisture class 4. Mean distance to unburned tree ranges from 38 to 185 m.

Plots with sparse to low tree regeneration (classified as lodgepole pine/aspen/

mixed conifer and whitebark pine/subalpine fir) are strongly associated with high elevation (> 2400 m), prefire abundance of whitebark pine (frequency = 77 percent), exposed mineral soil, dry topographic positions, and long distance to unburned trees. These plots are predominantly in the 1988 HUCK and 1987 DAH fires, hence are correlated with late fire year. Maximum stand values for percent bare mineral soil (52 percent) and distance to closest unburned tree (477 m) are associated with the group. Sixty-seven percent occur on topographic moisture class 2 (more xeric). The lodgepole/aspen/mixed conifer type tends to be on relatively more moist topographic positions, with greater importance of lodgepole pine before fire, compared with plots classified as whitebark pine/subalpine fir, which are typically on drier situations with higher prefire abundance of whitebark pine (Doyle 1997).

As observed in WFC fire stands (Fig. 11.5), successful reestablishment of spruce and fir soon after fire is common in the Rocky Mountains (Habeck and Mutch 1973, Loope and Gruell 1973, Noble and Alexander 1977, Whipple and Dix 1979, Romme and Knight 1981, Peet 1981, McCune and Allen 1985a, Veblen 1986, Aplet and others 1988, Doyle and others 1998). As I observed, Engelmann spruce is generally a better postfire invader than subalpine fir (Despain 1973, Habeck and Mutch 1973, Whipple and Dix 1979, Peet 1978, Doyle and others 1998). Successful regeneration of Engelmann spruce and subalpine fir, however, can be highly localized and is only partly explained by the variables I measured. On most high-elevation spruce/fir stands and some low-elevation stands with a limited seed source, postfire stand development is likely to be uneven and very slow (as seen in stands in the MYS, DAH, and HUCK fires, Stahelin 1943). In contrast, recruitment of Engelmann spruce and subalpine fir may occur within a window of ten years after fire on some favorable sites in the Rocky Mountains (such as in the WFC fire), but typically recruitment is much slower and may occur over a period of forty to seventy years (Peet 1988). A dense herb layer in some areas (such as an early successional stand classified as subalpine fir after the HUCK fire) may greatly impede tree recruitment and may favor subalpine fir over Engelmann spruce (Knapp and Smith 1981). This observation suggests that programs initiated by the USFS to seed grasses and forbs after fire could retard forest regeneration. Lodgepole pine stands at low elevations will be invaded gradually by subalpine fir, and Engelmann spruce on mesic sites (Loope and Gruell 1973); subalpine fir and whitebark pine will invade on drier high-elevation sites. Early postfire Engelmann spruce and subalpine fir stands should persist over long periods of time (Aplet and others 1988, Bradley and others 1992).

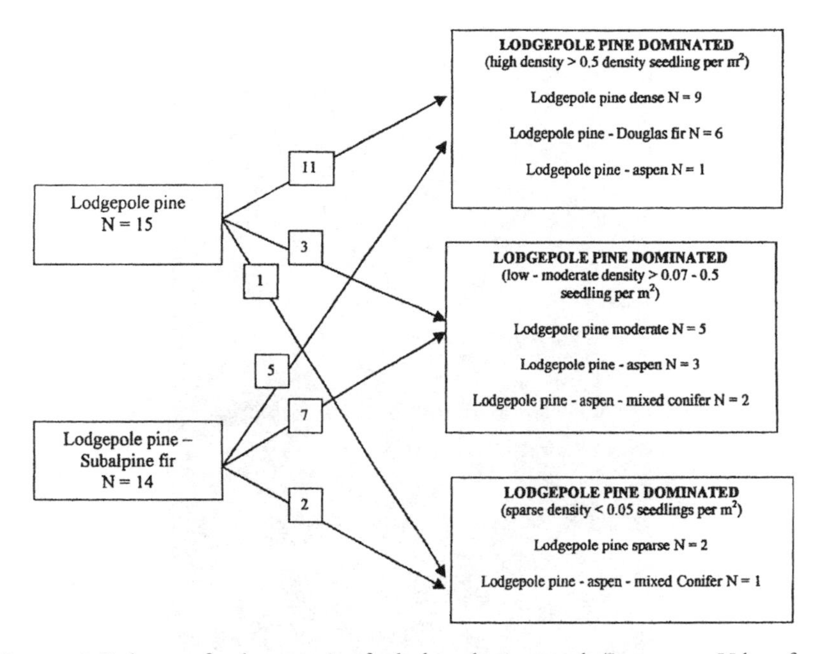

Figure 11.8. Pathways of early succession for lodgepole pine stands (Importance Value of lodgepole pine >38 percent). The postfire forest types (boxes on the right) that emerged from each prefire forest type (boxes on the left) are shown. N = number of stands.

Lodgepole Pine Stands

Lodgepole pine is the most widespread and successful early successional species in the Teton region; no species reaches greater density (Table 11.3). All prefire lodgepole pine and lodgepole pine/subalpine fir stands (prefire IV of lodgepole pine > 38 percent) are dominated by lodgepole pine after fire (Fig. 11.8). Most stands dominated solely by lodgepole pine before fire have high lodgepole pine tree density after fire (> 0.05 trees m^{-2}, Fig. 11.9), whereas mixed lodgepole pine/subalpine fir stands have lower lodgepole tree density after fire (Fig. 11.10). Subalpine fir regeneration is always sparse after fire but is more frequent in stands where subalpine fir was codominant with lodgepole pine before fire (64 percent versus 33 percent). Aspen had regenerated well in some burned lodgepole pine stands, reaching densities as high as 1.22 trees m^{-2} (Table 11.3).

Stands dominated by lodgepole pine after fire are differentiated from other postfire types by their positive correlation with prefire lodgepole pine, distance to seed source, lower cations, and lower elevation. However, the variability in the density of lodgepole pine trees and the presence of co-occurring species (Fig. 11.8) is not well accounted for by the variables included in the ordination (Doyle

Figure 11.9. Dense regeneration of lodgepole pine, eleven years after a severe burn in the Mystic Isle fire. The prefire forest was classified as lodgepole pine. Although percent serotiny was not measured for this stand, it was undoubtedly quite high, as observed for other stands in the Mystic Isle fire (as high as 42 percent). Prefire importance of lodgepole pine was found to be the best predictor of percent serotiny and postfire density of lodgepole pine. Photo by Kathleen M. Doyle.

1997). Measured variables explain only 41 percent of the variance in postfire tree composition (compared with 51 percent and 84.5 percent of the variance explained for Engelmann spruce/subalpine fir and Douglas fir stands, respectively, Doyle 1997). There is considerable variability in postfire density of lodgepole pine between stands (even adjacent stands within the same fire) (Figs. 11.9, 11.10). Dense lodgepole pine emerged after fire in four of the fires (MYS, BC, HUCK, and HUNT). Abundance ranges from low to dense in the MYS fire and sparse to dense in both the BC and HUCK fires.

Prefire importance of lodgepole pine and estimated percent serotiny are the best predictors of postfire lodgepole pine density (Fig. 11.11). Percent cone serotiny ranged from 0–51 for the twenty stands I sampled for this variable. Percent serotiny

Figure 11.10. Moderate regeneration of lodgepole pine, eleven years after a severe burn in the Mystic Isle fire. The prefire forest was classified as lodgepole pine/subalpine fir. This stand was adjacent to a stand with dense regeneration of lodgepole pine. The small-scale heterogeneity in the density of lodgepole pine regeneration was poorly explained by physical factors such as topography or soils. In general, stands where subalpine fir was codominant with lodgepole pine before fire had lower postfire lodgepole pine density than did nearby stands where lodgepole pine was dominant before fire. Photo by Kathleen M. Doyle.

tended to be higher for some fires, such as the MYS fire, and quite low in others (for example, DAH and WFC, Table 11.1). All stands with relatively high percent serotiny (> 25 percent) and relatively high prefire importance of lodgepole pine (> 45 percent) yielded dense stands of lodgepole pine (> 1.0 tree m^{-2}) after fire. However, the density of lodgepole pine after fire is quite variable for stands where lodgepole pine is a prefire dominant (IV > 40 percent). This is in part because prefire importance of lodgepole pine does not take into consideration the absolute density of lodgepole pine. For example, one stand dominated solely by lodgepole pine before fire (IV = 100 percent) had widely spaced trees before fire and had sparse postfire density of lodgepole pine after fire (0.01 trees m^{-2}).

Although lodgepole pine is the most successful postfire species in the Teton region in terms of abundance and extent, climax lodgepole pine, as described by Bradley and others (1992) and Despain (1990) does not appear to be common in the Tetons. All stands dominated by lodgepole pine before fire had some other

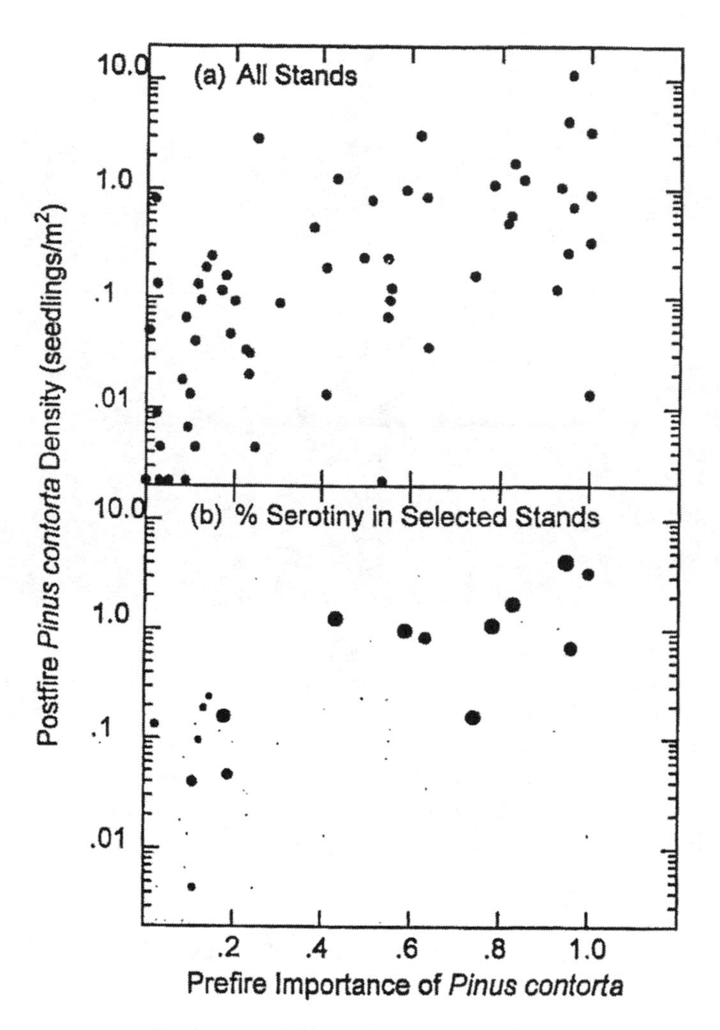

Figure 11.11. Mean postfire density of lodgepole pine in each stand by prefire importance of lodgepole pine: (a) data from each of eighty-one stands are indicated by a circle; (b) the size of the circle indicates relative estimated percent serotiny (ranging from 0 to 51 percent) for each of the twenty stands where percent serotiny was estimated.

tree species (usually subalpine fir) in the understory. It is likely that more tolerant species, especially subalpine fir, will invade early successional lodgepole pine stands (Loope and Gruell 1973, Peet 1988). In the most mesic situations, Engelmann spruce is expected to establish along with subalpine fir. In some stands—especially those in the HUNT fire where cations are high and Douglas fir is present in the initial postfire stand with lodgepole pine—I expect Douglas fir to become a late successional dominant.

Loope and Gruell (1973) assert that serotiny is relatively unimportant in the Teton region. Much higher levels of serotiny have been documented for other areas in the Rocky Mountains (Muir and Lotan 1985, Tinker and others 1994). Although the maximum degree of serotiny I observed in the Teton region is comparable to that found by Anderson and others (Chapter 4) in Yellowstone, low percent serotiny, especially at low elevation, appears to be more prevalent in the Tetons. Moreover, the maximum density of lodgepole pine saplings after fire was less (31.4 trees m^{-2} averaged over a 30-m^2 plot in the Tetons) compared with much higher densities observed in Yellowstone (Chapter 4). Nonetheless, as has been frequently noted (Muir and Lotan 1985, Tinker and others 1994, Anderson and others, Chapter 4), stands with a relatively high proportion of serotinous lodgepole pine tend to yield high densities of lodgepole pine after fire. Turner and others (1997) found that serotiny in Yellowstone National Park is an important predictor of postfire lodgepole pine density when the abundance of serotinous trees is above a certain threshold (Chapter 14). When serotiny is low, other factors are better predictors. Prefire serotiny was high for some of the fires (such as the MYS and HUNT) where lodgepole pine is more likely to be an important pioneer species, even when other species were dominant before fire. In contrast, in the WFC or DAH fires, where estimated percent serotiny was less than 21 percent for all stands sampled, postfire lodgepole pine tree density is comparatively low, and lodgepole pine was not an important dominant before fire.

Serotiny is not well correlated with measured environmental variables, consistent with the results of Muir and Lotan (1985). Stands initiated by fire are more likely to be highly serotinous, and thus the variation in the level of serotiny may be best explained by fire regime, geography, and broad topographic features (Muir and Lotan 1985, McCune and Allen 1985a, Peet 1988, Gauthier and others 1996). Fire return is likely to be shorter in areas where fire can spread rapidly than in landscapes that are dissected and have more barriers to fire, such as the WFC area. Correlation between serotiny and drier, low-elevation sites, as demonstrated in Yellowstone National Park (Tinker and others 1994, Chapter 4), may be due to the consistently shorter fire interval on such sites.

Aspen and Whitebark Pine Stands

Because they made up a small portion of the burned landscape, only six aspen and two whitebark pine stands were examined in this study. Similar to lodgepole pine stands, all aspen and whitebark pine stands remained dominated by the same species after fire. Whereas lodgepole pine, Douglas fir, Engelmann spruce, and subalpine fir infrequently were observed in stands where they were not observed in the prefire canopy, whitebark pine and aspen appear to have become more widespread after fire. Aspen established in forty stands where it was not observed in the prefire canopy, but typically aspen was a minor component in these stands after fire. Romme and others (1997) note that establishment of aspen from seed after the 1988 fires is a rare event, facilitated by available seed and good conditions for seedling establishment (Chapter 14). Whitebark pine established in sixteen stands where I did not observe it in the prefire canopy.

All stands dominated by aspen before fire quickly reestablished from root sprouts after fire. Density of sprouts is high (0.47–4.41 sprouts m^{-2}) and growth in height has been rapid (for example, aspen sprouts had reached 3 m in height within four years after the 1988 HUNT fire). Prefire aspen is the factor most strongly associated with postfire aspen. Aspen stands are also associated with soils high in cations and are generally a long distance from conifer seed source.

It is expected that all of the aspen stands sampled will develop into mature stands of aspen. Over the long term, in the absence of fire, mature aspen stands on both wet and dry sites are frequently invaded by more shade-tolerant tree species (Peet 1988). However, aspen may remain dominant on some sites if a well-developed herb layer impedes conifer regeneration (Peet 1981, 1988). It is plausible that the high herb cover observed for many aspen stands could restrict the establishment of conifers.

With the exception of one mesic stand in the HUCK fire, where aspen density reached 1.06 trees m^{-2}, the density of aspen that have established from seed is low ($< 0.14\ m^{-2}$), and browsing is evident. It is unlikely, therefore, that many of the new aspen seedlings will develop into mature aspen trees. Kay (1993) concluded that it is improbable that aspen that had established (0.09–0.27 m^{-2}) after recent fires within the BC fire in GTNP and in Yellowstone National Park would develop into mature aspen because of heavy browsing by elk combined with competition from lodgepole pines. However, in the one small early successional stand classified as lodgepole pine/aspen in the HUCK fire, aspen is dense and is growing more rapidly than lodgepole pine, so the development of aspen to maturity may be possible (Bradley and others 1992).

Stands with whitebark pine after fire are associated with high elevations and high prefire whitebark pine. Typically postfire tree density of whitebark pine is sparse, and only in 15 percent of stands does it reach tree densities greater than 0.05 trees m^{-2}. Fire has provided an opportunity for whitebark pine to increase in both extent and dominance at high elevations. It is expected that whitebark pine seedlings will continue to become established for many decades in the high-elevation burns, as the stands are mostly still quite open (Tomback and others

1993). As the canopy closes, subalpine fir is expected to increase in importance (Murray 1996). Although the highest-elevation stands may develop into climax whitebark pine (Loope and Gruell 1973), the long-term viability of these stands is uncertain because of white pine blister rust (Keane and others 1994).

RELATIVE IMPORTANCE OF FACTORS INFLUENCING EARLY SUCCESSION

When examining all stands, prefire tree composition—especially prefire aspen and prefire lodgepole pine—accounts for a large portion of the explained variation in the distribution of trees after fire, although much of the variance is jointly attributed to other variables (Table 11.4). For example, variation ascribed to an elevation gradient may be partly explained by the abundance of prefire whitebark pine (associated with high elevation) and prefire lodgepole pine (more abundant at lower elevations). Topographic variables, especially elevation, explain the next-largest portion of variation in postfire tree composition when considering all stands. When only groups of stands of a similar prefire forest type are considered, however, the portion of the variance accounted for by prefire tree composition decreases, and topographic variables increase in relative importance. For each major forest type, much of the variation explained by topographic variables can also be attributed to other factors, with a considerable portion ($>$ 7 percent) explained solely by topographic factors (Table 11.4).

Second to topographic variables, prefire vegetation accounts for the largest portion of the variance in postfire vegetation for Douglas fir and spruce/fir stands. Much of the variance explained by prefire vegetation is shared by other variables. The conditions associated with a given fire year account for a relatively large portion of the variation in postfire tree composition, especially for Douglas fir and Engelmann spruce/subalpine fir stands. Interestingly, the variance attributed to fire year in Douglas fir and lodgepole pine stands is almost entirely accounted for by other factors. In contrast, after fire in spruce/fir stands, a more sizable portion of the variance is attributed solely to fire year, indicating that important differences in postfire tree composition between the older and more recent fires (or different geographic locations) are accounted for by factors not included in the ordination, such as weather and climatic patterns, percent serotiny, or landscape context. Fire effects, such as distance from unburned forest, are important in explaining a portion of the variance in postfire composition not explained by other factors, especially for Douglas fir stands and lodgepole pine stands. Soil parameters, especially cations, account for a substantial portion of

Table 11.4. Percent variance in postfire tree composition based on partial CCA and CCA analyses

Set of variables	Prefire vegetation types				
	All Stands	Conifer Stands	Douglas Fir Stands	Engelmann Spruce/ Subalpine Fir Stands	Lodgepole Pine Stands
Prefire vegetation (prefire-abla, prefire-pial, prefire-pien, prefire-pico, prefire-potr; prefire-psme)	**53.0** (20.1)	**41.0** (13.0)	**35.3** (4.2)	**23.4** (5.0)	**12.0** (5.8)
Topography (elevation, slope, moisture index, topographic moisture scalar, topographic position scalar, distance to ridge, distance to stream)	**24.2** (4.6)	**26.9** (5.6)	**59.0** (9.7)	**28.1** (7.1)	**15.7** (7.3)
Soils (cations, texture [silt/clay]; phosphorus, nitrogen)	**11.8** (1.9)	**11.7** (2.4)	**26.8** (6.8)	**7.9** (1.7)	**13.9** (7.4)
Fire effects (distance to live tree, distance to unburned forest, seed survival scalar, seed source scalar, fire severity)	**11.2** (1.2)	**7.2** (1.7)	**25.6** (7.0)	**10.3** (3.2)	**15.5** (6.2)
Year of fire	**9.7** (2.1)	**11.6** (2.6)	**24.8** (1.0)	**16.0** (4.7)	**8.7** (2.8)
Percent variance explained by all variables	**66.7**	**58.7**	**85.8**	**54.8**	**47.3**

Note: Variance accounted for by each set of variables is shown in bold. The portion of the variance that is unique to a set of variables is shown in parentheses. The variables included in each set are listed following the italicized name of the variable set.

the explained variance in Douglas fir and lodgepole pine stands, but they are less important for explaining the postfire variability in spruce/fir stands (Table 11.4).

Prefire Vegetation

Species traits help to explain why prefire vegetation is a predictor of postfire tree establishment. Aspen and lodgepole pine both have mechanisms for establishing from propagules that survive fire—aspen from surviving roots and lodgepole pine from seed stored in serotinous cones. The two species that expanded their distribution after fire—whitebark pine and aspen—are both capable of long-distance seed dispersal: whitebark pine by Clark's nutcrackers *(Nucifraga columbiana)* and aspen seeds by wind. The reestablishment of Douglas fir where it was important before fire can be attributed to the tendency of Douglas fir to survive fire as a result of thick bark and open canopy structure (Peterson and Arbaugh 1986). In contrast, the thin bark, low and dense branching habit, and dense canopy of subalpine fir and Engelmann spruce make them highly susceptible to fire (Starker 1934), and seed dispersal distances are often limited (McCaughey and others 1985). Thus it is not surprising that the presence of Engelmann spruce and subalpine fir is not well predicted by prefire vegetation and that neither spruce nor subalpine fir was observed in stands where it was not part of the prefire canopy. The composition of the surrounding vegetation undoubtedly has a greater influence on postfire vegetation for stands dominated by spruce and fir than for the other tree species.

Topography

The influence of topography on patterns of mature vegetation is well documented (Whittaker 1956, Loope and Gruell 1973, Peet 1981, Wentworth 1981, Smith 1985, Allen and Peet 1990, Busing and others 1993). In early postfire stands in the Oregon Cascades, Chappell and Agee (1996) found that a topographic moisture index had the highest correlations with the abundance of red fir *(Abies magnifica)* seedlings. In the Teton region, my results suggest that topographic variables explain a large portion of the variance in postfire tree composition not accounted for by prefire species composition. For example, even though mature forests dominated by Douglas fir, Engelmann spruce, or subalpine fir occur over a range of topographic situations, each species is abundant soon after fire only on relatively moist topographic situations at lower elevations. Similarly, Tesch (1981) found rapid development of an even-aged Douglas fir stand on a mesic north-facing slope in Montana, whereas postfire establishment of Douglas fir was slow in an uneven-aged stand on an adjacent south-facing slope.

The establishment of Engelmann spruce and subalpine fir may be limited because of less efficient water conservation compared with lodgepole pine (Knapp and Smith 1981). The more successful reestablishment of subalpine fir—compared with Engelmann spruce—on drier, shaded, or high-elevation sites (subalpine fir was present after fire on 73 percent of sampled stands where it was found before fire, compared with 56 percent for Engelmann spruce) may be due to a longer taproot and comparatively greater water use (Day 1964, Knapp and

Smith 1981). Alternatively, subalpine fir seed may be relatively more available, because subalpine fir is the most common species across the GTNP region, and cone crops are more consistent than for spruce (Doug Eggers, silviculturalist, BTNF, personal communication). Mortality of subalpine fir seedlings can be quite high (Cui and Smith 1991), however, and this may contribute to the low density of fir seedlings I observed in most stands.

With few exceptions, early postfire tree density is sparse at the highest elevations for all prefire vegetation types. Postfire vegetation is highly correlated with elevation in Douglas fir and spruce/fir stands. Stands higher than 2400 m have limited or no regeneration of spruce, regardless of its prefire importance. As a whole, high-elevation sites have more exposed mineral soil, are relatively dry, and are often on slopes and far from the closest live tree. Still, the precise reasons why early tree establishment may be low at high elevations are not clear. Slow postfire tree regeneration at high elevations has been attributed to a short growing season, insufficient snow cover, harsh conditions, scarcity of a lodgepole pine seed source, low percent cone serotiny, the often xeric nature of high-elevation slopes, or competition with dense herb layers (Stahelin 1943, Habeck and Mutch 1973, Loope and Gruell 1973, Peet 1988, Little and others 1994, Tinker and others 1994). A combination of factors is undoubtedly important. Greater distance from unburned trees at high elevations in the HUCK fire contributes to the low density of trees. Whitebark pine, which can be dispersed by Clark's nutcrackers as far as 3 km from mature cone-bearing trees (Tomback and others 1990), is an exception to the pattern and is most abundant at high elevations.

Soils

Cation availability appears to be the most important soil factor influencing postfire tree establishment patterns in the Teton region. Cations are positively associated with the prefire abundance of aspen and Douglas fir, moderately associated with the prefire abundance of Engelmann spruce, and negatively associated with lodgepole pine. Lodgepole pine is often associated with relatively infertile, coarse-textured soils in the Rocky Mountains (Stahelin 1943, Despain 1973, Allen and Peet 1990, Despain 1990), but lodgepole pine also can be important on fine-textured, base-rich soils as well, such as in the HUNT fire. For Douglas fir stands, soil variables (cations in particular) help to explain 7 percent of the variation in postfire tree composition not explained by other variables. For Engelmann spruce/subalpine fir stands, variability explained by soils can be largely accounted for by other measured factors, such as prefire vegetation, topographic position, and geographic location. It has been suggested that the most important influence of soils on vegetation in the Rockies is due to variation in soil texture, which regulates moisture availability (Peet 1988); however, within the range of soils sampled in my study, soil texture does not have a strong influence on postfire vegetation patterns.

Distance to Seed Source and Other Fire Effects

Distance to seed source and burn severity have been identified as critical to explaining variability in early forest succession. For example, McClanahan (1986) observed distance to seed source to be the best predictor of the abundance of late successional species on mined sites, and Fastie (1995) identified distance to seed source as the most important factor contributing to different pathways of primary succession at Glacier Bay, Alaska. Moreover, fire severity and distance to seed source influence the density of lodgepole pine seedlings after fire in Yellowstone National Park (Ellis and others 1994, Turner and others 1997, Chapters 4, 14). In the Oregon Cascades, fire severity is a major factor determining the density of postfire red fir seedlings (Chappell and Agee 1996). In other studies, however, the importance of distance to seed source is less straightforward. Archibold (1980) attributed the greater abundance of spruce seed in severely burned areas to higher seed predation in moderate burns. Following the eruption of Mount St. Helens, Dale (1989) found that distance to seed source was not significantly correlated with seed abundance or seedling density. Beland and Bergeron (1993) noted that distance to seed source is only weakly associated with white spruce and balsam fir regeneration, and not correlated with the abundance of jack pine or black spruce.

Different species respond to the effects of fire in different ways (Rowe 1983, Agee 1993). I found sparse regeneration of Douglas fir in stands that were more severely burned and farther from live trees. However, because of the wide spacing and thick bark of Douglas fir trees, a relatively small proportion of Douglas fir stands were in the most severe burn class (21 percent, compared with 50 percent for lodgepole pine stands and 55 percent for Engelmann spruce/subalpine fir stands).

In the Teton region, aspen and lodgepole pine postfire forest types were more likely, compared with other postfire types, to be found a long distance from live trees that provide a seed source. Both aspen and lodgepole pine don't require seed dispersal from adjacent conifer forests for establishment. Lodgepole pine forests tend to be perpetuated in areas where large fires are likely (Peet 1988), and aspen stands are common in landscapes where forest patches are widely spaced in a matrix of shrublands and grasslands. The ability of aspen and lodgepole pine to occur far from patches of live trees reflects the ability of both species to reestablish where they were present before fire.

Although lodgepole pine stands in the Teton region tend to be found in more severely burned areas and a longer distance from unburned trees, the role of distance to unburned trees and fire severity in explaining the postfire density of lodgepole pine seedlings is equivocal. In Yellowstone National Park, lodgepole pine seedling density is higher near surface burns (Turner and others 1997) or moderately burned or unburned forest (Ellis and others 1994, Chapter 4), especially when fires were severe. However, in the Tetons I found the density of regenerating lodgepole pine trees to range widely, from sparse to very dense in stands located a long distance from unburned forest (> 300 m). Anderson and others (Chapter 4) found densities of lodgepole pine to be four to twenty-four times as high in moderately burned plots in Yellowstone as in paired severely burned plots. I did not observe a similar trend. Seedling density was highest in the more severely burned areas. I observed a stand average of less than 0.16 lodgepole pine trees m^{-2} in the ten moderately burned stands I sampled, and up to 4.15 lodgepole pine trees m^{-2} in the thirty-four severely burned stands. Undoubtedly, differences in fire severity may be particularly important when comparing similar sites (Chapter 4).

The influence of burn severity or proximity to a live seed source on the reestablishment of Engelmann spruce/subalpine fir after fire is not straightforward. Both Engelmann spruce and subalpine fir require a nearby seed source for successful recruitment after fire (Noble and Alexander 1977, McCaughey and others 1985). Notably, spruce/fir stands at high elevation often were a longer distance from the nearest live tree, and overall tree density after fire is often sparse. However, although there is a tendency for greater establishment of spruce and fir where the distance to live trees is relatively short, proximity to live trees did not ensure successful seedling establishment. In addition, burn severity did not account for variation in postfire tree establishment. In general, fire effects do not account for much variation in tree establishment in spruce/fir stands not already accounted for by other factors, such as elevation, moisture, and geographic location.

Fire Year and Geographic Location

Differences in successional patterns between sites in a given region have often been noted (Abrams and others 1985, McCune and Allen 1985b, Gagnon and Bradfield 1987, Halpern 1989, Turner and others 1997). However, it is difficult to determine the environmental and historical interactions that contribute to site differences (Pickett and White 1985, Motkin and others 1999). For lodgepole pine and Douglas fir stands, most of the variation in early succession between the different years of the fires is explained by site variables, including prefire vegetation, soils, topography, and fire severity (Table 11.4). However, measured variables fail to adequately explain why Engelmann spruce and subalpine fir have reestablished so successfully in the first few years after the oldest fire (Fig. 11.5, the 1974 WFC fire, Doyle and others 1998), but not on stands in the other, more recent, fires. Even when soils, topography, prefire vegetation, and distance to seed source are similar to the WFC fire, spruce and fir regeneration is typically sparse or absent in the more recent fires (Fig. 11.6). Grubb (1977) suggests that overlooked factors that affect regeneration may often be responsible for unexplained variation in species composition along ordination axes. The availability of seed, early postfire conditions, climatic patterns, and differences in fire history may account for the successful reestablishment of spruce and fir. Above-normal precipitation and cool temperatures during the growing season have been attributed to successful regeneration of Engelmann spruce and subalpine fir (Habeck and Mutch 1973, Noble and Alexander 1977, Little and others 1994). Notably, during the first growing season after the 1974 WFC fire, during which substantial spruce and fir establishment occurred (Doyle and others 1998), weather conditions were unusually cool and moist (Fig. 11.2). Moreover, because of its proximity to the Teton Range, the annual precipitation is considerably greater than at other sites at a similar elevation (Farnes and others 1998, Chapter 10). Engelmann spruce and subalpine fir seed probably were relatively more abundant after the WFC, as both species are dominant and appear vigorous in unburned patches, which are common within the perimeter of the WFC fire. Moreover, percent serotiny of lodgepole pine was apparently low throughout the WFC fire, and thus lodgepole pine seed may have been less available than after other fires. Differences in fire history can lead to different patterns of vegetation and levels of serotiny between sites. Fire frequency is likely to be lower at higher elevations and in sheltered locations, such as the WFC fire, that are more humid (Loope and Gruell 1973, Romme and Knight 1981, Martin 1982, Chapters 6, 8, 14). Limited data suggest that the WFC area probably had not burned for more than two hundred years (Barmore, Taylor, and Hayden, Grand Teton National Park, unpublished manuscript). In contrast, much of the area in the MYS and lower-elevation portions of the HUCK fires were dominated by lodgepole pine before and after fire, and a map of burned areas from 1898 indicates that large portions of these fires burned in the late 1800s (Brandegee 1899).

Engelmann spruce and subalpine fir reestablished quickly after the WFC fire apparently because of available seed and favorable conditions for establishment—cool weather and abundant precipitation coupled with moist topographic situations and fertile soils. McCune and Allen (1985a) suggest that insular sites (such as the WFC fire tucked between the slopes of the Tetons and Jackson Lake), the effects of history, and local anomalies may result in postfire patterns that are not typical for the region.

Some other differences in postfire vegetation patterns between locations may be related to the timing of conditions associated with specific fires. For example, new aspen regeneration was common only after the fires in 1987 and 1988. Notably, the above-normal temperature and precipitation in 1989 have been identified as contributing to an unusually good year for aspen regeneration in the Rocky Mountains (Romme and others 1997). Similarly, Tomback and others (1993) note that seedling establishment of whitebark pine is associated with slightly higher precipitation coincident with available seed. Seed production was abundant for whitebark pine in 1989 in the Greater Yellowstone Ecosystem (Tomback and others 1990), and several years of above-average precipitation occurred in the early 1990s (Fig. 11.2) that contributed to a relatively high level of establishment. The highest densities of whitebark pine seedlings that I observed were in the 1987 DAH and 1988 HUCK fires.

In conclusion, forests are establishing in all areas that have burned in the past twenty years, and all tree species present before fire have reestablished, although rates and patterns of establishment differ across the landscape. Overall, prefire vegetation and topography appear to be the most important factors influencing vegetation patterns, but other factors—such as soil cations, seed availability, fire severity, percent serotiny, weather, and site history—appear critical as well. The importance of the variables I measured varies among vegetation types. In the Teton region, if aspen, whitebark pine, or lodgepole pine dominates the prefire vegetation, the dominant species replace themselves after fire. Dense stands of lodgepole pine have developed where levels of cone serotiny are high. At higher elevations, postfire tree establishment is likely to be slow for all species, but whitebark pine increases in importance. At low elevations, moist, cation-rich soil and seed production by surviving Douglas fir are clearly related to successful reestablishment of Douglas fir. For spruce/fir stands, reestablishment appears to be a function of elevation, moisture, site history, the species composition of surrounding forests, distance to seed source, climate, and weather conditions after the fire.

The Teton region is striking in its geology and topography, which vary over short distances. Superimposed on this dramatic landscape are mosaics of vegetation, soil, climate, and fire history. After fire, the resulting patterns of forest development are complex and clearly cannot be predicted by considering only a few factors. Moreover, although descriptions of postfire patterns in adjacent Yellowstone National Park help us to understand patterns that emerge after fire in the Teton region, observations from Yellowstone or other areas cannot be extrapolated automatically to the Teton landscape. Large-scale fires in the region are unreplicated, are sometimes unpopular, and present problems for analysis, but they do provide information that could not be obtained in any other way and offer a fascinating look into the forces that shape our natural world.

Acknowledgments

The National Park Service, Rocky Mountain Regional Office, the University of Wyoming–National Park Service Research Center, and the Wyoming Native Plant Society provided generous funding for this research. Grand Teton National Park, the Bridger Teton National Forest, and the U.W.–N.P.S. Research Center provided technical and logistical support. My deep appreciation goes to Dennis Knight, William Baker, Jim Graves, Dan Tinker, Monica Turner, Bill Romme, Susan Wiser, and Diana Tomback for their advice and support in designing and carrying out this research. Dennis Knight, Linda Wallace, Bill Reiners, Ron Beiswenger, Larry Munn, Greg Brown, Jim Graves, Joan Doyle, Shawn White, and anonymous reviewers provided critical reviews of the manuscript. Special appreciation goes to Laura Archer, Joan Doyle, James Krumm, Kevin Vaughn, and numerous volunteers for their assistance with data collection. Thanks goes to Irene Thien, who created the map. I am grateful to Dave Love, who gave me access to unpublished geologic maps, and Bryce Frost, who helped with the identification of rock specimens.

References

Abrams, M. D., D. G. Sprugel, and D. I. Dickman. 1985. Multiple successional pathways on recently disturbed Jack pine sites in Michigan. Forest Ecology and Management 10:31–48.

Agee, J. K. 1993. Fire ecology of Pacific Northwest forests. Island, Washington, D.C.

Allen, R. B., and R. K. Peet. 1990. Gradient analysis of forests of the Sangre de Cristo Range, Colorado. Canadian Journal of Botany 68:193–201.

Aplet, G., R. D. Laven, and F. W. Smith. 1988. Patterns of community dynamics in Colorado Engelmann spruce–subalpine fir forests. Ecology 69:312–319.

Archibold, O. W. 1980. Seed input into a postfire forest site in northern Saskatchewan. Canadian Journal of Forest Research 10:129–134.

Austin, M. P. 1972. Models and analysis of descriptive vegetation data. Pages 61–86 in J. N. R. Jeffers, ed., Mathematical models in ecology. Blackwell Scientific, Oxford.

Baker, W. L., and G. M. Walford. 1995. Multiple stable states and models of riparian vegetation succession on the Animas River, Colorado. Annals of the Association of American Geographers 85:320–338.

Barton, A. M. 1993. Factors controlling plant distributions: Drought, competition, and fire in montane pines in Arizona. Ecological Monographs 63:367–397.

———. 1994. Gradient analysis of relationships among fire, environment, and vegetation in a southwestern USA mountain range. Bulletin of the Torrey Botanical Club 121:251–265.

Beland, M., and Y. Bergeron. 1993. Ecological factors affecting abundance of advanced growth in jack pine (*Pinus banksiana* Lamb.) stands of the boreal forest of northwestern Quebec. Forestry Chronicle 69:561–568.

Borcard, D., P. Legendre, and P. Drapeau. 1992. Partialling out the spatial component of ecological variation. Ecology 73:1045–1055.

Borgegård, S. O. 1990. Vegetation development in abandoned gravel pits: Effects of surrounding vegetation, substrate, and regionality. Journal of Vegetation Science 1:675–682.

Bradley, A. F., W. C. Fischer, and N. V. Noste. 1992. Fire ecology of the forest habitat types of eastern Idaho and western Wyoming. U.S. Department of Agriculture Forest Service General Technical Report INT-290.

Brandegee, T. S. 1899. Teton forest reserve. House document no. 5, 55th Congress, 34th session. Serial 3763, Washington, D.C.

Brown, J. R., and D. Warncke. 1988. Recommended cation tests and measures of cation exchange capacity. Pages 15–17 in NCR-13 Soil Testing Committee, eds., Recommended chemical soil test procedures for the North Central Region. North Central Regional Publication no. 221, revised. North Dakota Agricultural Experiment Station Bulletin no. 499, revised. North Dakota State University, Fargo.

Busing, R. T., P. S. White, and M. D. MacKenzie. 1993. Gradient analysis of old spruce–fir forests of the Great Smoky Mountains circa 1935. Canadian Journal of Botany 71:951–958.

Chappell, C. B., and J. K. Agee. 1996. Fire severity and tree seedling establishment in *Abies magnifica* forests, Southern Cascades, Oregon. Ecological Applications 6:628–640.

Cui, M., and W. K. Smith. 1991. Photosynthesis, water relations, and mortality in *Abies lasiocarpa* seedlings during natural establishment. Tree Physiology 8:37–46.

Dale, V. H. 1989. Wind dispersed seeds and plant recovery on the Mount St. Helens debris avalanche. Canadian Journal of Botany 67:1434–1441.

Daubenmire, R. 1943. Vegetation zonation in the Rocky Mountains. Botanical Review 9:325–393.

Day, R. J. 1964. The microenvironments occupied by spruce and fir regeneration in the Rocky Mountains. Department of Forestry Publication 1037, Ottawa, Ont.

De Ronde, C. 1990. Impact of prescribed fire on soil properties: Comparison with wildfire effects. Pages 127–136 in J. G. Goldammer and M. J. Jenkins, eds., Fire in ecosystem dynamics. Proceedings of the Third International Symposium on Fire Ecology, Freiburn, West Germany.

Despain, D. G. 1973. Vegetation of the Big Horn Mountains, Wyoming, in relation to substrate and climate. Ecological Monographs 43:329–355.

———. 1990. Yellowstone vegetation: Consequences of environment and history in a natural setting. Roberts Rinehart, Boulder, Colo.

Despain, D. G., D. Houston, M. Meagher, and P. Schullery. 1986. Wildlife in transition. Roberts Rinehart, Boulder, Colo.

Dole, N. E., M. H. Mitchell, H. E. Bailey, and W. D. Thomas. 1936. Vegetation type map of Grand Teton National Park. U.S. Department of the Interior, Grand Teton National Park.

Doyle, K. M. 1997. Fire, environment, and early forest succession in a heterogeneous Rocky Mountain landscape, northwestern Wyoming. Ph.D. diss., University of Wyoming, Laramie.

Doyle, K. M., D. H. Knight, D. L. Taylor, W. J. Barmore, and J. M. Benedict. 1998. Seventeen years of forest succession following the Waterfalls Canyon Fire in Grand Teton National Park, Wyoming. International Journal of Wildland Fire 8:45–55.

Eckert, D. J. 1988. Recommended pH and lime requirement tests. Pages 6–8 in NCR-13 Soil Testing Committee, eds., Recommended chemical soil test procedures for the North Central Region, North Central Regional Publication no. 221, revised. North Dakota Agricultural Experiment Station Bulletin no. 499, revised. North Dakota State University, Fargo.

Ellis, M., C. D. von Dohlen, J. E. Anderson, and W. H. Romme. 1994. Some important factors affecting density of lodgepole pine seedlings following the 1988 Yellowstone fires. Pages 139–150 in D. G. Despain, ed., Plants and their environments: Proceedings of the first biennial scientific conference on the Greater Yellowstone Ecosystem. Technical Report NPS/NRYELL/NRTR-93/XX. U.S. Department of the Interior, National Park Service, Natural Resources Publication Office, Denver.

Farnes, P., C. Heydon, and K. Hansen. 1998. Snowpack distribution in Grand Teton National Park. Draft Annual Report submitted to Grand Teton National Park, Moose, Wyoming.

Fastie, C. L. 1995. Causes and ecosystem consequences of multiple pathways of primary succession at Glacier Bay, Alaska. Ecology 76:1899–1916.

Frank, E. C., and R. Lee. 1966. Potential solar beam irradiation on slopes: Tables for 30° to 50° latitudes. United States Forest Service, Rocky Mountain Forest and Range Experiment Station Research Paper RM-18.

Gagnon, D., and G. E. Bradfield. 1987. Gradient analysis of west central Vancouver Island forests. Canadian Journal of Botany 65:822–833.

Gauthier, S., Y. Bergeron, and J. Simon. 1996. Effects of fire regime on the serotiny level of jack pine. Journal of Ecology 84:539–548.

Gee G. W., and J. W. Bauder. 1986. Particle-size analysis. Pages 383–411 in A. Klute, ed., Methods of soil analysis, part 1, Physical and mineralogical methods. American Society of Agronomy, Soil Science Society of America, Madison, Wis.

Gelderman, R. H., and P. E. Fixen. 1988. Recommended nitrate-N tests. Pages 10–12 in NCR-13 Soil Testing Committee, eds., Recommended chemical soil test procedures for the North Central Region. North Central Regional Publication no. 221, revised. North Dakota Agricultural Experiment Station Bulletin no. 499, revised. North Dakota State University, Fargo.

Glenn-Lewin, D. C. 1980. The individualistic nature of plant community development. Vegetatio 43:141–146.

Greig-Smith, P. 1982. Quantitative Plant Ecology. Blackwell Scientific, Oxford.

Grubb, P. J. 1977. The maintenance of species-richness in plant communities: The importance of the regeneration niche. Biological Reviews 52:107–145.

Habeck, J. R., and R. W. Mutch. 1973. Fire-dependent forests in the Northern Rocky Mountains. Quaternary Research 3:408–424.

Halpern, C. B. 1989. Early successional patterns of forest species: Interactions of life history traits and disturbance. Ecology 70:704–720.

Hill, M. O. 1979. TWINSPAN: A FORTRAN program for arranging multivariate data in an ordered two-way table by classification of the individuals and attributes. Section of Ecology and Systematics, Cornell University, Ithaca, N.Y.

Jackson, W. D. 1968. Fire, air, water, and earth: An elemental ecology of Tasmania. Proceedings of the Ecological Society of Australia 3:9–16.

Kay, C. E. 1993. Aspen seedlings in recently burned areas of Grand Teton and Yellowstone National Parks. Northwest Science 67:94–103.

Keane, R. E., P. Morgan, and J. P. Menakis. 1994. Landscape assessment of the decline of whitebark pine *(Pinus albicaulis)* in the Bob Marshall Wilderness Complex, Montana, USA. Northwest Science 68:213–229.

Knapp, A. F., and W. K. Smith. 1981. Water relations and succession in subalpine conifers in southeastern Wyoming. Botanical Gazette 142:502–511.

Knight, D. H. 1997. Natural catastrophes as opportunities for ecosystem research. Biogeomon: Journal of Conference Abstracts 2:218.

Langenheim, J. H. 1962. Vegetation and environmental patterns in the Crested Butte area, Gunnison County, Colorado. Ecological Monographs 32:249–285.

Little, R. L., D. L. Peterson, and L. L. Conquest. 1994. Regeneration of subalpine fir *(Abies lasiocarpa)* following fire: Effects of climate and other factors. Canadian Journal of Forest Research 24:934–944.

Loope, L., and G. E. Gruell. 1973. The ecological role of fire in the Jackson Hole area, northwestern Wyoming. Quaternary Research 3:425–443.

Loucks, O. L. 1962. Ordinating forest communities by means of environmental scalars and phytosociological indices. Ecological Monographs 32:137–166.

Love, J. D., J. C. Reed, and A. C. Christiansen. 1992. Geologic map of Grand Teton National Park, Teton County, Wyoming. U.S. Geological Survey, U.S. Department of the Interior.

Martin, R. E. 1982. Fire history and its role in succession. Pages 92–99 in J. E. Means, ed., Forest succession and stand development research in the Northwest: Proceedings of a symposium. United States Forest Service Research Laboratory, Oregon State University, Corvallis.

Martner, B. E. 1986. Wyoming climate atlas. University of Nebraska Press, Lincoln.

McCaughey, W. W., W. C. Schmidt, and R. C. Shearer. 1985. Seed dispersal characteristics of conifers of the Inland Mountain West. Pages 50–61 in R. C. Shearer, ed., Conifer seed in Inland Mountain West, symposium proceedings, Missoula, Mont.

McClanahan, T. R. 1986. The effect of a seed source on primary succession in a forest ecosystem. Vegetatio 65:175–178.

McCune, B., and T. F. H. Allen. 1985a. Forest dynamics in the Bitterroot Canyons, Montana. Canadian Journal of Botany 63:377–383.

———. 1985b. Will similar forests develop on similar sites? Canadian Journal of Botany 63:367–376.

McCune, B., and M. J. Mefford. 1995. PC-ORD. Multivariate analysis of ecological data. Version 2.05. MjM software, Gleneden Beach, Ore.

Mehlich, A. 1984. Mehlich 3 soil test extractant: A modification of Mehlich 2 extractant. Communications in Soil Science and Plant Analysis 15:1409–1416.

Motkin, G., P. Wilson, D. R. Foster, and A. Allen. 1999. Vegetation patterns in heterogeneous landscapes: The importance of history and environment. Journal of Vegetation Science 10: 903–920.

Mueller-Dombois, D., and H. Ellenberg. 1974. Aims and methods of vegetation ecology. Wiley, New York.

Muir, P. S., and J. E. Lotan. 1985. Disturbance history and serotiny of *Pinus contorta* in western Montana. Ecology 66:1658–1668.

Murray, M. P. 1996. Landscape dynamics of an island range: Interrelationships of fire and whitebark pine *(Pinus albicaulis)*. Ph.D. diss., University of Idaho, Moscow.

Noble, D. I., and R. R. Alexander. 1977. Environmental factors affecting natural regeneration of Engelmann spruce in the Central Rocky Mountains. Forest Science 23:420–429.

Noy-Meir, I. 1973. Desert ecosystems: Environment and producers. Annual Review of Ecology and Systematics 4:25–51.

Parker, A. J. 1982. The topographic relative moisture index: An approach to soil-moisture assessment in mountain terrain. Physical Geography 3:160–168.

Peet, R. K. 1978. Latitudinal variation in southern Rocky Mountain forests. Journal of Biogeography 5:275–289.

———. 1981. Forest vegetation of the Colorado Front Range. Vegetatio 45:3–58.

———. 1988. Forests of the Rocky Mountains. Pages 63–101 in M. G. Barbour and W. D. Billings, eds., North American terrestrial vegetation. Cambridge University Press, New York.

Peterson, D. L., and M. J. Arbaugh. 1986. Postfire survival of Douglas-fir and lodgepole pine: Comparing the effects of crown and bole damage. Canadian Journal of Forest Research 16:1175–1179.

Pickett, S. T. A., and P. S. White. 1985. The ecology of natural disturbance and patch dynamics. Academic Press, Orlando, Fla.

Rodman, A., D. Thorma, and H. Shovic. 1992. John D. Rockefeller, Jr., Memorial Parkway draft soil survey. Soils and Watershed Section, Division of Research, Yellowstone National Park, Wyo.

Romme, W. H., and D. H. Knight. 1981. Fire frequency and subalpine forest succession along a topographic gradient in Wyoming. Ecology 62:319–326.

Romme, W. H., M. G. Turner, R. H. Gardner, W. W. Hargrove, G. A. Tuskan, D. G. Despain, and R. A. Renkin. 1997. A rare episode of sexual reproduction in aspen (*Populus tremuloides* Michx.) following the 1988 Yellowstone fires. Natural Areas Journal 17:17–25.

Rowe, J. S. 1983. Concepts of fire effects on plant individuals and species. Pages 135–154 in R. W. Wein and D. A. MacLean, eds., The role of fire in northern circumpolar ecosystems. Wiley, New York.

Schulte, E. E. 1988. Recommended soil organic matter tests. Pages 29–32 in NCR-13 Soil Testing Committee, eds., Recommended chemical soil test procedures for the North Central Region, North Central Regional Publication no. 221, revised. North Dakota Agricultural Experiment Station Bulletin no. 499, revised. North Dakota State University, Fargo.

Smith, W. K. 1985. Western montane forests. Pages 95–126 in B. F. Chabot and H .A. Mooney, eds., Physiological ecology of North American plant communities. Chapman and Hall, New York.

SPSS Inc. 1996. Systat 6.0 for Windows: Statistics. SPSS Inc., Chicago.

Stahelin, R. 1943. Factors influencing the natural restocking of high altitude burns by coniferous trees in the central Rocky Mountains. Ecology 24:19–30.

Starker, T. J. 1934. Fire resistance in the forest. Journal of Forestry 32:462–467.

Steele, R., S. V. Cooper, D. M. Ondov, D. W. Roberts, and R. D. Pfister. 1983. Forest habitat types of eastern Idaho–western Wyoming. United States Forest Service General Technical Report INT-144.

ter Brak, C. J. F. 1991. CANOCO version 3.12: A FORTRAN program for canonical community ordination. Microcomputer Power, Ithaca, N.Y.

Tesch, S. D. 1981. Comparative stand development in an old-growth Douglas-fir *(Pseudotsuga menziesii* var. *glauca)* forest in western Montana. Canadian Journal of Forest Research 11:82–89.

Thornbury, W. D. 1965. Regional geomorphology of the United States. Wiley, New York.

Tinker, D. B., W. H. Romme, W. W. Hargrove, R. H. Gardner, and M. G. Turner. 1994. Landscape-scale heterogeneity in lodgepole pine serotiny. Canadian Journal of Forest Research 24:897–903.

Tomback, D. F., L. A. Hoffmann, and S. K. Sund. 1990. Coevolution of whitebark pine and nutcrackers: Implications for forest regeneration. Pages 118–129 in W. C. Schmidt and K. McDonald, eds., Proceedings, symposium on whitebark pine ecosystems: Ecology and management of a high-mountain resource, Bozeman, Montana. United States Forest Service General Technical Report INT-270.

Tomback, D. F., S. K. Sund, and L. A. Hoffman. 1993. Post-fire regeneration of *Pinus albicaulis:* Height-age relationships, age structure, and microsite characteristics. Canadian Journal of Forest Research 23:113–119.

Turner, M. G., W. W. Hargrove, R. H. Gardner, and W. W. Romme. 1994. Effects of fire on landscape heterogeneity in Yellowstone National Park, Wyoming. Journal of Vegetation Science 5:731–742.

Turner, M. G., W. H. Romme, R. H. Gardner, and W. W. Hargrove. 1997. Effects of fire size and pattern on early post-fire succession in subalpine forests of Yellowstone National Park, Wyoming. Ecological Monographs 67:411–433.

Veblen, T. T. 1986. Age and size structure of subalpine forests in the Colorado Front Range. Bulletin of the Torrey Botanical Club 113:225–240.

Wentworth, T. R. 1981. Vegetation on limestone and granite in the Mule Mountains, Arizona. Ecology 62:469–482.

Whipple, S. A., and R. L. Dix. 1979. Age structure and successional dynamics of a Colorado subalpine forest. American Midland Naturalist 101:142–158.

Whittaker, R. H. 1956. Vegetation of the Great Smoky Mountains. Ecological Monographs 26:1–80.

Young, J. F. 1982. Soil survey of Teton County, Wyoming. United States Department of Agriculture Soil Conservation Service.

Chapter 12 Snags and Coarse Woody Debris: An Important Legacy of Forests in the Greater Yellowstone Ecosystem

Daniel B. Tinker and Dennis H. Knight

Severe forest fires, such as those that occurred in the Greater Yellowstone Ecosystem during the summer of 1988, create ephemeral forests of dead trees. For many people the trees are both an eyesore and a waste of salvageable wood. Harvesting the wood of burned trees is an option in many areas, but ecological processes in national parks are allowed to proceed whenever possible with minimal human intervention. The standing dead trees, commonly known as snags, have been falling to the ground and decomposing for millennia in most forest ecosystems. On the ground, the logs and large branches (> 7 cm diameter) are collectively known as coarse woody debris (CWD) and, together with the snags (standing dead trees), are remnants—a legacy—of the previous forest that have important influences on biological diversity, nutrient cycling, and soil development. Removing snags and CWD changes the forest ecosystem in significant ways (Harmon and others 1986, Harvey and Neuenschwander 1991, Edmonds 1991, Bull and others 1997).

The 1988 fires in Yellowstone National Park (YNP) killed millions of trees, mostly lodgepole pine (*Pinus contorta* var. *latifolia* [Engelm. ex Wats.] Critchfield), creating an estimated 25 million metric tons of

CWD (Tinker and Knight, unpublished data). During the first year or two after the fires, only a few trees fell to the ground, but later, as the relatively shallow root systems began to decay, the rate of treefall increased. Thirteen years after the fires, some stands had very few snags while others still had many. An earlier study of lodgepole pine treefall rates in Montana estimated that approximately 90 percent of the trees in a fire-killed lodgepole pine stand fall to the ground within thirty years (Lyon 1984). Thus most trees killed by the 1988 fires will have fallen to the ground and will have begun their new role in Yellowstone's forests as CWD by 2018.

The additions of CWD to terrestrial and aquatic ecosystems in Yellowstone occur continually as a result of the regular mortality of trees, whether caused by fire or by some other disturbance. As forests develop, individual trees die periodically for a variety of reasons. In very dense stands of lodgepole pine, mortality may begin after only ten to fifteen years because of intense competition for limited water and nutrients. More of the trees in less dense stands grow to maturity, but eventually they become susceptible to various pathogenic root and stem fungi, which often result in the weakening and death of the tree.

A more important class of causes of tree mortality in western coniferous forests comprises such large, infrequent disturbances as fires, insect outbreaks, windstorms, and avalanches. The effects of fires are discussed frequently, as are the effects of two native insects, namely, the spruce budworm *(Choristoneura occidentali)* and the mountain pine beetle *(Dendroctonus ponderosae)*. Both insects are present in Yellowstone's forests, usually at low, endemic population sizes that have little effect on forest structure; but many trees die when their populations increase to epidemic proportions, as happened during the 1970s and 1980s in the Greater Yellowstone Area (Despain 1990). Other disturbances such as avalanches and windstorms are less publicized than insect outbreaks and fires, but they also create large amounts of CWD. For example, the avalanche that occurred on Factory Hill in southern YNP during 1997 swept most of the trees from the side of a 350-m-high slope down to a meadow near the base. Obviously, this event changed the environment of both the meadow and the hillside for decades.

WOOD DECOMPOSITION

The transition from dead trees to downed logs to decayed logs to soil takes many decades, if not a century or more, where cool, dry summers and long, cold winters slow decomposition rates of CWD (Yavitt and Fahey 1982, Fahey 1983), as in the Yellowstone region. A variety of organisms are responsible for the breakdown and decay of logs on the forest floor. Initially, weathering cracks begin to appear as the wood dries, providing access for insects and other macroinvertebrates (Frankland and others 1982). These pioneer decomposers, known as reducers, fragment the CWD and create a favorable substrate for many species of fungi and bacteria (Harmon and others 1986). After one hundred years or more, depending on the size of the logs, most of the more easily decomposed portions of the logs have been utilized by the decomposers, leaving only the most recalcitrant residues. Eventually, even they can be mineralized by white-rot fungi.

An important result of log decay is the accumulation and release of plant nutrients and organic acids (for example, humic and fulvic acids). As logs decompose, bacteria and fungi immobilize important nutrients such as nitrate and ammonium, with the CWD accumulating nutrients for ten to thirty years (Fahey 1983, Hart 1999). As the decay process continues, however, the decomposer organisms eventually die and are decomposed themselves, releasing nutrients and organic acids into the soil. Because microbial decomposition takes so long, mineralization by fire is an important cause of nutrient release from organic materials.

THE IMPORTANCE OF SNAGS AND CWD

The snags that remain for twenty to thirty years after a fire serve as perching, feeding, or nesting habitat for birds and mammals (Bull and others 1997). Many cavity-nesting birds, such as woodpeckers and sapsuckers, create nest cavities in older snags (Davis and others 1983, Harmon and others 1986). The black-backed three-toed woodpecker, which prefers recently burned forests for nesting and feeding habitat, is only rarely seen in Yellowstone, yet was relatively abundant following the 1988 fires. Many small mammals, such as squirrels and chipmunks, also occupy cavities in the dead trees. Notably, we have observed snags in Yellowstone that have been standing for more than one hundred years, providing important habitat for many organisms far beyond the normal "life expectancy" of dead trees.

Similarly, logs on the forest floor provide habitat for a wide variety of insects, bacteria, fungi, and invertebrates that inhabit rotting logs until the decay process is complete. Ants are common inhabitants of CWD, which, along with other insects, are food for foraging bears. Similarly, small mammals such as chipmunks, voles, and shrews use CWD as habitat and are the prey for predatory birds and mammals, such as the boreal owl and pine marten (Hayward and Verner 1994).

Although more commonly thought of as habitat for small mammals, insects, and microbial organisms, CWD is used by plants as well (Grier and others 1981, Little and others 1994, Jurgensen and others 1997). For example, in some areas decaying wood provides an excellent site for the establishment of tree seedlings (Little and others 1994). Seedling establishment on "nurse logs" happens occasionally with lodgepole pine in Yellowstone, though not to the extent as in other areas, where the logs remain moist for a longer time during the growing season. In the more humid Pacific Coast region, more than 90 percent of the Douglas fir and Sitka spruce seedlings have been found growing on CWD in some forests (McKee and others 1982). We have frequently observed aspen and lodgepole pine seedlings in the shade of a downed log, suggesting that CWD provides a favorable microenvironment even when the seedlings are not growing on the log itself. (However, see Chapter 4 for additional information on lodgepole seedling sites.)

We have also observed roots of trees and herbaceous plants growing into rotting logs on the forest floor in YNP, apparently taking advantage of the water and nutrients in the decaying wood, a growth pattern that would be adaptive in an environment that is often dry and infertile (Fig. 12.1). Little evidence suggests that plants require CWD (Harmon and others 1986), but CWD appears to improve the growth of some individuals, and decayed wood may facilitate the formation of mycorrhizae (Harvey and Neuenschwander 1991, Jurgensen and others 1997).

Not as widely recognized, but perhaps just as important, is the role of CWD in affecting soil structure and chemical composition. CWD in the form of snags, logs, and stumps covers as much as 24 percent of the forest floor immediately following a fire, and this percentage increases to almost 60 percent by fifty years postfire as the snags fall (Tinker and Knight 2001). So many logs on the forest floor surely influence the soil that is formed under these logs (Anna Kryszowska-Waitkus and George Vance, personal communication). Decaying logs, roots, and stumps constitute one of the primary sources of organic matter inputs to forest soils (Edmonds 1991), which may help maintain higher levels of productivity on Yellowstone's generally infertile, volcanic soils. A plausible hypothesis is that forest soils overlain by wood for long periods of time will be more fertile and have a higher water holding capacity than where wood has been absent from the surface for the same time. Because the forest floor is a shifting mosaic of fine litter and CWD, it is important to determine how long it takes for the entire forest floor to be covered by pieces of large wood, a topic that we address in the next section.

Given the proximity of forests to streams, rivers, and lakes in the Yellowstone area, a relatively large amount of CWD eventually ends up in aquatic ecosystems where it creates debris jams that influence fluvial and ecological processes

Figure 12.1. Photograph of a linear depression (log trench) where a decaying log had previously occupied the forest floor. The log was consumed during the 1996 Pelican Creek fire in Yellowstone National Park. Note that an unburned woody lateral root (denoted by an arrow) from a nearby lodgepole pine had grown into the decaying log, making a right-angle turn and growing up the length of the log, apparently using it as a source of water and nutrients. Photo by Dennis H. Knight.

(Harmon and others 1986, Bisson and others 1987, Bragg and Kershner 1999). Because of the extreme reduction in wood decay while in water, CWD resides in streams and lakes for centuries (Swanson and Lienkaemper 1978), providing habitat for fish and other aquatic organisms that can become limiting after timber harvesting in riparian zones or well-intentioned attempts to clean up a channel or beach (Bisson and others 1987; Young and others 1994).

HETEROGENEITY IN CWD DISTRIBUTION ACROSS THE LANDSCAPE

Although CWD is ubiquitous across Yellowstone's forested landscapes, the amount of CWD varies tremendously from place to place. Its mass and distribution are controlled largely by the age or successional stage of a particular stand (Fig. 12.2). Young stands from ten to fifty years of age typically contain large amounts of CWD, created by the last stand-replacing fire. As the stand continues to develop and the logs and stumps decompose, and are incorporated into the soil, the amount of visible CWD declines. Then, as the stand reaches maturity, trees of the current generation begin to die and fall to the ground, adding new CWD to the forest floor (Fig. 12.2). Our earlier study of CWD dynamics in Yellowstone revealed that CWD biomass might range from 41 to 284 Mg/ha (one megagram equals one metric ton) in burned and unburned stands. We also found that it varies significantly with time since the last disturbance (Tinker and Knight 2000).

The effects of varying amounts of CWD on the microenvironment are poorly understood, but some relationships seem clear. For example, in stands that were relatively dense before 1988, a tangle of suspended logs now covers much of the forest floor. This vertical stratification of CWD occurs because many of the fallen trees are resting above the ground, supported by trees that fell earlier. The suspended logs shade the ground below, thereby reducing soil temperature. The lower soil temperature could slow decomposition and mineralization. Similarly, shade from CWD may reduce evaporation from the soil, conserving moisture for plant growth. Logs that are elevated above the forest floor persist much longer than if they were in contact with the ground, where the onset of decomposition occurs more rapidly.

QUANTIFYING CWD GAINS AND LOSSES FOLLOWING DISTURBANCE

In contrast to natural forest disturbances, human disturbances such as timber harvesting and some forms of slash treatment remove much of the wood destined to

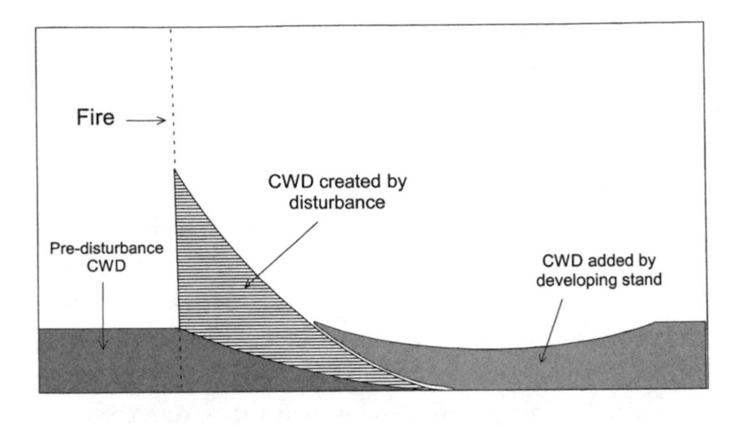

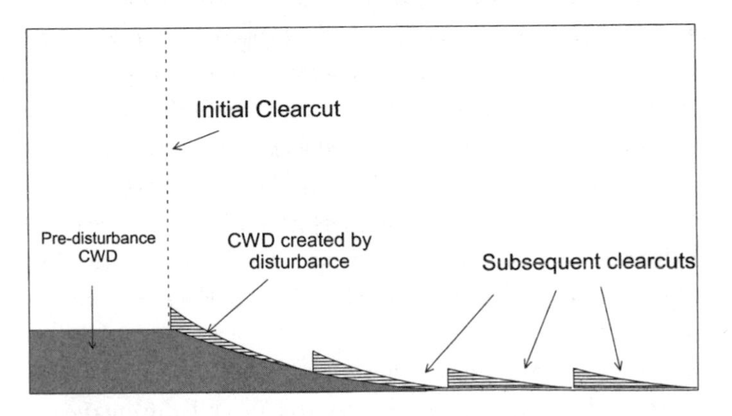

Figure 12.2. Hypothetical trends of CWD (> 7.5 cm) biomass in lodgepole pine forests under (a) a natural fire regime and (b) a 100-year rotation clear-cut harvest regime. Predisturbance CWD is the baseline amount of CWD present in mature stands, assuming that inputs are approximately balanced by decomposition. CWD created by each disturbance is from fire-killed snags for the fire regime and CWD (> 7.5 cm) left as postharvest slash for the timber harvesting simulation. As depicted, there would be essentially no CWD originating from the developing stand under the clear-cut regime, since each successive harvest removes the trees before natural mortality occurs.

become CWD (Jurgensen and others 1997). There is a large amount of inherited wood present on the forest floor before an initial harvest, a legacy of the trees that were killed during the last stand-replacing fire (Maser and others 1979, Wei and others 1997), as well as trees from the present stand that died as the stand matured (Gore and Patterson 1986, Franklin and others 1987). The amount of this wood, and therefore total CWD, is reduced with repeated timber harvesting.

Considering the recent interest in whether or not some forms of timber harvesting such as clear-cutting mimic natural disturbances such as fire (Hammond 1991, Keenan and Kimmins 1993), we compared CWD amounts following fire and logging. To make such a comparison, it is necessary to know how much CWD is produced and eliminated by each disturbance type. Estimating the amount of CWD removed and created by harvesting is relatively straightforward (for example, Brown 1974). However, such estimates are more problematic in stands burned by wildfires, where some of the wood is consumed and some is converted to charcoal. We recently estimated the net loss or gain of CWD following each disturbance type in the Greater Yellowstone Area (GYA) and southeastern Wyoming (Tinker and Knight 2000). As is customary in the ecological literature, CWD was defined as all downed woody material > 7.5 cm diameter, including root crowns (stumps) and woody lateral roots > 7.5 cm. The two belowground components comprised approximately 20 percent of the total CWD and were included because we assumed the belowground wood provides habitat and influences soil structure in a manner similar to the aboveground wood. Data were collected from lodgepole pine stands in Yellowstone National Park and the Medicine Bow National Forest (MBNF) in southeastern Wyoming. Yellowstone provides a unique opportunity to investigate CWD dynamics in forests that are essentially undisturbed by human activity. In contrast, many parts of the MBNF have been heavily influenced by timber harvesting and salvaging for half a century or more.

Our estimates of wood consumption and conversion to charcoal were made in a single stand that burned near Pelican Creek in YNP during the summer of 1996. There are no generally accepted methods for estimating wood loss during natural wildfires, but we proceeded by adding our estimates of (1) the mass of wood converted to charcoal, yet still present and measurable, and (2) the mass of logs completely consumed by the fire. We attempted unsuccessfully to estimate the biomass of logs partially consumed by the fire (Tinker 1999), which, when it occurs, typically creates cupped depressions on the log surface.

Our approach to estimating the conversion of downed CWD to charcoal was based on the volume of a tapered cylinder (Harmon and Sexton 1996). The maximum diameters at three locations along each of fifty downed logs that had been charred or partially burned in 1996 were measured using tree calipers. This provided diameters of the log that included the charred exterior. Each log was then cut at the three points of measurement, exposing the inner, unburned portion of the wood. The diameter measurements were then repeated on the unburned portion of the logs. With these two sets of measurements, we calculated the volumes of two tapered cones: one outer cone that included the charcoal and one smaller cone for the unburned portion of the log. An estimate of charcoal volume was calculated as the difference between the two cone volumes. The charcoal volume was then converted to CWD mass lost using prefire wood bulk density values for the unburned wood of each individual log (Harmon and Sexton 1996). Charcoal that had sloughed to the ground was not estimated; therefore our estimates are conservative.

Estimating the volume of wood that was completely combusted was more problematic, but we used two types of evidence on the forest floor—log shadows and log trenches. Log shadows are linear, light gray patches on the otherwise blackened soil surface. The high temperatures of glowing combustion on logs touching or in close proximity to the forest floor create these linear patterns. Log trenches are elongated depressions caused by the nearly complete combustion of logs that had been partially buried in the forest floor. Log shadows and trenches were mapped, along with CWD that was still present in the stand, in three 20 × 20–m plots. The maps were then overlaid with digital maps of the unburned CWD using ARC/INFO (ESRI 1995). The overlays linked the shadows and trenches with unburned portions of logs of identifiable size and decay classes. The shadow-log associations were used to estimate the probable dimensions and decay stage of the "missing" logs prior to the fire. The mass of missing logs in log trenches not associated with adjacent, unburned logs were estimated using regression models developed in unburned forests in YNP, which predicts log diameter from trench width. The estimated volumes of each missing log were multiplied by wood density values for the appropriate decay class (Harmon and Sexton 1996) to calculate the biomass of the logs that previously occupied the trenches.

The log shadow/log trench method also provided a conservative estimate of wood consumption because some of the wood < 7.5 cm diameter that was consumed leaves no obvious trace of its previous existence. Without prefire measurements, no alternative for estimating wood consumption during natural fires exists at the present time. Notably, our method can be used only in the first few months after a fire because the log shadows soon disappear and the log trenches become less conspicuous.

The amount of CWD converted to charcoal during the Pelican Creek fire in 1996 was estimated to be at least 6.4 Mg/ha (approximately 8 percent of CWD present in the stand we studied). The log-shadow and log-trench evidence revealed that an additional 6.4 Mg/ha (about 8 percent) of downed CWD was completely consumed. When both losses are added, approximately 12.8 Mg/ha (about 16 percent of total) of downed CWD was either consumed or converted to charcoal by the Pelican Creek fire.

We found no significant differences between downed CWD or total downed wood (CWD plus fine woody debris < 7.5 cm) when comparing burned stands with recently clear-cut stands (Fig. 12.3, $p > 0.2$). However, there are significantly more snags in burned stands. When snags are considered in the calculation of total CWD, there is more than double the biomass of CWD > 7.5 cm in burned stands than in clear-cut stands (Fig. 12.3, $p < 0.05$). Clear-cut stands, on the other hand, had five times more fine woody debris (less than 7.5 cm) than burned stands ($p < 0.05$; 20.6 and 4.5 Mg/ha, respectively), largely because fine debris is left as slash following clear-cuts but is mostly consumed during intense fires. Our comparisons were made in recent clear-cut stands that had been subjected to only a single harvest.

When comparing disturbance types (fire in YNP and timber harvest in the MBNF), clear-cut harvesting resulted in an average net CWD *loss* of 83 Mg/ha (137 Mg/ha harvested as live trees minus 54 Mg/ha gained from slash, Fig. 12.4). In contrast, assuming the 16 percent reduction in CWD due to consumption by fire and conversion to charcoal, lodgepole pine stands that burn still would have an average net CWD *increase* of 99.2 Mg/ha after all snags fall to the ground (112 Mg/ha gained minus 12.8 Mg/ha burned; Fig. 12.4). In our study areas, clear-cutting removed almost eleven times as much CWD biomass as a natural fire (137 Mg/ha removed by clear-cutting ÷ 12.8 Mg/ha removed by wood combustion; Fig. 12.4).

CWD DYNAMICS IN LODGEPOLE PINE FORESTS AFTER FIRE AND TIMBER HARVESTING

Single measurements of CWD in unmanaged or naturally developing forests are useful for establishing baseline information, but they fail to capture the dynamics of CWD in space and time. Downed log biomass and dead-standing trees vary temporally and spatially as new stands develop repeatedly after a series of disturbances spread out over hundreds of years. Previous studies of CWD biomass in various forest ecosystems (Harmon and others 1986, Comeau and Kimmins 1989, Arthur and Fahey 1990, Wei and others 1997, and others) have provided excellent data on CWD amounts but little information on the spatial and temporal dynamics of CWD. Several models of CWD dynamics exist (for example, Harmon and others 1996, Bragg 1997, McCarter 1997), but none address the changes in spatial distribution of CWD over long periods of time. We

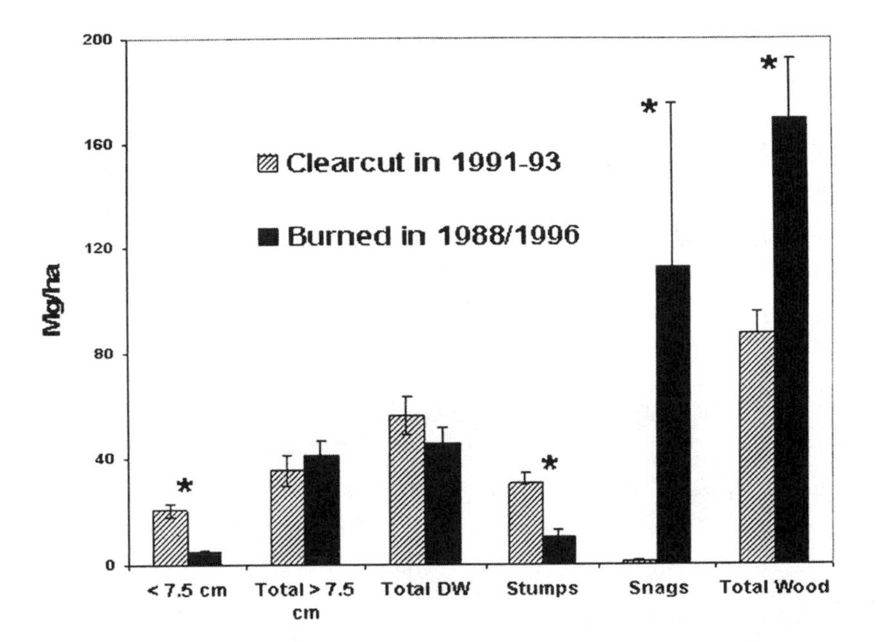

Figure 12.3. The mass of wood in snags, stumps, small wood (< 7.5 cm), and large wood (> 7.5 cm) in clear-cut stands (hatched bars; data from the Medicine Bow National Forest) and burned stands (solid bars; data from Yellowstone National Park). DW = downed wood; Total wood = total downed wood of all sizes + stumps + wood in snags. Error bars are ± 1 SE. Asterisks indicate significant differences between clear-cut and burned stands. Stump biomass is higher in clear-cuts than in burned stands only because bole wood removal creates stumps during timber harvest.

attempted to fill this void for Rocky Mountain coniferous forests by developing a spatially explicit simulation model called DEADWOOD that is parameterized with data from YNP and the MBNF. As the new paradigm of ecosystem management continues to evolve, understanding the historic range of variability (HRV) for key forest processes through simulation modeling becomes even more important. Simulation models such as DEADWOOD are currently the best if not the only alternative for estimating the HRV of most ecosystem variables. Our objective was to simulate different natural fire-return intervals to obtain an estimate of the HRV for CWD, within which forest managers might decide to maintain CWD amounts following timber harvesting.

As discussed previously, the presence of wood on the soil surface, with its associated biota and modified environment, probably has significant effects on soil structure and chemistry. This would be especially true if the wood remains on the forest floor for a sufficiently long time that it becomes incorporated into

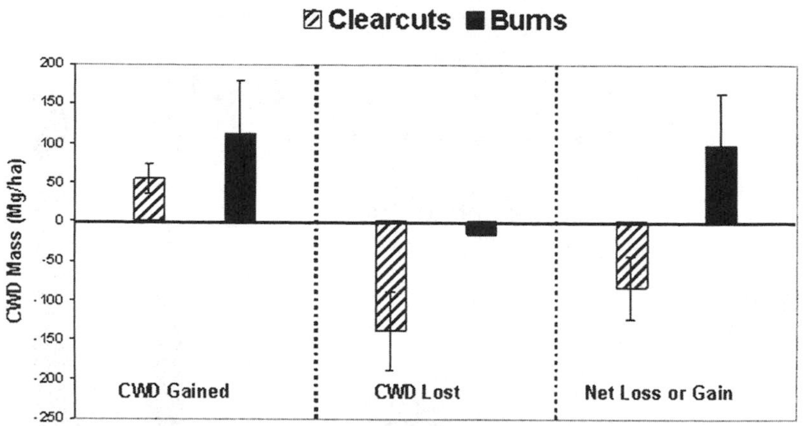

Figure 12.4. Average CWD mass (Mg/ha) gained, lost, and the net gain or loss for all clear-cut (hatched bars) and burned (solid bars) stands. In clear-cut stands, gain is from postharvest slash and loss is from the removal of live tree boles. In burned stands, gain is from fire-killed trees and loss is from wood consumption by fire and conversion to charcoal. Error bars are ± 1 SE. There is no error bar for the loss category for burns because this estimate is from a single stand.

the forest floor (forming what some scientists refer to as soil wood; Jurgensen and others 1997). Normally the wood remains in place for decades or even a century, as fire removes only a small proportion of the CWD (Tinker and Knight 2000). Thus the time required for every square decimeter of the forest floor to be occupied by a tree or covered with CWD, which we refer to as forest floor occupancy rate, would seem to be an important forest ecosystem variable. To explore this phenomenon, we tested two hypotheses: (1) the amount of time required for 100 percent of the forest floor to be affected by wood (logs, stumps, trees, snags, and root crowns) is longer under a clear-cut timber harvest regime than under natural fire regimes, regardless of fire-return interval; and (2) CWD biomass declines over time as a result of repeated clear-cut timber harvesting, compared with a natural fire regime. Here we present a summary of our results. The details of model construction and simulation results are available in Tinker and Knight (2001).

Three burn intervals (100-, 200-, and 300-year intervals) were simulated for a period of 1000 years. Similarly, five different clear-cut harvesting operations were simulated for 1000 years, assuming that clear-cutting was the adopted silvicultural system and that the stands had not previously been harvested. The CWD amounts for the five clear-cut operations spanned the range of slash left

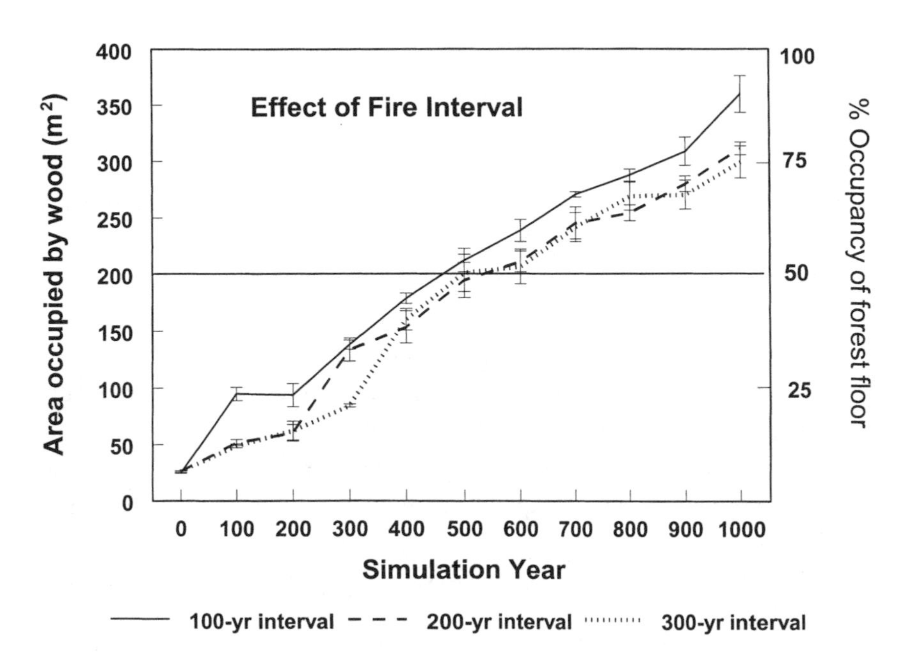

Figure 12.5. The 1000-year simulation results for 100-, 200-, and 300-year fire-return intervals. Values for each 100-year period indicate the amount of the forest floor directly affected by wood (includes CWD and live- and dead-standing trees), as predicted by the DEADWOOD model for the 400 m^2 plots. All values are averages of five replicate simulations for each treatment. Percent occupancy is the percent of the 400 m^2 plots occupied by wood. The horizontal line represents 50 percent occupancy of the forest floor by wood. Error bars are $\pm$ 1 SD.

on the sites that were measured in the field. Each of the five measured slash amounts was simulated at three different applications: 100 percent of slash (the amount measured); 50 percent of the measured amount; and 200 percent of the measured amount. This approach produced what can be considered as fifteen different treatments for our analysis—namely, five slash amounts estimated in the field at the three slash application levels. Different proportions of the actual slash amounts were simulated because forest managers have some control over how much slash will be left on a site after harvesting.

Using the DEADWOOD model, the 100-year fire-return interval produced more CWD after 1000 years than the 200- or 300-year intervals in lodgepole pine forests (Fig. 12.5), probably because most of the tree basal area (wood volume) of trees is accrued during the first 100 years. This suggests that more frequent inputs (every 100 years) of smaller diameter logs to the forest floor resulted in more forest floor coverage by CWD than less frequent inputs of slightly larger trees every 200 or 300 years (Fig. 12.5). In addition, when averaged over the 1000-year simulation period, the average decadal rate of CWD input was not significantly different between the 100- and 300-year intervals.

All of the burn simulations produced sufficient CWD to completely cover the forest floor within 1400 years (Fig. 12.6). However, only by doubling the amount of slash applied during the clear-cut simulations was 100 percent of the forest floor covered by CWD within the same time period, and only then for the three highest measured slash amounts (Fig. 12.6). None of the amounts of slash estimated on the MBNF would leave enough CWD to mimic CWD inputs by natural fires (Fig. 12.6). Perhaps more importantly, our analysis suggests that even reducing the current amount of CWD left on a site following clear-cutting by only 50 percent requires several thousand years more than our measured slash amounts to completely cover the forest floor with CWD (Fig. 12.6).

Notably, CWD on the forest floor was consistently higher during the first few hundred years for both fire and clear-cut simulations because of the large amount of inherited wood present prior to the first simulated disturbance. However, once the inherited wood had decomposed, CWD inputs to the forest floor declined in comparison with natural fires. Harmon and others (1996) observed a similar decline in CWD following timber harvesting in the 1920s using the HARVEST model. Sturtevant and others (1997) measured CWD abundance in a balsam/fir forest in Newfoundland and found that, if harvest rotation times coincided with periods of lowest CWD, then CWD amounts decreased through time. Sturtevant and others (1997) also found that structural diversity contributed by large dead and downed logs in old-growth forests was higher than in stands that had been clear-cut fifty to sixty years earlier.

Our simulations identified significant differences among the fifteen clear-cut simulation treatments. Notably, the divergence through time in forest floor CWD emphasizes the cumulative effect of reduced amounts of CWD input after repeated harvests. Deficiencies in the amount of CWD slash as small as 5 Mg ha^{-1} 100 yrs^{-1} may result in a cumulative reduction in forest floor occupancy by CWD after several hundred years of repeated clear-cutting, which could have adverse effects on both biological diversity and the soil characteristics on which current levels of wood production may depend.

If clear-cutting is to be used as a harvest technique that mimics natural fire in the region where our study was done, it is necessary that managers leave more CWD as slash on the site than current amounts. Obviously, broadcast burning or pile-and-burn slash treatments, which consume large amounts of

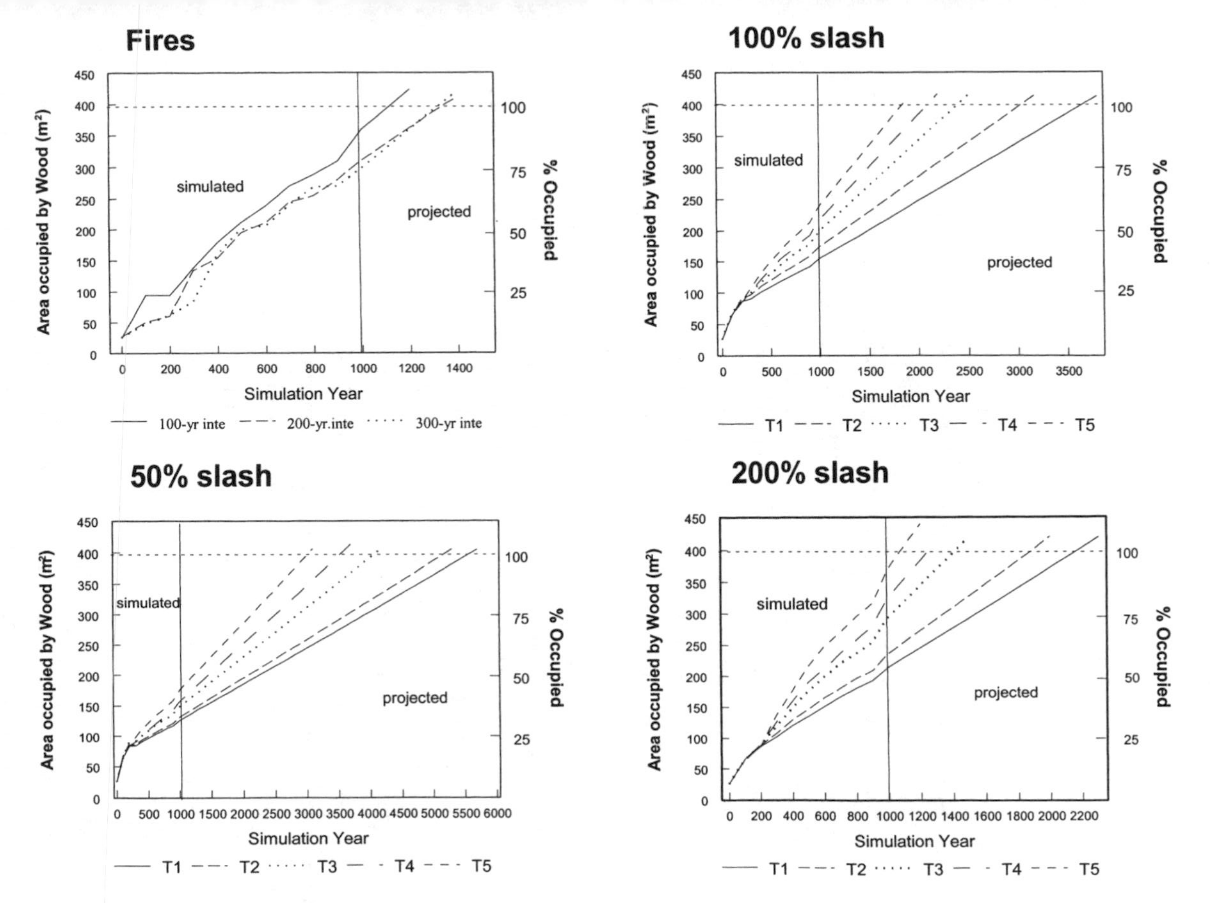

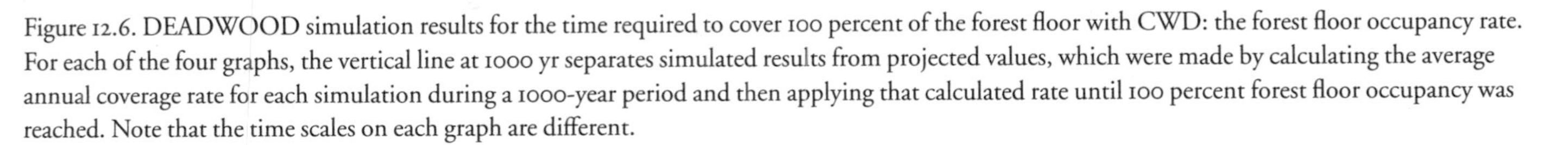
Figure 12.6. DEADWOOD simulation results for the time required to cover 100 percent of the forest floor with CWD: the forest floor occupancy rate. For each of the four graphs, the vertical line at 1000 yr separates simulated results from projected values, which were made by calculating the average annual coverage rate for each simulation during a 1000-year period and then applying that calculated rate until 100 percent forest floor occupancy was reached. Note that the time scales on each graph are different.

CWD, would leave less CWD than other treatments (such as now left by roller-chopping or tractor-walking). Indeed, our analysis suggests that only by doubling the amount of slash currently left on clear-cut harvest sites in our study area would CWD be within the historic range of variability of CWD inputs from fires (Fig. 12.7), undoubtedly due to the absence of inherited wood subsequent to the initial harvest. Stands subjected to natural fires without harvesting, such as occurred in 1988, produce regular and larger inputs of CWD. Bragg (1997) showed that CWD delivery to stream channels in the Bridger-Teton National Forest was significantly reduced following clear-cut timber harvesting, but that it increased following simulated spruce beetle outbreaks. Our results for the same area suggest that natural fires may create up to four times more CWD during a 100-year period than current harvesting and postharvest slash treatment practices, regardless of fire-return interval (see also Chapters 8, 9).

These data lead us to conclude that leaving the unsightly snags that were created in Yellowstone National Park by the 1988 fires was an ecologically important and appropriate decision. The innumerable ecosystem processes that occur on, near, or within snags and CWD, or are mediated by CWD, proceed only with periodic inputs of decaying wood through tree mortality and treefall. This wood is one of the most important legacies of the predisturbance forest. Yellowstone has again proven to be an extremely valuable natural laboratory, one where natural ecological functions occur with a minimum of human interference. As the snags continue to fall in YNP and as new forests continue to develop, eventually removing from sight all of the CWD created in 1988, most park visitors will soon forget the forests of blackened trees. However, the dead wood on the forest floor persists for many decades, occupied by a plethora of living organisms and influencing numerous ecological processes. Some of the CWD will burn in the next fire, but much of it slowly contributes to the soil on which future generations of trees will depend—a process that has been underway for millennia, and one that depends on a rich array of forest organisms that most people never see.

Acknowledgments

This research was supported by grants from the University of Wyoming–National Park Service Research Center and the U.S. Department of Agriculture (NRI 96-35101-3244). We wish to thank Mike Sanders and Dave Carr of the Medicine Bow National Forest, and Dave Phillips, Kathleen O'Leary, and John Varley of Yellowstone National Park, for their help and cooperation during this

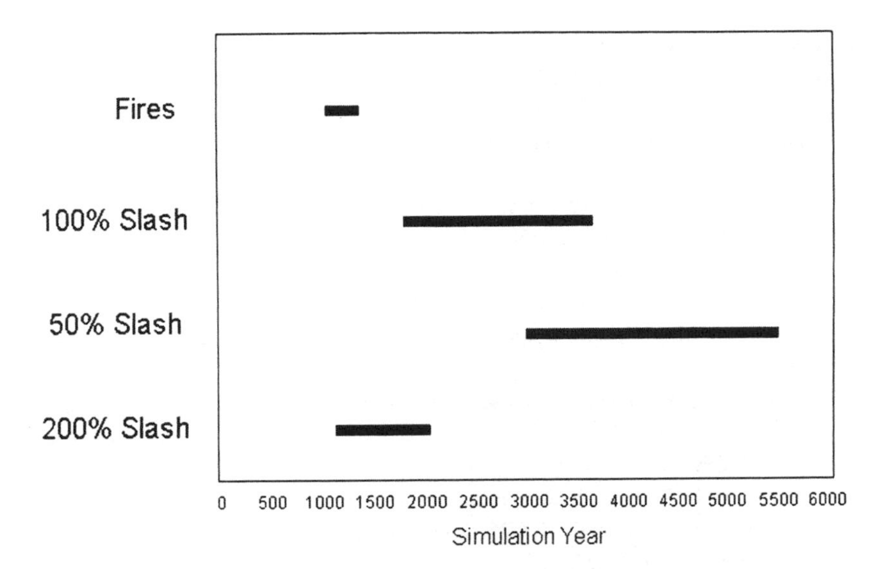

Figure 12.7. Horizontal bars indicate the range of years required to occupy 100 percent of the forest floor by fire and each slash treatment simulation (100 percent, 50 percent, and 200 percent), as predicted by DEADWOOD. The range of values for the burn simulations is considered to be the historic range of variability (HRV) in naturally developing stands. The time required for complete forest floor cover under the natural fire regimes ranged from 1125 to 1350 years; clear-cutting and 100 percent slash simulations required 1800–3600 years to completely cover the forest floor; 50 percent slash simulations required 3000–5600 years; and 200 percent slash simulations required only 1100–2150 years. The shorter forest floor CWD occupancy rate of the 200 percent slash simulations (double the observed amounts) overlap the HRV predicted by natural fire simulations.

study. The University of Wyoming and the UW-NPS Research Center provided valuable logistical support. Mark Harmon and William Pulliam provided helpful suggestions during the early phases of the development of DEADWOOD. We also thank Sharon Stewart , Kris Johnson, Donna Ehle, David Melkonian, Mark Lyford, and Sally Tinker for assistance with field data collection and analysis; and George Vance, Larry Munn, and Anna Krzyszowska-Waitkus for their work on the influences of CWD on soil characteristics in our study areas. William Hicks and Hua Chen provided helpful reviews of an earlier draft of the chapter manuscript.

References

Arthur, M. A., and T. J. Fahey. 1990. Mass and nutrient content of decaying boles in an Engelmann spruce–subalpine fir forest, Rocky Mountain National Park, Colorado. Can. J. For. Res. 20:730–737.

Bisson, P. A., R. E. Bilby, M. D. Bruant, C. A. Dolloff, G. B. Grette, R. A. House, M. L. Murphy, K. V. Koski, and J. R. Sedell. 1987. Large woody debris in forested streams in the Pacific Northwest: Past, present, and future. Pages 143–190 in E. O. Salo and T. W. Cundy, eds., Streamside management: Forestry and fishery interactions. University of Washington Press, Seattle.

Bragg, D. C. 1997. Simulating catastrophic disturbance effects on coarse woody debris production and delivery. Pages 148–156 in Proceedings: Forest vegetation simulator conference. U.S.D.A. Gen. Tech. Rep. INT-GTR-373. U.S. Department of Agriculture, Forest Service, Intermountain Research Station, Ogden, Utah.

Bragg, D. C., and J. L. Kershner. 1999. Coarse woody debris in riparian zones. Journal of Forestry 97:30–35.

Brown, J. K. 1974. Handbook for inventorying downed woody material. USDA Forest Service General Technical Report INT-16. Intermountain Forest and Range Experimental Station, Ogden, Utah.

Bull, E. L., C. G. Parks, and T. R. Torgersen. 1997. Trees and logs important to wildlife in the Interior Columbia River Basin. USDA Forest Service PNW-GTR-391. Pacific Northwest Research Station, Portland, Ore.

Busse, M. D. 1994. Downed bole-wood decomposition in lodgepole pine forests of Central Oregon. Soil Sci. Soc. Am. J. 58:221–227.

Comeau, P. G., and J. P. Kimmins. 1989. Above- and below-ground biomass and production of lodgepole pine on sites with differing soil moisture regimes. Can. J. For. Res. 19:447–454.

Davis, J. W., G. A. Goodwin, and R. A. Ockenfels. 1983. Snag habitat management. USDA For. Serv. Gen. Tech. Rep. RM-99.

Despain, D. G. 1990. Yellowstone vegetation: Consequences of environment and history in a natural setting. Roberts Rinehart, Boulder, Colo.

Edmonds, R. L. 1991. Organic matter decomposition in western United States forests. Pages 118–128 in A. E. Harvey and L. F. Neuenschwander, eds., Proceedings: Management and productivity of western-montane forest soils. USDA, Forest Service General Technical Report INT-280.

ESRI, Inc. 1995. Understanding GIS: The ARC/INFO method. Environmental Systems Research Institute, Inc., Redlands, Calif.

Fahey, T. J. 1983. Nutrient dynamics of aboveground detritus in lodgepole pine *(Pinus contorta* ssp. *latifolia)* ecosystems, southeastern Wyoming. Ecological Monographs 53:51–72.

Fahey, T. J., and D. H. Knight. 1986. The lodgepole pine ecosystem. BioScience 36:610–617.

Frankland, J. C., J. N. Hedger, and M. J. Swift. 1982. Decomposer basidiomycetes: Their biology and ecology. Cambridge University Press, London.

Franklin, J. F., H. H. Shugart, and M. E. Harmon. 1987. Tree death as an ecological process. BioScience 37:550–556.

Gore, J. A., and W. A. Patterson III. 1986. Mass of downed wood in northern hardwood forests in New Hampshire: Potential effects of forest management. Can. J. For. Res. 16:335–339.

Grier, C. C., K. A. Vogt, M. R. Keyes, and R. L. Edmonds. 1981. Biomass distribution and above- and below-ground production in young and mature *Abies amabilis* zone ecosystems of the Washington Cascades. Can. J. For. Res. 11:155–167.

Hammond, H. 1991. Seeing the forest among the trees: The case for holistic forest use. Polestar, Vancouver, B.C.

Harmon, M. E., J. F. Franklin, F. J. Swanson, P. Sollins, S. V. Gregory, J. D. Lattin, N. H. Anderson, S. P. Cline, N. G. Aumen, J. R. Sedel, G. W. Lienkaemper, K. Cromack Jr., and K. W. Cummins. 1986. Ecology of coarse woody debris in temperate ecosystems. Adv. Ecol. Res. 15:133–302.

Harmon, M. E., and J. Sexton. 1996. Guidelines for measurements of woody detritus in forest ecosystems. Publication no. 20. U.S. LTER Network Office. University of Washington, Seattle.

Hart, S. C. 1999. Nitrogen transformations in fallen tree boles and mineral soil of an old-growth forest. Ecology 80:1385–1394.

Harvey, A. E. 1982. The importance of residual organic debris in site preparation and amelioration for reforestation. Pages 75–85 in D. M. Baumgartner, ed., Proceedings: Site preparation and fuel management on steep terrain. Washington State University, Cooperative Extension, Pullman.

Harvey, A. E., and L. F. Neuenschwander, eds. 1991. Proceedings: Management and productivity of western-montane forest soils. USDA, Forest Service, Gen. Tech. Rep. INT-280.

Hayward, G. D., and J. Verner. 1994. Flammulated, boreal, and great gray owls in the United States: A technical conservation assessment. USFS Gen. Tech. Report RM-253. U.S. Dept. of Agriculture, Forest Service, Rocky Mountain Forest and Range Experiment Station, Rocky Mountain Region.

Jurgensen, M. F., A. E. Harvey, R. T. Graham, D. S. Page-Dumroese, J. R. Tonn, M. J. Larsen, and T. B. Jain. 1997. Impacts of timber harvesting on soil organic matter, nitrogen, productivity, and health of inland northwest forests. Forest Science 43:234–251.

Keenan, R. J., and J. P. Kimmins. 1993. The ecological effects of clearcutting. Environ. Rev. 1:121–144.

Koch, P. 1987. Gross characteristics of lodgepole pine trees in North America. Gen. Tech. Rep. INT-227. U.S. Department of Agriculture, Forest Service, Intermountain Research Station, Ogden, Utah.

Koch, P., and J. Schlieter. 1991. Spiral grain and annual ring width in natural unthinned stands of lodgepole pine in North America. Res. Pap. INT-449. U.S. Department of Agriculture, Forest Service, Intermountain Research Station. Ogden, Utah.

Little, R. L., D. L. Peterson, and L. L. Conquest. 1994. Regeneration of subalpine fir *(Abies lasiocarpa)* following fire: Effects of climate and other factors. Canadian Journal of Forest Research 24:934–944.

Lotan, J. E., and W. B. Critchfield. 1990. *Pinus contorta* Dougl. ex Loud. Pages 302–315 in R. M. Burns, B. H. Honkala, tech. coords., Silvics of North America, vol. 1, Conifers. Agric. Handb. 654. U.S. Department of Agriculture, Washington, D.C.

Lyon, L. J. 1984. The Sleeping Child burn: 21 years of postfire change. Res. Pap. INT-330. U.S. Department of Agriculture, Forest Service, Intermountain Research Station, Ogden, Utah.

Maser, C., R. G. Anderson, K. Cromack, Jr., J. T. Williams, and R. E. Martin. 1979. Dead and down woody material. Pages 78–85 in J. W. Thomas, ed., Wildlife habitats in managed forests: The Blue Mountains of Oregon and Washington. USDA Forest Service Agricultural Handbook no. 553.

McCarter, J. B. 1997. Integrating forest inventory, growth and yield, and computer visualization into a landscape management system. Pages 159–167 in Proceedings: Forest vegetation simulator conference. U.S.D.A. Gen. Tech. Rep. INT-GTR-373. U.S. Department of Agriculture, Forest Service, Intermountain Research Station, Ogden, Utah.

McKee, A., G. LeRoi, and J. F. Franklin. 1982. Structure, composition, and reproductive behavior of terrace forests, South Fork Hoh River, Olympic National Park. Pages 22–29 in E. E. Starker, J. F. Franklin, and J. W. Matthews, eds., Ecological research in national parks of the Pacific Northwest. Natl. Park Serv. Coop. Stud. Unit, Corvallis, Ore.

Pearson, J. A., T. J. Fahey, and D. H. Knight. 1984. Biomass and leaf area in contrasting lodgepole pine forests. Can. J. For. Res. 14:259–265.

Romme, W. H., and D. G. Despain. 1989. Historical perspective on the Yellowstone fires of 1988. BioScience 39:695–699.

Spies, T. A., J. F. Franklin, and T. B. Thomas. 1988. Coarse woody debris in Douglas-fir forests of western Oregon and Washington. Ecology 69:1689–1702.

Stone, J. N., A. MacKinnon, J. V. Parminter, and K. P. Lertzman. 1998. Coarse woody debris decomposition documented over 65 years on southern Vancouver Island. Can. J. For. Res. 28:788–793.

Sturtevant, B. R., J. A. Bissonette, J. N. Long, and D. W. Roberts. 1997. Coarse woody debris as a function of age, stand structure, and disturbance in boreal Newfoundland. Ecological Applications 7(2):702–712.

Swanson, F. J., and G. W. Lienkaemper. 1978. Physical consequences of large organic debris in Pacific Northwest streams. USDA For. Serv. Gen. Tech. Rep. PNW-69.

Tinker, D. B. 1999. Coarse woody debris in Wyoming lodgepole pine forests. Ph.D. diss., University of Wyoming, Laramie.

Tinker, D. B., and D. H. Knight. 2000. Coarse woody debris following fire and logging in Wyoming lodgepole pine forests. Ecosystems 3:472–483.

———. 2001. Temporal and spatial dynamics of coarse woody debris in harvested and unharvested lodgepole pine forests. Ecological Modelling 141:125–149.

United States Department of Agriculture. 1985. Land and resource management plan. Medicine Bow National Forest and Thunder Basin National Grassland. U.S.F.S., Laramie, Wyo.

Wei, X., J. P. Kimmins, K. Peel, and O. Steen. 1997. Mass and nutrients in woody debris in harvested and wildfire-killed lodgepole pine forests in the central interior of British Columbia. Can. J. For. Res. 27:148–155.

Yavitt, J. B., and T. J. Fahey. 1982. Loss of mass and nutrient changes of decaying woody roots in lodgepole pine forests, southeastern Wyoming. Can. J. For. Res. 12:745–752.

Young, M. K., D. Haire, and M. A. Bozek. 1994. The effect and extent of railroad tie drives in streams of southeastern Wyoming. Western Journal of Applied Forestry 9:125–130.

Chapter 13 Fire Patterns and Ungulate Survival in Northern Yellowstone Park: The Results of Two Independent Models

Linda L. Wallace, Michael B. Coughenour,
Monica G. Turner, and William H. Romme

Yellowstone National Park (YNP) is renowned for its natural beauty. The original park boundaries were established to contain the geyser basins and hot springs, but they were later altered to include more of the ecosystem, such as watersheds and vegetational communities, within natural boundaries (Haines 1977, Schullery 1997). This was done in an effort to protect habitat for the abundant wildlife found in the park. According to current surveys, one of the reasons park visitors give for their decision to see Yellowstone is the chance to view wildlife in its natural surroundings. Therefore it is not surprising that when the 1988 fires burned a great proportion of the YNP landscape, one concern was how fires affect the wild animals, particularly populations of elk *(Cervus elaphus)* and bison *(Bison bison)*. Concern was so great that in the winter of 1988–1989, feeding programs for elk were proposed, but these were never implemented by the park or the Forest Service (Christensen and others 1989).

Given the large areas utilized by native ungulates and the heterogeneity of habitats, burned areas, and winter conditions across the park landscape, a unique set of tools was needed to explore the effects of the

1988 fires on the elk and bison of YNP. For obvious logistical and ethical reasons, this question could not be addressed using field manipulations of fire and snow. Therefore, computer simulation models needed to be developed that accurately portrayed vegetation, fire, weather, and ungulate population dynamics. These models could then be manipulated to determine what effect fire had on ungulate survival in the park. This was a question that, given its spatial and temporal nature, incorporated issues from population, ecosystem, and landscape ecology. Therefore models needed to recognize the complex nature of this problem.

Two groups independently addressed this issue. Michael Coughenour and Francis Singer of the Natural Resources Ecology Lab at Colorado State University developed a landscape carrying capacity model (LCCM) based on nutritional and energetic requirements of elk. Elk were not treated as individuals in this model, but rather the response of the entire population was assumed based on proportional responses of the landscape to different fire and winter severity scenarios (Coughenour 1994, Coughenour and Singer 1996a). Monica Turner, William Romme, and Linda Wallace developed an individually based, spatially explicit, energetically based model (NOYELP) (Turner and others 1993, 1994, Wallace and others 1995, Wu and others 1996). Both models had a strong landscape component, using spatially explicit and realistic distributions of vegetation, slope, aspect, and fire patterns.

In this chapter, we first compare the assumptions and structures of the two models (Tables 13.1, 13.2). Then we compare the outcomes of the models and the modeling experiments that were conducted and discuss the implications of those results to future fire scenarios in YNP and elsewhere where large populations of native ungulates occur.

MODEL ASSUMPTIONS AND STRUCTURES

The LCCM is a spatially explicit forage-based carrying capacity model, consisting of linked models of forage production, snow cover, elk forage intake, dietary mixing, and elk energy and nitrogen. The model estimates forage biomass based on typical values for a vegetation cover type and the previous year's precipitation. Precipitation at a location is estimated by elevation-corrected spatial interpolation from weather station data (Coughenour 1991). Snow cover is dynamically simulated based on precipitation input and rates of snowmelt derived from temperature. Snow input is corrected for slope and aspect after Farnes and Hartman (1989). Temperature varies with elevation according to an adiabatic lapse rate. Forage intake rate is estimated from a functional response equation

Table 13.1. Comparison of model structure and function between the Landscape Carrying Capacity Model (LCCM) and the Northern Yellowstone Park (NOYELP) models

Model factor	LCCM	NOYELP
Pixel size (grain)	25 ha	1 ha
Time step	2 week	Daily
Duration	365 days	180 days
Area modeled	Winter and summer range	Winter range
Vegetation classes	Eleven classes, including 5 forest types	Twelve classes, including 2 forest types
Density dependent effects	No	Yes
Migration outside of park?	Yes	No
Probabilistic responses of ungulates	No	Yes
Normal distribution of forage biomass within different vegetation types	No	Yes

in which intake is a function of forage biomass. The forage biomass that is actually available, and used in the functional response equation, is affected by snow cover (Table 13.2).

Energy requirements are derived from calculations of the costs of basal metabolism, lactation, pregnancy, travel in snow, and thermoregulation as in Hobbs (1989). An energetic reserve is considered by diminishing the over-winter energy requirements by the amount of energy reserves that can be drawn upon before a certain level of mortality occurs. For summer runs, the energy requirements are increased by the amount of reserves that must be accumulated. After Hobbs (1989), it was assumed that a loss down to 67 percent of labile fat reserves results in mortality (Table 13.2). Furthermore, 70 percent of the energy reserves come from fat. Available reserves are thus fat energy times 0.67 divided by 0.7. Using a normal distribution of animal energy reserves with a CV of 0.21, it can be shown that if 74 percent of the available reserves are utilized, then 10 percent of the population will have overused the available reserves and will die. When the LCCM is run, the user specifies the energy reserve level that is to be subtracted from forage energy requirements as a fraction of available reserves. This value is set with a certain level of mortality in mind. Normally, the model was run under an assumption of 10 percent mortality and thus a 74 percent level of reserve use.

Table 13.2. Comparisons of key assumptions made by the two models with respect to population structure and factors limiting population growth

Assumption Factor	LCCM	NOYELP
Population Structure (elk)	12 percent bulls, 77 percent cows	19 percent bulls, 65 percent cows, 20 percent calves,[a] 16 percent calves[b]
Population Structure (bison)	Not modeled	42 percent bulls, 38 percent cows, 18 percent calves[c]
Interaction between bison and elk?	No	No
Landscape factor limiting ungulate survival[d]	Forage on winter range	Forage on winter range
Physiological factor(s) limiting survival	Energy and nitrogen[e]	Energy[f]
Snow limitations on ungulate foraging	[g]$For_{avl} = For_{tot} x \dfrac{(1)}{1 + 90\, e^{-0.12Sdep}}$	Snow depth at brisket height, therefore movement of each age class affected differently, SWE over 10 cm starts to limit foraging with a linear feedback
Winter forage availability	Estimated from fall forage values	Refuge of ~15 percent initial forage is unavailable, measured values from fall forage
Maintenance energy requirement	$BW^{0.75}$[h]	$BW^{0.75}$[h]
Rates of vegetation recovery to prefire biomass levels	Sagebrush grassland to recover in 25 years, forested to recover in 11 years, meadows recovered in 3 years[a]	Measured biomass levels used from Fall 1990, 1991
Ungulate death[i]	67 percent loss of fat weight	70 percent of fat weight

Notes: LCCM is the landscape carrying capacity model and NOYELP is the Northern Yellowstone Park model.

[a]Coughenour and Singer 1996b
[b]Lemke and Singer 1989
[c]Barmore 1980
[d]Houston 1982
[e]Hobbs and Swift 1985
[f]Turner and others 1994
[g]Cassirer 1990
[h]Thompson and others 1973, Parker and others 1984
[i]Hobbs 1989

Given the energy intake rate, the frequency distributions of forage energy, and nitrogen contents, the model calculates what fraction of the forage is of sufficient energy or nitrogen content to meet an animal's requirements at the calculated rate of forage intake (Hobbs and others 1982, Hobbs and Swift 1985). As intake rate goes down due to low forage mass or deep snow, the forage must have a higher energy content to meet requirements. Thus as intake rate decreases, only a small fraction of the forage that is on the landscape would have sufficiently high energy content to be able to meet the animal's requirements. Based on the amount of forage that can meet requirements, the model calculated the number of animals that can be supported at the location and time. Mean animal days per ha is calculated as the minimum value of the following:

$$\frac{\text{Tfore}}{\text{Frate}_{\text{day}}}, \frac{\text{Tforn}}{\text{Frate}_{\text{day}}}$$

where Tfore = the total g/m^2 of forage required to meet daily energy requirements for elk at a given daily feeding rate ($\text{Frate}_{\text{day}}$) and Tforn = the total g/m^2 of forage required to meet nitrogen requirements. Tfore was always more limiting and therefore the nitrogen limitation did not come into play in the model.

The model is run over all of the 25-ha grid cells on the entire northern winter range, including areas outside the park. Although 100,000 ha are in the park, an additional 40,000 ha lie outside of the northern park boundary. For each grid cell, the model runs through time at two-week increments. This is important because snow cover varies over time and space, as do the animal's energy requirements due to stage of pregnancy or lactation, thermoregulatory costs, and costs of travel in snow. Thus for each grid cell and two-week period (Table 13.1), the model determines the number of elk-days that can be supported using the procedure described above. The results are integrated over space and time to derive a season-long estimate of elk-days that can be supported by the landscape in a given winter.

Fire effects on forage supply are simulated as a function of time since fire, using empirical data from each habitat type (Coughenour and Singer 1996a). In addition, it is assumed that the burning off of tree cover results in increased forage production for a number of years, until tree cover is once again established.

NOYELP is an individually based model layered on top of a spatially explicit landscape model. It follows the fate of individual elk (and bison) as they make foraging decisions based upon snow depth and density and forage availability. These decisions are probabilistic—that is, the animal has a greater probability of moving when snow depth or forage amount within a pixel becomes unfavorable. Hence multiple runs of the model could be followed during model experiments and the results could then be statistically analyzed. When an individual animal loses 30 percent of its lean body weight and 70 percent of its fat reserves, it dies (Hobbs 1989) (Table 13.2). This is very similar to the body weight that would result in death that was assumed in the LCCM model (for example, if an average elk bull weighed 292 kg, it would die at 260.56 kg using the LCCM calculations and at 259.3 kg using the NOYELP calculations). Movement is constrained to the winter range within park boundaries, and only the winter range conditions are considered in this model (Table 13.1). The model runs only from November through April, hence there is no opportunity for plant regrowth. However, between-year variation in forage availability can be represented easily by modifying the initial conditions of forage abundance across the landscape for a given simulation. Nutrients were not considered because forage analyses showed no nutritional differences among forage types available during the winter (Wallace and others 1995). Forage availability differs by ungulate age class and is determined by snow water equivalent (SWE), animal brisket height, and vegetation class. Most animals are limited by snow depths deeper than their knees, and movement is severely limited by snow up to the height of the brisket (Parker and others 1984). SWE is the amount of water that snow of a given depth and density would produce if it were melted. Hence the higher the SWE found in an area, the heavier the snow is, making it harder for animals to move about in or dig through the snow to reach forage. Snow conditions across the winter range are modeled based upon snow course data available from one location. Snow is then distributed across the landscape based upon aspect and slope (Turner and others 1994). For example, a flat location has the same amount of snow as would be found at the snow course site. However, on a south-facing steep slope of 15–30 degrees, only 70 percent of the snow found on a flat surface would accumulate. On a north-facing slope of 15–30 degrees, 120 percent of the snow found on a flat surface would accumulate. In unburned forests, 90 percent of the snow found on a flat surface would accumulate. Burned forest areas accumulated the same amount of snow as their nonforested counterparts. Regression equations corrected snow depth for the low-elevation sites of the northern range (areas around Mammoth Hot Springs and Gardiner, Montana).

Sensitivity analysis of the NOYELP model indicated that energy gain is the key process regulating ungulate survival (Turner and others 1994). Thus any factor governing the availability of forage, and the initial body size of the ungulates going into the winter, were critically important to model output. Similarly, energy content of forage and forage availability were critical factors in the LCCM model (Coughenour and Singer 1996a).

Partially because of its reduced spatial and temporal scope (winter season and winter range only), the NOYELP model is built at a much finer resolution than the LCCM. However, the reduced spatial scope of NOYELP results in some assumptions about ungulate survival outside of the park that differ from those of the LCCM. If ungulates were forced by snow conditions to the park boundaries, they were assumed to die. However, in reality, some elk survive leaving the park and are not shot by hunters. During the time period these models were built, any bison leaving the park were shot. Therefore this assumption results in an overestimate of elk mortality by NOYELP (Wu and others 1996).

The coarser resolution of the LCCM (25 ha versus 1 ha for NOYELP) results in some potential overestimates of elk mortality as well, with the assumption of 50 percent mortality if the population mean fat reserves are depleted to 67 percent. This reduces the effect of spatial heterogeneity, whereby pockets of resources could result in some elk doing quite well or some elk being "trapped" by winter conditions and doing very poorly.

Forage response to fire is different in the two models, as well. In the LCCM, forage regrowth back to prefire levels takes three years, whereas regrowth to prefire levels can occur in one year in the NOYELP model. This difference could result in an overestimate of ungulate mortality in LCCM if succeeding winters after a fire were mild and would result in an underestimate of ungulate mortality in NOYELP if postfire winters were severe. Both models used different forage amounts for different vegetation classes, with both showing increased forage amounts in burned forests the first year after fire.

This naturally leads to the question, "How do the predictions of these models compare with each other?" Do they identify similar controls on ungulate population size? Both models underwent a validation process where their results were compared to what actually occurred in the system (Table 13.3). Both models simulated elk survival well for years when actual weather data were available for the model run.

MODEL EXPERIMENTATION AND RESULTS

Following validation, both models were used in simulation experiments. The NOYELP model was used in a series of factorial experiments designed to determine the relative importance of winter severity, proportion of the landscape burned, patch structure of the burned areas, and elk population size at the on-

Table 13.3. Model validation results for elk population size during different winters

Year	Observed Mortality	NOYELP mortality	Comments
1987–1988	5%	0%	Prefire winter, mild winter
1988–1989	38–43%	40%	Postfire winter, severe winter
1990–1991	5%	4%	Later postfire winter, mild winter

Year	Observed Population	LCCM Population	Comments
1987	21,000	19,000	Prefire, model had actual weather data
1988	25,000	19,000	Postfire, model had actual weather data
1993	19,000	25,000	Postfire, model run on average weather data

Note: The NOYELP model (Turner and others 1994) examined percent mortality; the LCCM model (Coughenour and Singer 1996) examined total population size.

set of winter. The LCCM model was used in experiments to determine the relative effects of fire, drought, and snow on carrying capacity.

In the LCCM, snow alone in the severe winter of 1988–1989 was found to reduce carrying capacity by 50 percent (Coughenour and Singer 1996a). Drought alone reduced carrying capacity to 66–68 percent of its value for 1988–1989 by reducing forage production during the summer. Similarly the fire alone reduced carrying capacity to 80–84 percent of its value for the same year by consuming plant biomass.

Factorial simulation experiments with the NOYELP model showed that the strongest effect on elk survival the first year after a fire was winter severity, followed by fire size and fire pattern (clumped patches versus dispersed patches) (Turner and others 1994, Wu and others 1996). In these simulations, the initial sizes of the ungulate populations were set at sizes comparable to those observed at the beginning of the 1988–1989 winter. When winter conditions were very mild, even large fires (60 percent of the landscape) had no effect on ungulate population size at the end of the first winter following fire. Effects of fire on ungulate survival became important when winter conditions were average to severe, and effects were observed in both the initial and later postfire winters (Turner and others 1994). In average to severe winters, the spatial pattern of fire also was important: small dispersed burned patches reduced elk survival more than large clumped burned patches for any given total area of fire.

Weather Role in Both Models

Both models focused on forage biomass production as a critical element in elk survival. The LCCM model found that simulated forage biomass was closely tied to annual precipitation and that burning decreased sagebrush/grassland production during the first two postfire years. Alternatively, the NOYELP model did not examine forage production during the summer months, but used forage biomass measurements at the end of summer to create a normal distribution of forage biomass within a vegetation class. Forage in all vegetation classes increased the first year following a fire rather than having a lag period shown by the data used to parameterize the LCCM model (Singer and Harter 1996). Because no differences were found in wintertime forage quality after the fires (Wallace and others 1995), the NOYELP model focused on forage energy available to ungulates. The LCCM model used a normal distribution of forage protein contents (Singer and Harter 1996) to calculate carrying capacity because during the summer, there were differences in forage quantity and quality in the years following the fires. Both models showed that a large fraction of the winter range lost its forage in the fires, resulting in reduced forage supply the following winter. Therefore the weather effect on elk mortality noted by both models was due to a combination of several factors. For the LCCM model, weather affected forage availability in two ways: forage production tied to summer precipitation and forage availability tied to snow depth and duration in the winter. In the NOYELP model, the forage-related component tied to weather was primarily the availability of forage biomass due to snow depth, density, and duration.

Weather also had direct effects on elk energy expenditure in both models. As snow accumulated, more energy was required for locomotion, digging through the snow to reach forage, and for thermal regulation by elk. Also, elk typically needed to move farther to locate adequate forage supplies. Both models used a forage intake rate that decreased with increasing snow depth. The LCCM model intake rate was based on a type II functional response, whereas the NOYELP model used a logistic response to snow water equivalent with upper and lower thresholds.

Interaction of Fire and Weather with Ungulate Population Size

An extensive factorial simulation experiment was conducted with the NOYELP model to explore the interaction of ungulate population size with fire size and pattern and winter severity. Three winter severities were compared. The most mild (1976–1977) and most severe (1975–1976) winters recorded during the twentieth century with respect to snow conditions were identified from weather records. Conditions during these winters bracket known weather extremes. In addition, the NOYELP model simulated an "average" winter in which the mean monthly snow conditions were computed across all winters for which data were available. Nine levels were specified for the size of simulated fires by specifying the percentage, p, of the study area burned. In this experiment, p ranged from 10 to 90 in intervals of 10 (the 1988 fires burned ca. 22 percent of the northern range). In addition, two alternative fire patterns—fragmented or clumped in a single patch—were generated for each of the p values to bracket the extremes in spatial patterns of fire. Finally, the initial number of elk in the northern range was simulated at five levels: 6000, 12,000, 18,000, 24,000, and 30,000. Approximately 18,000 elk and 600 bison were present on the northern range at the beginning of the winter of 1988–1989 (Singer and others 1989), and we included both larger and smaller population sizes. Bison population sizes were adjusted simultaneously with the elk, for initial population sizes of 200, 400, 600, 800, and 1000. Simulations were run for both the initial postfire winter, when burned areas contained no forage, and the later postfire winter, when burned areas had enhanced forage quantities. Three replicates with different random number seeds were run for each scenario. Results were analyzed separately for initial and later postfire years using ANOVA (SAS 1990).

Density-dependent thresholds for the effect of fire size on the proportion of elk surviving the winter occurred during the initial postfire winter (Fig. 13.1). When the winter was mild, ungulate survival dropped precipitously from 100 percent to zero or very low levels at different fire sizes, depending on population density. Smaller population sizes survived larger fires, and for the same elk population and fire size, survival was usually greater with clumped fires compared with fragmented fires. For bison, the effects of fire pattern were statistically but not biologically significant (Wallace and others 1994, Fig. 13.2). Note that the difference in percent survival for all age groups was very small. Thresholds of survival were also observed in the average and mild winters and were most pronounced for the smaller population sizes and larger fires. When population densities were high (> 18,000 elk and 800 bison), survival was very low even with the smaller fires if winter conditions were average to severe (Fig. 13.1, Wallace and others 1995). ANOVAs revealed that most of the variation in elk survival during the initial postfire winter was due to variation in winter severity, with ungulate population size and fire size having secondary effects. During later postfire winters, when burned areas generally provided more abundant forage than

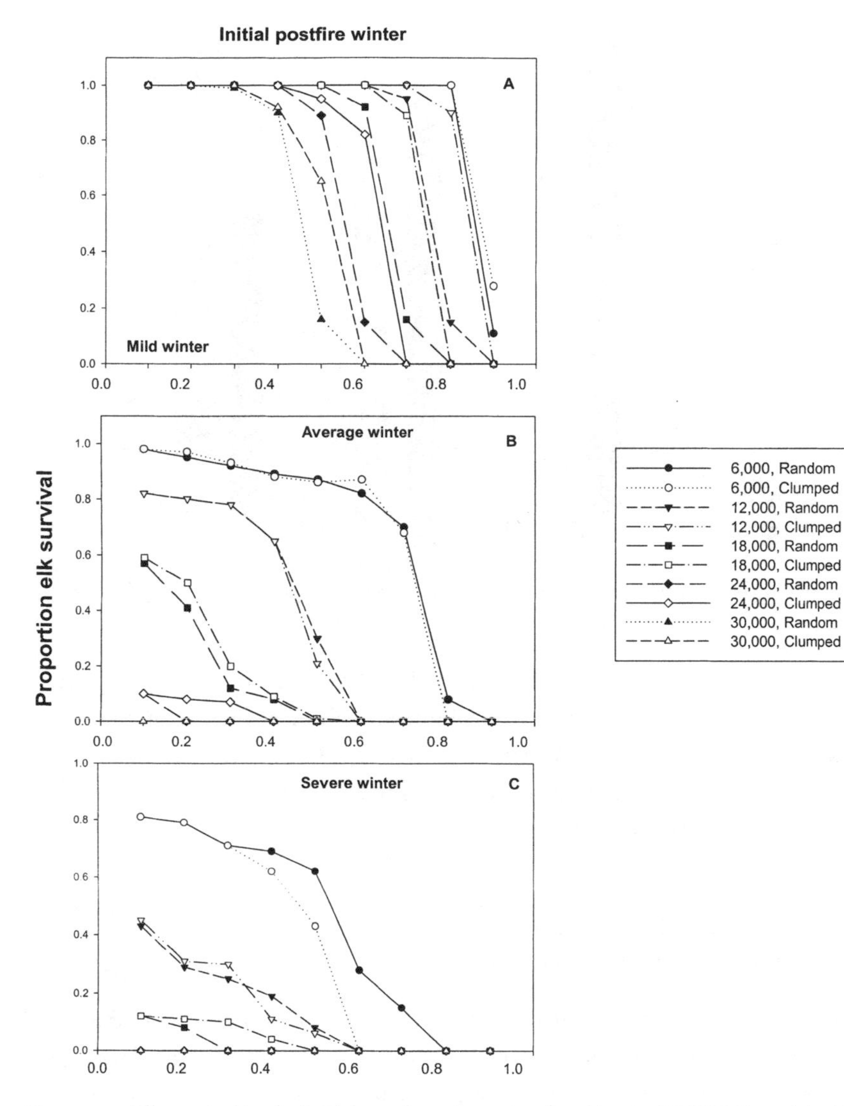

Figure 13.1. Elk survival in the initial postfire winter simulated by the NOYELP model with mild, average, and severe winter conditions under either random or clumped fire patterns with increasing proportions of the winter range landscape being burned indicated on the x-axis.

comparable unburned sites, elk survival was 100 percent at the lower population densities. When postfire winter conditions were average or severe, elk survival increased with fire size (Fig. 13.3). The increase in survival attributable to fire size (and forage enhancement) was most pronounced for the highest population sizes (for example, 24,000 and 30,000 elk and 800 or 1,000 bison during

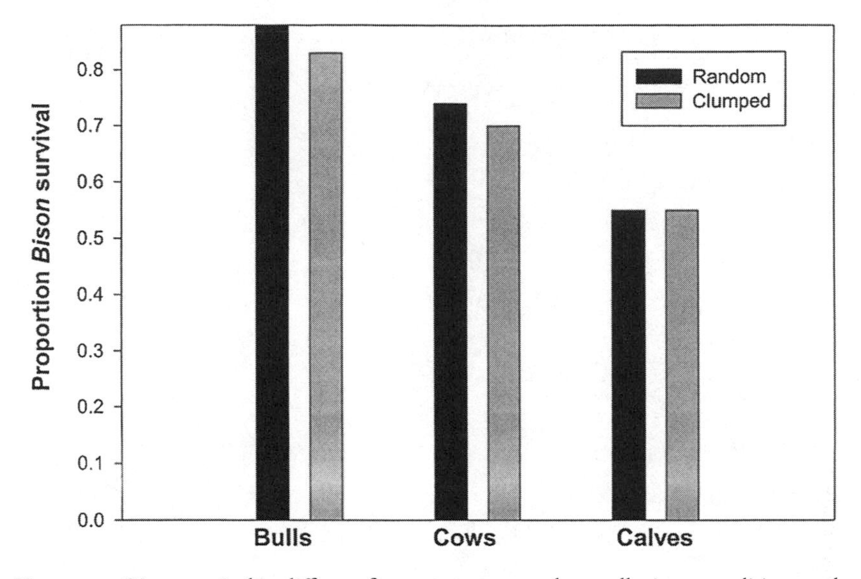

Figure 13.2. Bison survival in different fire patterns averaged over all winter conditions and years.

the average winter). Threshold effects with population sizes were not apparent during the later postfire winters, but there was substantial variation in survival associated with population density (Fig. 13.3). For bison, threshold effects were also minimized, but dramatic effects of winter severity were seen on survival, particularly for calves (Fig. 13.4).

WHAT THESE MODEL RESULTS MEAN FOR FIRE ECOLOGISTS AND LAND MANAGERS

There are substantial differences between the two models in terms of their structure and the objectives for constructing them. The NOYELP model was based on an elk and bison energetics, movement, and spatial variation in winter forage availability and was constructed primarily to investigate the relative importance of fire size and pattern in determining winter survival of ungulates on the northern range. The LCCM model looked at the carrying capacity of the environment (both nutrients and energy) across the entire summer and winter range. Its primary objective was to predict how the carrying capacity of the environment relative to elk changed over time and how quickly the system would return to prefire levels. Given that these models were developed in isolation from each other, it is important to note that both show that weather is the single most important factor governing elk survival in the Greater Yellowstone Ecosystem.

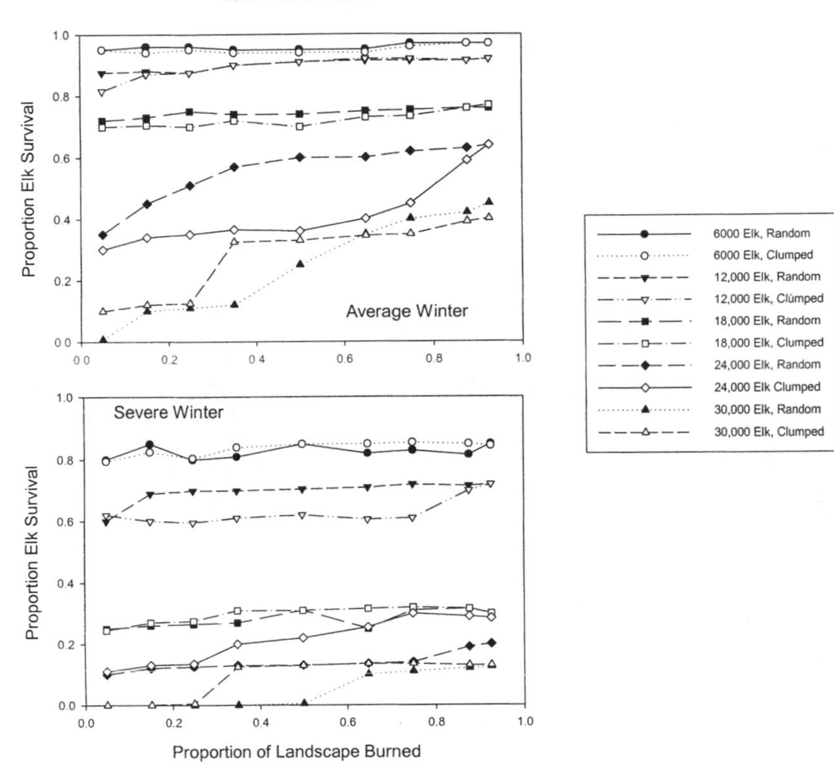

Figure 13.3. Elk survival in the second winter postfire simulated by the NOYELP model with average or severe winter conditions under either random or clumped fire patterns with increasing proportions of the winter range landscape being burned indicated on the x-axis.

Unfortunately, it is also the factor that is the least predictable and under the least control by land managers. Although the fires of 1988 were not under the control of land managers, a future prescribed burn would most likely have little, if any, effects on elk and bison survival in this landscape—unless it occurred under conditions of high ungulate populations and was followed by a severe winter (Coughenour and Singer 1996a, Turner and others 1994, Wallace and others 1994).

Fire patterns had no effects on ungulate survival when fire size was large in NOYELP simulations (Turner and others 1994) with no responses measured in the LCCM model. Fire pattern can be controlled in prescribed fires, which presumably would be much smaller than the large wildfires of 1988. When fire size is small to moderate, more ungulate survival is noted for clumped rather than

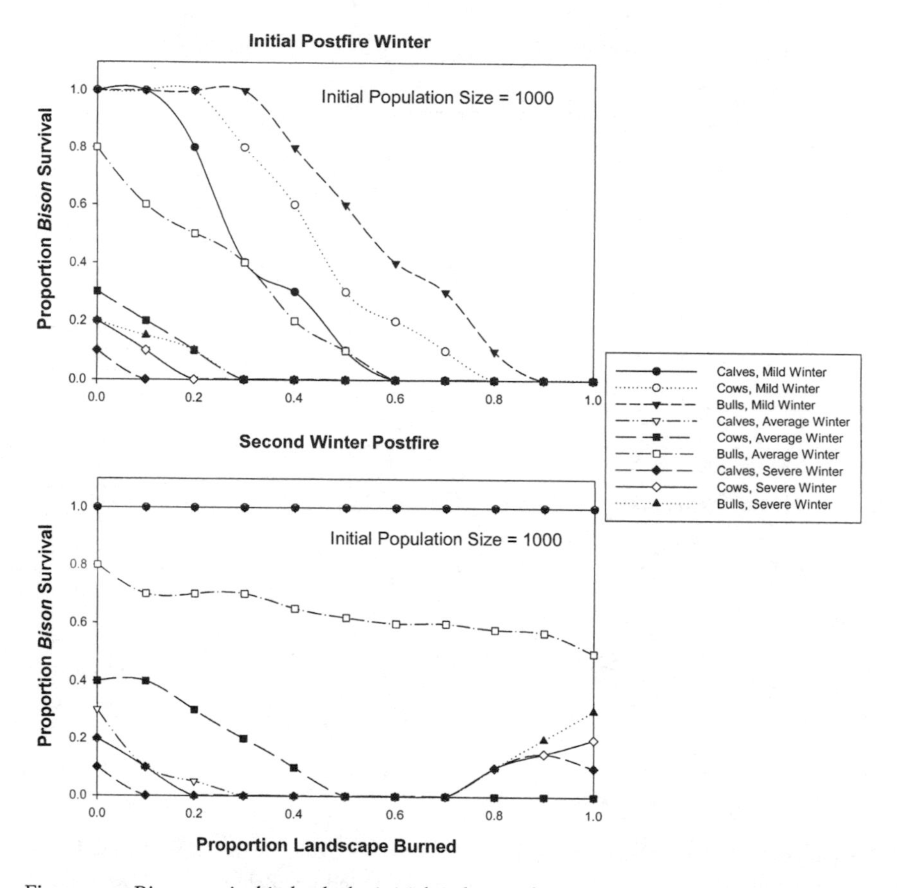

Figure 13.4. Bison survival in both the initial and second winter postfire. Initial population size for these analyses was 1000. Population responses to decreasing initial population size were similar to what was seen for elk.

dispersed fire patterns (Turner and others 1994). Thus managers may enhance survival the first winter postfire as well as subsequent winters by burning clumped rather than dispersed areas. This is more evident for elk than for bison (Wallace and others 1994).

The effects of weather can be readily seen in the LCCM model (Coughenour and Singer 1996a). Coughenour and Singer predicted elk population dynamics under a stochastic set of weather scenarios, by running the model with alternative stochastic weather patterns beginning in 1991 and going through the year 2011 using 1972–1991 weather data. Given the overwhelming importance of weather in determining elk and bison (Wallace and others 1994, Coughenour and Singer 1996a) survival in this ecosystem, it is not surprising that the predicted results of these stochastic runs are very different from the actual observed population sizes for the years 1991–2000. Another factor that cannot be predicted is the success of annual hunting efforts just outside of the park, which were also great during this time frame. When one compares the model results to the actual elk counts obtained by scientists in Yellowstone Park (Lemke and others 1998) for some of these years after the model was run, we see that the model estimates of carrying capacity are larger than observed elk numbers by 20,000 or more animals. However, in some other years, carrying capacity estimates are far below observed numbers of elk. It is important to point out that the LCCM is not predicting population dynamics; it is only predicting forage-based carrying capacity. During 1972–1991, another reason for the seeming discrepancy between model predictions and observed elk populations is that Yellowstone experienced a series of mild winters, a pattern that has not been seen since (Farnes and Hartman 1989, Chapter 10).

Additional simulations with the NOYELP model that explored the implications of spatial heterogeneity at different scales on winter survival of elk and bison demonstrated the critical role of habitat heterogeneity in determining forage availability (Turner and O'Neill 1995). Even though the total amount of forage available on the landscape remained constant in these simulations, removal of spatial variation in forage substantially reduced simulated elk survival. Thus the landscape-level distribution of areas with high and low forage biomass can enhance ungulate survival. Similarly, spatial heterogeneity in snow conditions—especially the maintenance of snow-free or low-snow areas—was demonstrated as being critical for winter ungulate survival. Thus managers should be attentive to maintaining the ecological processes responsible for producing and maintaining spatial heterogeneity in forage resources.

The effects of fire or winter weather on ungulate population size will be modified by the population size at the beginning of the winter. The proportion of animals that survive is density dependent, with survival being higher for lower population densities when conditions produce a reduction in the available forage. Migration out of the park may occur, however, and this was not explicitly simulated in either model.

When predicting ungulate survival in wild land ecosystems in the Northern Rocky Mountains, land managers need to realize that those factors over which they have the most control will be of the least help in their predictions. Weather is the most limiting factor to survival in this environment: summer weather strongly influences forage production, and winter weather largely controls forage availability. Therefore weather conditions will fundamentally determine whether the environment can support elk and bison well. If it cannot, then those individual animals that find areas of low snow accumulation with adequate forage will survive. Animals elsewhere in the ecosystem will be at risk. Things that managers can control, such as hunting pressure outside protected reserves and prescribed fire within reserves, also may influence elk and bison populations—but the effects of these other controls are constrained by the overarching effects of weather.

In both models, the factor ultimately limiting ungulate survival was energy (Coughenour and Singer 1996a, Turner and others 1994). Why would this limitation yield the responses seen here, the overwhelming effects of climate on survival rather than the effects of fire? As discussed above, the spatial heterogeneity provided by the patch structure of fire may be crucial in forage availability, provided that the total proportion of the landscape burned is not too great (Turner and others 1994, Wallace and others 1994, Turner and O'Neill 1995). The effects of fire in reducing forage availability (energy) are much more localized than the effects of climate in reducing forage availability (energy). Thus it makes sense that both models found energy to be limiting to survival during the winter and that the effects of climate would supercede those of fire.

If land managers wish to predict what will occur in wild land systems in the face of global climate change, they need to consider several factors. The observed values of elk and bison population growth show that both species are capable of rapid population growth, if they experience a series of mild winters. Global change models (Bartlein and others 1997) indicate that conditions will vary across the park but that generally warmer, wetter winters and warmer summers will occur. No change in summer precipitation is called for, a condition that could lead to greater frequency and extent of summer fires. Even then, it would appear, given the results of these separate modeling exercises, that winter weather will be more critical to elk survival than would be the occurrence of fire in this system. However, the survival of elk and bison in this landscape will be difficult to predict if winters are warmer with deeper snows and there is a more widespread occurrence of low-elevation vegetation types (Romme and Turner 1991, Bartlein and others 1997).

Sala and others (2000) suggest that the single greatest factor affecting boreal systems (where Yellowstone Park is classified in their global classification system) will be changes in land use patterns. However, if the park continues to exist, this factor will be of less importance, making climate change the single greatest driver of biodiversity change in this system, followed by nitrogen deposition, biotic exchange and atmospheric CO_2. This assumes that tremendous exurban landscape development does not occur in the areas immediately outside of the park. If such development does occur, then land use change will be a critical factor along with climate change. This corroborates the predictions of Bartlein and others (1997) and again points to how critical it is to understand the effects of weather on ungulate survival in this landscape.

Little work has analyzed how global change will affect fire cycles directly (see Chapter 3). However, if, as Bartlein and others (1997) suggest, we have greater landscape heterogeneity, we may expect the relative severity of fire to be greater when both mesic and xeric elements of the landscape are more juxtaposed (He and Mladenoff 1999). Ungulate survival may be more influenced by changes in plant community structure, phenology, winter weather, and possible changes in land use outside of the park than by changes in fire severity (Price and Waster 1998, Post and Stenseth 1999).

In summary, fire is a critical and recurring element in the ecology of the Greater Yellowstone Ecosystem. Large-scale fires have occurred in the past (Chapters 2, 3) and will probably occur again in the future despite efforts at human intervention (Christensen and others 1989, Pyne 2000). Concern over fires in this landscape will continue to be great as long as humans occupy it. However, in terms of ungulate survival, the scenario popularized by the Disney film *Bambi* is not completely valid. Fire is not the primary driver of ungulate survival. As the landscape is currently structured, weather is the single most important element driving ungulate survival. If landscape use changes as described by Sala and others (2000), then ungulate survival may become more closely tied to habitat loss as we have seen occur in numerous other boreal environments. If Yellowstone Park continues to exist with some buffer zone surrounding it, then the current and long-term variability of winter weather will remain the key to elk and bison survival in this extraordinary environment.

References

Barmore, W. J. 1980. Population characteristics, distribution, and habitat relationships of six ungulates in northern Yellowstone National Park. Final Report, Yellowstone National Park files.

Bartlein, P. J., C. Whitlock, and S. L. Shafer. 1997. Future climate in the Yellowstone National Park region and its potential impact on vegetation. Conservation Biology 11:782–792.

Cassirer, E. F. 1990. Responses of elk to disturbance by cross-country skiers in northern Yellowstone National Park. M.S. thesis, University of Idaho, Moscow.

Christensen, N. L., J. K. Agee, P. F. Brussard, J. Hughes, D. H. Knight, G. W. Minshall, J. M. Peek, S. J. Pyne, F. J. Swanson, J. W. Thomas, S. Wells, S. E. Williams, and H. A. Wright. 1989. Interpreting the Yellowstone fires of 1988. BioScience 39:678–685.

Coughenour, M. B. 1989. Grazing responses of upland steppe on Yellowstone's northern winter range. Final Report to the National Park Service.

Coughenour, M. B. 1991. Biomass and nitrogen responses to grazing of upland steppe on Yellowstone's northern winter range. J. Appl. Ecol. 28:71–82.

———. 1994. Elk carrying capacity on Yellowstone's northern elk winter range: Preliminary modeling to integrate climate, landscape, and elk nutritional requirements. Pages 97–122 in D. Despain, ed., Plants and their environments: Proceedings of the First Biennial Scientific Conference on the Greater Yellowstone Ecosystem, Mammoth Hot Springs, September 1991. Technical Report NPS/NRYELL.NRTR-93/XX. USDI/NPS, Denver.

Coughenour, M. B., and F. J. Singer. 1996a. Yellowstone elk population responses to fire: A comparison of landscape carrying capacity and spatial-dynamic ecosystem modeling approaches. Pages 169–179 in J. M. Greenlee, ed., Ecological implications of fire in Greater Yellowstone. International Association of Wildland Fire, Fairfield, Wash.

Coughenour, M. B., and F. J. Singer. 1996b. Elk population processes in Yellowstone National Park under the policy of natural regulation. Ecol. Appl. 6:573–593.

Farnes, P. E., and R. K. Hartman. 1989. Estimating effects of wildfire on water supplies in Northern Rocky Mountains. Pages 90–99 in Proceedings: Western Snow Conference, April 18–20, 1989, Fort Collins, Colo.

Haines, A. L. 1977. The Yellowstone story: A history of our first national park. Yellowstone Library and Museum Assoc. and Colorado Assoc. Univ. Press, Yellowstone National Park, Wyoming.

He, H. S., and D. J. Mladenoff. 1999. Spatially explicit and stochastic simulation of forest-landscape fire disturbance and succession. Ecology 80:81–99.

Hobbs, N. T. 1989. Linking energy balance to survival in mule deer: Development and testing of a simulation model. Wildlife Monographs no. 101.

Hobbs, N. T., D. L. Baker, J. E. Ellis, D. M. Swift, and R. A. Green. 1982. Energy- and nitrogen-based estimates of elk winter range carrying capacity. J. Wildl. Manage. 46:12–21.

Hobbs, N. T., and D. M. Swift. 1985. Estimates of habitat carrying capacity incorporating explicit nutritional constraints. J. Wildl. Manage. 49:814–822.

Houston, D. B. 1982. The Northern Yellowstone elk: Ecology and management. Macmillan, New York.

Lemke, T., J. A. Mack, and D. B. Houston. 1998. Winter range expansion by the Northern Yellowstone elk herd. Intermountain Journal of Sciences 4:1–9.

Lemke, T., and F. J. Singer. 1989. Northern Yellowstone elk: The big herd. Bugle, Fall 1989, 113–121.

Parker, K. L., C. T. Robbins, and T. A. Hanley. 1984. Energy expenditures for locomotion by mule deer and elk. J. Wildl. Manage. 48:474–488.

Post, E., and N. C. Stenseth. 1999. Climatic variability: Plant phenology and northern ungulates. Ecology 80:1322–1339.

Price, M. V., and N. M. Waser. 1998. Effects of experimental warming on plant reproductive phenology in a subalpine meadow. Ecology 79:1261–1271.

Pyne, S. J. 2000. Green skies of Montana. Forest History Today, Spring 2000, 37–38.

Romme, W. H., and M.G. Turner. 1991. Implications of global climate change for biogeographic patterns in the Greater Yellowstone Ecosystem. Conservation Biology 5:373–386.

Sala, O. E., F. S. Chapin, J. J. Armesto, E. Berlow, J. Bloomfield, R. Dirzo, E. Huber-Sanwald, L. F. Huenneke, R. B. Jackson, A. Kinzig, R. Leemans, D. M. Lodge, H. A. Mooney, M. Oesterheld, N. L. Poff, M. T. Sykes, B. H. Walker, M. Walker, and D. H. Wall. 2000. Global biodiversity scenarios for the year 2100. Science 287:1770–1774.

SAS 1990. SAS procedures guide, Version 6. SAS Institute, Cary, N.C.

Schullery, P. 1997. Searching for Yellowstone: Ecology and wonder in the last wilderness. Houghton Mifflin, New York.

Singer, F. J., and M. K. Harter. 1996. Comparative effects of elk herbivory and 1988 fires on Northern Yellowstone National Park grasslands. Ecol. Appl. 6:185–199.

Singer, F. J., W. Schreier, J. Oppenheim, and E. O. Garton. 1989. Drought, fire, and large mammals. BioScience 39:716–722.

Thompson, C. B., J. B. Holter, H. H. Hayes, H. Silver, W. E. Urban. 1973. Nutrition of white-tailed deer, part 1, Energy requirements of fawns. J. Wildl. Manage. 37:301–311.

Turner, M. G., and R. V. O'Neill. 1995. Exploring aggregation in space and time. Pages 194–208 in C. G. Jones and J. H. Lawton, eds., Linking species and ecosystems. Chapman and Hall, New York.

Turner, M. G., Y. Wu, W. H. Romme, and L. L. Wallace. 1993. A landscape simulation model of winter foraging by large ungulates. Ecol. Modelling 69:163–184.

Turner, M. G., Y. Wu, L. L. Wallace, W. H. Romme, and A. Brenkert. 1994. Simulating winter interactions among ungulates, vegetation, and fire in Northern Yellowstone National Park. Ecol. Appl. 4:472–496.

Wallace, L. L., M.G. Turner, W. H. Romme, R. V. O'Neill, and Y. Wu. 1995. Scale of heterogeneity of forage production and winter foraging by elk and bison. Landscape Ecology 10:75–83.

Wallace, L. L., M. G. Turner, W. H. Romme, Y. Wu. 1994. Bison and fire: Landscape analysis of ungulate response to Yellowstone's fires. Pages 79–120 in Proceedings: North American Public Bison Herds Symposium, July 27–29, 1993, LaCrosse, Wis.

Wu, Y., M. G. Turner, L. L. Wallace, and W. H. Romme. 1996. Elk survival following the 1988 Yellowstone fires: A simulation experiment. Natural Areas Journal 16:198–207.

Chapter 14 Ten Years After the 1988 Yellowstone Fires: Is Restoration Needed?

William H. Romme and Monica G. Turner

Coniferous forests in many parts of western North America have been substantially altered by human activities in the past 150 years and are now at risk of abnormally severe disturbance, impaired ecosystem function, and loss of biodiversity. For example, ponderosa pine forests in many locations today are characterized by unusually high densities of small-diameter trees, a paucity of large trees and snags, severely suppressed herbaceous plants, and excessive quantities of well-connected live and dead fuels. These changes are the result primarily of fire exclusion, excessive livestock grazing, and unsustainable logging activities conducted in the past. Efforts are now under way to restore missing ecological structure, processes, and species in these degraded ecosystems via mechanical thinning, prescribed fire, or a combination of thinning and fire (for example, Covington and Moore 1994, Hardy and Arno 1996, Fule and others 1997, Lynch and others 2000, Allen and others 2002, Friederici 2003).

In the wake of the past several devastating fire years, the public and policymakers have called for accelerated efforts to reduce the risk of uncontrollable wildfires in western coniferous forests by means of logging, prescribed fire, and other restoration techniques. However, it is critical that restoration efforts focus on ecosystems where structure and processes really have been altered by human activities. Although ponderosa pine forests appropriately have received much attention, some other types of coniferous forest in western North America have not been much altered from their natural range of structure, composition, and ecological function. Mechanical thinning, prescribed burning, and other attempts to "restore" these systems would be misguided at best and potentially destructive at worst (Cole and Landres 1996, Cole 2000).

The fires of 1988 were the largest fires in recorded history in the Yellowstone region. The size and severity of the 1988 fires led many people at the time to conclude that the park had been "destroyed" and that the National Park Service had been remiss in not actively intervening to prevent such fires prior to 1988—for example, by conducting programs of mechanical thinning or manager-ignited prescribed burning. Our objective in this chapter is to evaluate these assertions—in other words, to test the proposition that the 1988 Yellowstone fires were an abnormal event, and that they severely damaged park ecosystems. Based on an examination of Yellowstone's fire history during the past several centuries (Chapters 2, 3), plus the postfire responses that we have measured since 1988, we argue that Yellowstone's coniferous forest ecosystems were not degraded or altered significantly in any abnormal ways, either by the 1988 fires or by pre-1988 management activities. We further suggest that many other high-elevation wilderness areas in western North America are similar, in other words, they remain within their range of natural variation (Landres and others 1999) and do not require active intervention to restore natural conditions.

It is important to note that we restrict our analysis to the high-elevation forested plateaus and mountains that cover approximately 80 percent of Yellowstone National Park plus extensive contiguous areas outside the park boundaries (which we refer to as the Yellowstone Plateau). We do not deal with lower-elevation ecosystems in Yellowstone's northern winter ungulate range, where climate, vegetation, and ecological history are very different from the high-elevation systems upon which we focus. Nor do we challenge the interpretations of ecological change and need for restoration in ponderosa pine and some other low-elevation forest types in western North America. On the contrary, we emphasize that assessments of ecosystem "health" must explicitly consider the ecological characteristics and histories of different forest types, and that management goals and methods must be tailored to unique local conditions (Dahms and Geils 1997).

As a framework for our assessment, we pose a set of general criteria by which

Table 14.1. Criteria for evaluating the need for active restoration

1. Is the current disturbance regime within the historical or natural range of variability?
2. Are current stand structure and landscape structure within the historical or natural range of variability?
3. Are any species or communities extinct or threatened with extinction because of alterations in the disturbance regime?
4. Have recent disturbances been accompanied by normal return of community structure and composition?
5. Have recent disturbances been accompanied by normal return of ecosystem function, for example, energy flow and material cycling?
6. Have recent disturbances been associated with any novel or unexpected effects that are regarded as undesirable?

we can evaluate the need for active intervention to restore degraded ecological conditions (Table 14.1). These criteria may also be useful in other ecosystems to determine whether restoration efforts are needed. Harig and Bain (1998), Moore and others (1999), and others have utilized similar approaches in both aquatic and terrestrial ecosystems.

IS THE CURRENT DISTURBANCE REGIME WITHIN THE RANGE OF NATURAL VARIATION?

Fire is the most important natural disturbance in high-elevation forested landscapes of Yellowstone National Park (Despain 1990). Most lightning-caused fires are small (< 1 ha) and produce little ecological change (Renkin and Despain 1992). However, infrequent large, severe fires create and maintain a mosaic of patches of different stand ages and stages of postfire recovery. This mosaic produces a landscape pattern that is very striking when viewed from the air and that probably influences a variety of ecological characteristics and processes (Romme and Knight 1982, Knight and Wallace 1989, Foster and others 1998). Tree-ring evidence shows that a fire regime dominated by infrequent but large, severe fires has shaped the high-elevation Yellowstone landscape for at least the past three centuries (Romme and Despain 1989). Charcoal and other remains in lake sediments further reveal that infrequent large fires occurred for many centuries prior to the earliest fires documented in tree rings (Millspaugh and others 2000, Chapters 2, 3).

The occurrence of large fires is controlled primarily by regional weather patterns, though fuel conditions also are important under certain weather conditions (Renkin and Despain 1992, Turner and Romme 1994). Large fires occur only in years with prolonged summer droughts, especially when accompanied by many windy days (Johnson and Wowchuk 1993). The year 1988 was such a time: summer precipitation in Yellowstone Park was 20 percent of normal in June, 79 percent in July, and 10 percent in August. Overall, 1988 was the driest year on record in the park (Christensen and others 1989). In addition to drought, winds exceeding 40 mph were recorded day after day (Rothermel 1991). This distinctive fire regime, which characterizes the high-elevation landscapes of Yellowstone Park, is similar to the fire regimes of boreal forests and other high-elevation or very moist temperate coniferous forest ecosystems (for example, Johnson 1992, Hemstrom and Franklin 1982).

Was Yellowstone's fire history significantly different during the time of park management than during the preceding period without substantial Euro-American influence (Chapter 3)? The period of park management extends from the late nineteenth century through the twentieth century. Although Yellowstone was designated a National Park in 1872, it received little or no fire protection until the U.S. Army assumed management responsibilities in 1886 (Schullery 1997). We regard the century prior to the army's arrival as a reference period characterized by "natural" ecological conditions—at least with respect to fire. This period, from the late eighteenth through late nineteenth century, is a suitable reference period for evaluating the current fire regime because we have reasonably detailed fire history information, and environmental conditions, notably climate, were generally similar (but see Pickett and Parker 1994 for important caveats about this approach).

This is not an easy question to answer. In many southwestern ponderosa pine forests, researchers have applied rigorous statistical models to high-resolution fire history data obtained from tree rings, and have demonstrated that twentieth-century fire intervals have been significantly longer than fire intervals of the eighteenth and nineteenth centuries (for example, Grissino-Mayer 1999, Fule and others 1997). However, for two reasons a rigorous statistical analysis of this kind is problematical in systems—like Yellowstone—characterized by very long fire intervals. First, fire intervals at the scale of individual stands are on the order of decades or centuries even under natural conditions. Therefore even though we have a reasonably detailed 350-year fire history record based on tree rings (Romme and Despain 1989), this record contains too few fire intervals to permit a robust statistical analysis of central tendencies and variance. Second, if we attempt to solve the first problem by extending the fire history record farther back in time based on fire events recorded in lake bottom sediments, we lose

Table 14.2. Occurrence of large fires, by decade since the 1790s, within the 130,000-ha study area in central Yellowstone National Park where Romme and Despain (1989) reconstructed fire history

Decade	Did a Large Fire (>400 ha) Occur in One or More Years?
Period of park management	
1980s	Yes (in 1981 and 1988)
1970s	No
1960s	No
1950s	No
1940s	Yes (in 1940 and 1949)
1930s	Yes (in 1931)
1920s	No
1910s	Yes (in 1910 and 1919)
1900s	No
1890s	No
Premanagement period	
1880s	Yes
1870s	Yes
1860s	Yes
1850s	Yes
1840s	Yes
1830s	No
1820s	No
1810s	Yes
1800s	Yes
1790s	Yes

Note: Large fires are defined as those that burned >400 ha within the study area. Individual fire years are listed for the twentieth century, but precise years are uncertain for earlier fires (see text). The period from the 1890s through the 1980s is regarded as the period of park management; the previous century from the 1790s through the 1880s is the premanagement reference period.

temporal resolution and also encounter significant changes in fire frequency associated with century-scale, regional climatic variation (Millspaugh and others 2000, Chapters 2, 3).

Given these constraints on statistical analysis of Yellowstone's fire history, we made a conservative, semiquantitative assessment based on the frequency of decades in which large fires occurred during the period of park management compared with the preceding reference period (Table 14.2). The focus was on large fires rather than all fires because a few large fires generally account for most of the area burned in any century (Johnson 1992), and our fire history reconstructions based on tree rings were focused on detecting large fire events rather than all past ignitions. We chose 400 ha as a minimum fire size for this analysis, because nearly all early fires of this extent would be readily detectable in the landscape mosaic, even if portions had been reburned by later fires. The analysis was restricted to Romme and Despain's (1989) 130,000-ha study area in central Yellowstone National Park, for this is the only portion of the Yellowstone Plateau where we have detailed information on fire occurrence and fire size in the period before written records were kept. (See Houston 1973 and Barrett 1994 and Chapters 2 and 3 for fire history data from other physiographic regions of the park.) Although Romme and Despain (1989 and unpublished data) detected fires as long ago as the 1400s, our analysis here deals only with fires since 1780 because some of the area burned in earlier fires was reburned by later fires, thus obscuring the actual extent of those earlier fires. We identified decades in which large fires occurred rather than individual fire years, because this reflects the resolution of our pre-1900 fire history data: fire scars are uncommon in Yellowstone's subalpine forests, so Romme and Despain (1989) mapped the spatial extent of pre-1900 fires primarily on the basis of the current age structure of dominant trees, tying those age structure data to precise fire-scar dates wherever available. Where we have only the age of the dominant canopy trees, we can determine actual fire dates to within about a decade but cannot distinguish between two or more separate fires within a single year or fires that burned within a few years of each other.

Large fires (greater than 400 ha within the boundaries of the study area) occurred in four of the ten decades between 1890 and 1989 and in eight of the decades from 1790 through 1889 (Table 14.2). The longest intervals without any large fires were three decades in the late twentieth century (1950s–1970s), two decades at the turn of the twentieth century (1890s–1900s), and two decades in the early nineteenth century (1820s–1830s). The three-decade period without large fires in the twentieth century coincides with a period of consistently effective fire suppression in Yellowstone Park. This period began after World War II, when dramatic improvements were made in fire detection and firefighting technologies, and ended in the 1970s when the park implemented a natural fire management program (Schullery 1989). However, the lack of large fires from 1890 through 1909 probably was more a result of wet summer weather conditions than active fire suppression, since the technology for detecting and suppressing fire was not well developed at that time (Schullery 1989, 1997). The relatively fire-free period in the 1820s–1830s was undoubtedly attributable mainly to regional climatic conditions of the time. A paucity of large fires has been documented for this same time period in the American Southwest (Swetnam and Betancourt 1998)—further supporting the idea of regional climate as a primary control on the occurrence of large fires in western North America.

From this analysis we see that the frequency of decades with large fires during the century of park management was about half the frequency during the previous, premanagement century. Especially in the 1950s–1970s, effective fire control undoubtedly prevented some fires that would have been relatively large in the absence of suppression, even if the impact of fire suppression earlier in the century is equivocal. But do those "missing" fires from the 1950s–1970s really make much difference in the long-term history of this landscape? We think not. The 1988 fires probably encompassed some of the areas that would have burned in the 1950s–1970s without suppression. And although burning some of those areas a few decades earlier, rather than all in 1988, might have led to somewhat greater heterogeneity in successional stages and stand structures across the burned landscape, the post-1988 landscape is remarkably heterogeneous (see below). Moreover, large fires (greater than 400 ha) also have occurred in the park in the 1990s and in 2000 and 2001. Finally, although the 1988 fires were seemingly unprecedented in sheer size—they were more extensive than the fires in any other year of the preceding two centuries—the tree-ring data reveal very large fires in the early 1700s that may have been comparable to the 1988 fires (Romme and Despain 1989). We really cannot trace the exact extent of those early-1700s fires because an unknown number of the trees they scarred and of the forests that originated in their aftermath were destroyed by later fires.

We conclude, therefore, that Yellowstone's high-elevation fire regime remains within or nearly within its range of natural variation, and that twentieth-century fire suppression has produced no seriously adverse ecological changes. Thus no attempt at restoration of the fire regime is needed. In fact, it is hard to imagine how we even could "manage" Yellowstone's natural fire regime, controlled as it is mainly by regional climatic patterns rather than local ignitions or fuel conditions. However, we caution that anticipated global climate changes could lead to significant shifts in the disturbance regimes of Yellowstone and many other areas during the new century (Street 1989, Overpeck and others 1990, Balling and others 1992, Flannigan and Van Wagner 1991, Gardner and others 1996, Chapter 2). The question of whether Yellowstone's disturbance regime remains within its range of natural variation should be addressed again in a few decades.

ARE STAND AND LANDSCAPE STRUCTURE WITHIN THE RANGE OF NATURAL VARIABILITY?

Recent studies in southwestern ponderosa pine forests have shown that tree densities and size class distributions in many stands are far outside the range of natural variability in stand structure because of past grazing, logging, and fire exclusion (for example, Fule and others 1997, Friederici 2003). Although we lack detailed data on past stand structures in Yellowstone's subalpine forests, we have observed nothing here that would suggest abnormal conditions in the twentieth century. For example, photographs taken in the late nineteenth century depict lodgepole pine stands that generally resemble the stands one sees today (Meagher and Houston 1998). A central concern in southwestern ponderosa pine forests is excessive tree density in current stands (for example, Covington and Moore 1994). Many stands in Yellowstone are very dense and have tremendous fuel loads, but these are normal conditions in forests that naturally persist for very long times without major disturbance. We also know that such stand conditions are not unique to the twentieth century. Early European explorers in Yellowstone described similar conditions more than a century ago. For example, after traveling from the Lower to the Upper Geyser Basin in 1875, General W. E. Strong wrote, "We had to pick our way over fallen timber and through dense thickets of pine. . . . We emerged from this labyrinth of down timber." (Strong 1968:72–73). Therefore at the scale of individual subalpine forest stands, current stand structure in Yellowstone apparently is within the range of natural variability.

But what can be said about landscape-scale patterns? Did fire suppression in the twentieth century permit an abnormal area of older forest to develop before the fires of 1988? Is the area of recently burned forest now excessive? We address these questions by examining the mosaic of forest successional stages that existed at various times during the past three hundred years within a 130,000-ha study area in central Yellowstone Park (fig. 2 in Romme and Despain 1989). This historical reconstruction indicates substantial natural variation in proportions of early, middle, and late stages in forest succession. Yellowstone's subalpine landscape is a nonequilibrium system with high variance in landscape structure (Turner and others 1993); no single equilibrium point can serve as a reference condition. Instead, studies of ecological history can help us identify a range of natural variability in landscape structure, and that range can then serve as the reference condition for evaluating current conditions (Pickett and Parker 1994, Cole and Landres 1996, Swetnam and others 1999).

In the mid-1980s, the area occupied by old forests was as high as it had been at any time during the previous three hundred years (Romme and Despain 1989). However, this pattern cannot be explained entirely or even primarily by twentieth-century fire suppression. A gradual increase in area of old forests began in the late 1800s, long before there was effective fire suppression on Yellowstone's remote high plateaus, because of natural successional processes occurring after very extensive fires that had burned in the early 1700s.

The extent of old forests has been substantially reduced since the 1988 fires, and the area covered by young forests has greatly increased. However, the current proportions of successional stages are similar to the proportions that existed in the mid-1700s, following the extensive fires of the early 1700s (Romme and Despain 1989). Specific locations of old versus young forest patches obviously differ, but the overall landscape mosaic of the year 2000 probably was very similar to the mosaic that existed in 1750. Therefore we conclude that current and recent landscape structures in Yellowstone's subalpine zone are within the range of natural variation for this system.

This interpretation could change during the next century if global climate change leads to increased fire activity, as is predicted by some climate models (Overpeck and others 1990, Chapters 2, 13). Shorter fire intervals and larger fire sizes would produce a landscape dominated by younger forests; old forests might become rare compared to their extent during the past two hundred years (Gardner and others 1996).

ARE ANY SPECIES OR COMMUNITIES EXTINCT OR THREATENED WITH EXTINCTION BECAUSE OF ALTERATIONS IN THE DISTURBANCE REGIME?

Altered disturbance regimes or novel disturbances to ecosystems may lead to local extinctions of species or substantially altered community structure. For example, many marine ecosystems have lost numerous species and are now functionally different than in the past, because of overfishing, pollution, climatic changes, and other factors (Dayton and others 1998). Some southwestern forests may have lost obligate cavity nesting birds because of fire exclusion combined with overharvest of large trees and snags (Balda 1975). Do any of Yellowstone's plant or animal species appear threatened because of changes in fire frequency or severity?

The answer to this question is no—at least for the immediate future and with regard to fire. With the reintroduction of the wolf in 1995, Yellowstone now has its complete complement of pre-Columbian species. There is no evidence that the 1988 fires, or the fire suppression that occurred prior to 1988, impaired long-term viability of any population of native species in Yellowstone either through direct mortality or indirect habitat effects (Singer and others 1989). The extent of suitable habitat may be reduced for some obligate old-growth forest species—for example, pine marten *(Martes americana),* but mature or old-growth forests still form large patches encompassing thousands of hectares. The extent of habitat for species that thrive mainly in recently burned forests—for example, Bicknell's geranium *(Geranium bicknellii),* dragonhead *(Dracocephalum parviflorum),* and three-toed woodpeckers *(Picoides arcticus* and *P. tridactylus)*—has increased since 1988, and many of these species became locally abundant during the first three years after the fires (Turner and others 1997 and personal observations). The robust response of these previously rare species to the 1988 fires indicates that they can persist for many decades without large fires, and all of the expected successional stages and habitat types are still well represented on the Yellowstone Plateau. We conclude, therefore, that neither the large fires in 1988 nor the preceding fire suppression has threatened any native species or communities in the park. Indeed, Yellowstone's subalpine landscape demonstrates remarkable ecological integrity (see also Dayton and others 1998).

Despite this optimistic assessment of community response to fire, we must note the potential for other kinds of disturbances to threaten long-term ecological integrity of the Yellowstone subalpine system. Blister rust *(Cronartium ribicola),* an alien fungus disease from Eurasia, has decimated whitebark pine forests in Montana and Idaho (Tomback and others 2001). Although blister rust has not yet caused extensive mortality in whitebark pine in and around Yellowstone Park, the disease is present and could become more serious in the future. Loss of whitebark pine, or even a significant decline in its population, could have cascading effects on the ecosystem. For example, grizzly bears depend heavily on whitebark pine nuts during some portions of the year (Tomback and others 2001). Whitebark pine also may be threatened over longer time frames by global climate change (Bartlein and others 1997, Chapter 2).

HAVE RECENT DISTURBANCES BEEN ACCOMPANIED BY NORMAL RETURN OF COMMUNITY STRUCTURE AND COMPOSITION?

The 1988 Yellowstone fires were not only very large, but also strikingly heterogeneous in burn severity (Turner and others 1994). Even the largest expanses of severely burned forest contained patches of less severely burned or even unburned forest, and were surrounded by a zone of lower severity. In 1990 we initiated a long-term study of the effects of fire size and severity on early postfire succession, a study that is still ongoing (Turner and others 2003b). We reported our findings as of 1993 in Turner and others (1997), and we report here the trends through 1996.

This research was designed to address two general questions that are directly relevant to the question being addressed in this chapter: (1) What is the relative importance of burned patch size, fire severity, prefire vegetation, and local abiotic conditions (climate and soils) in predicting plant community composition and structure during the first few years after fire? (2) Do initial differences persist for many years, or do community structure and composition converge among sites having different fire size, severity, prefire vegetation, and abiotic conditions?

Methods

In our initial appraisal of the heterogeneity of fire effects in 1989 (Turner and others 2000), we recognized three classes of fire severity that could be readily identified and mapped either in the field or remotely (Table 14.3). We subsequently mapped all patches of crown fire in Yellowstone Park from satellite imagery and then selected three patches in each of three geographic locations within the burned portions of the Yellowstone Plateau (fig. 1 in Turner and others 1997). The three geographic locations encompass a range of elevation and substrate conditions representative of Yellowstone's subalpine plateaus (Table 14.4). In general, the Cougar Creek area is at low elevation on relatively infertile substrates; the Fern Cascades area is at higher elevation on very infertile substrates, and the Yellowstone Lake area is at the highest elevation on relatively fertile substrates. See Turner and others (1997) for additional details.

In each patch, we established four transects radiating from the center in subcardinal directions. Along each transect we placed permanent sampling points at variable distances. Points were generally 100 m apart in large tracts of uniform fire severity but were more closely spaced (5–20 m) in areas of transition between fire severity classes (for example, from crown fire to severe surface fire). Each sampling point was a 50-m^2 circular plot, the center of which was permanently marked with a stake or rock cairn and located with GPS. Within the 50-m^2 plot, we measured percent cover of plants and abiotic components in eight 0.25-m^2 point intercept frames (Floyd and Anderson 1987), and we listed all vascular plant species present within a 10 m × 1 m quadrat. We also determined the pro-

Table 14.3. Classification of fire severity

Crown Fire	Severe Surface Fire	Light Surface Fire
Canopy needles consumed	Canopy needles killed but not consumed	Canopy needles unharmed
Canopy trees killed	Canopy trees killed	Canopy trees not killed
Organic litter on the forest floor completely consumed	Organic litter on the forest floor mostly consumed; scorched needles added from the canopy	Organic litter on the forest floor consumed only in patches; mostly intact
Aboveground portions of all shrubs and herbs killed	Aboveground portions of most shrubs and herbs killed	Aboveground portions of shrubs and herbs killed only in patches
Stand-replacing	Stand-replacing	Not stand-replacing

Table 14.4. Description of the three geographic locations and burned patches where sampling was conducted from 1990 through 1996 in Yellowstone National Park

Attribute	Cougar Creek	Fern Cascades	Yellowstone Lake
Patch sizes and sampling points			
Large patch	500 ha (84 points)	480 ha (103 points)	3698 ha (59 points)
Moderate patch	91 ha (46 points)	200 ha (83 points)	74 ha (67 points)
Small patch	1 ha (34 points)	1 ha (37 points)	1 ha (40 points)
Elevation range	2150–2300 m	2270–2500 m	2400–2700 m
Geologic substrate	Infertile rhyolite and tuff	Infertile rhyolite	Moderately fertile andesite and lake sediments
Prefire vegetation	Dense, ca. 130-yr-old *Pinus contorta* forests, with *Ceanothus velutinus*, *Calamagrostis rubescens*, and *Carex rossii*	Ca. 290-yr-old *Pinus contorta* forests, with *Vaccinium scoparium*, *Lupinus argenteus*, and *Carex geyeri*	Ca. 250–400+-yr-old forests of *Pinus contorta* and *Picea engelmannii*, with *Vaccinium scoparium*, *Arnica cordifolia*, and *Carex geyeri*

Note: See Turner and others 1997 for additional details.

portion of serotinous versus open cone–bearing lodgepole pine trees that had been present at the time of the fire, using methods developed by Tinker and others (1994). See Turner and others (1997) and Chapter 4 for additional sampling details.

Results

Total plant cover was significantly different among *fire severity* classes (Table 14.3) in 1990, with mean values of 14 percent in crown fire sites, 32 percent in severe surface burns, and 47 percent in light surface fires (Fig. 14.1e). Total cover increased from 1990 to 1993 in all severity classes. We then saw an apparent decrease in cover, especially of herbs, from 1993 to 1996 (Fig. 14.1), which could reflect the onset of resource depletion, competition, or other processes. However, we believe that this seeming decline is actually an artifact of the cool, late spring in 1996 that slowed the phenological development of plants throughout our study area. Continued sampling over the next decade(s) will reveal whether the drop in plant cover from 1993 to 1996 is a phenological anomaly or a genuine trend (cf. Turner and others 2003b). What we believe to be the more significant pattern in 1996 is a convergence in total plant cover among all three burn severity classes: the large differences that were so conspicuous immediately after the fires had essentially disappeared within eight years. Trends in percent cover of forbs and graminoids (Fig. 14.1a,b) were generally similar to those for total biotic cover. In contrast, initial differences in tree cover appeared to become more pronounced over time, although confidence intervals generally overlapped (Fig. 14.1c). Shrub cover remained highest in light surface burns throughout the first eight years after fire (Fig. 14.1d).

Patterns in total plant cover related to *patch size* were similar to those of burn severity. Small patches initially had twice the cover of moderate and large patches (approximately 40 percent versus 20 percent), but by 1996 the differences had become smaller and cover appeared to be converging among patch sizes (Fig. 14.2c) (cf. Turner and others 2003b). Patterns for forb and graminoid cover (Fig. 14.2 a,b) largely mirrored the pattern for total biotic cover. However, some closely related species exhibited striking individualistic responses; for example, the highest cover of *Carex geyeri* was seen in small patches, while *Carex rossii* was most abundant in large patches (data not shown).

Differences in total plant cover among *geographic locations* were apparent throughout the period of measurement. The lowest total cover values (Fig. 14.3e) were measured in the Fern Cascades area, which has the most infertile soils. Total percent cover was similar in the Lake and Cougar Creek areas through 1993,

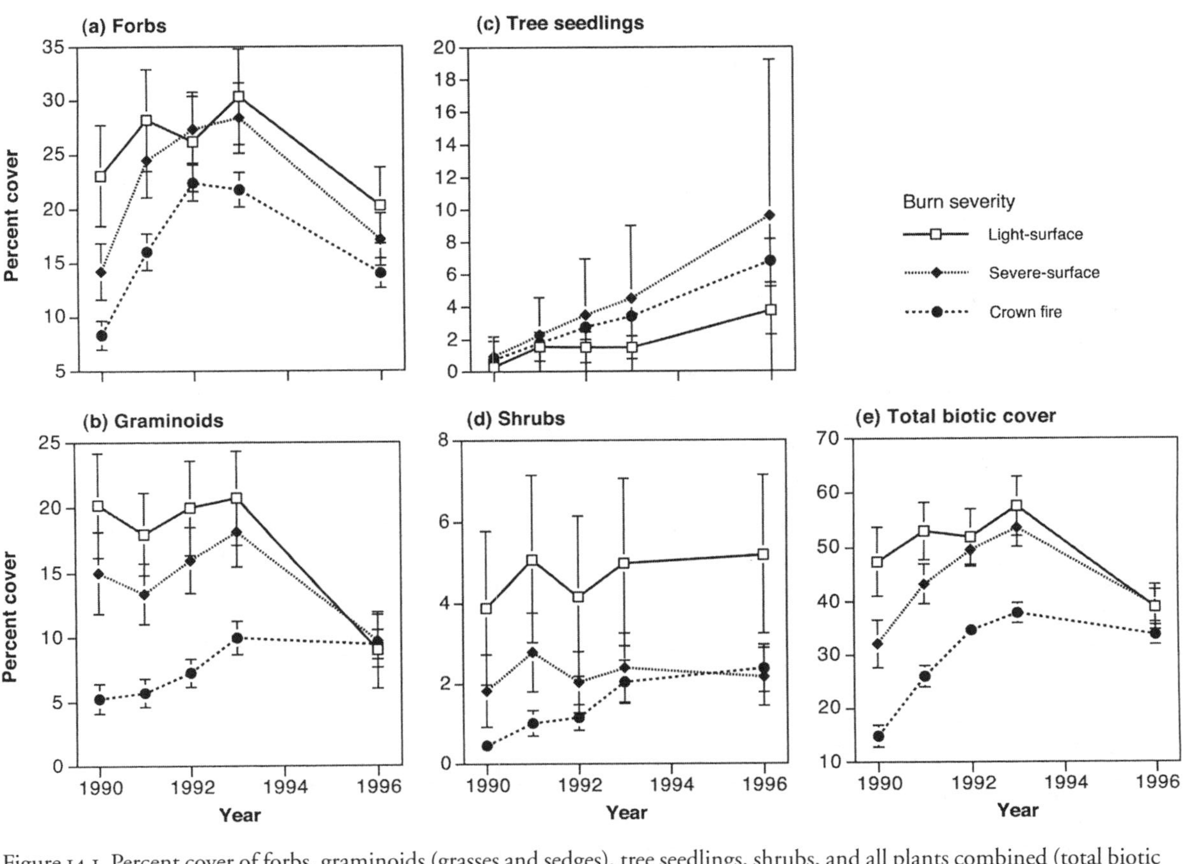

Figure 14.1. Percent cover of forbs, graminoids (grasses and sedges), tree seedlings, shrubs, and all plants combined (total biotic cover) in relation to burn severity in Yellowstone National Park, Wyoming.

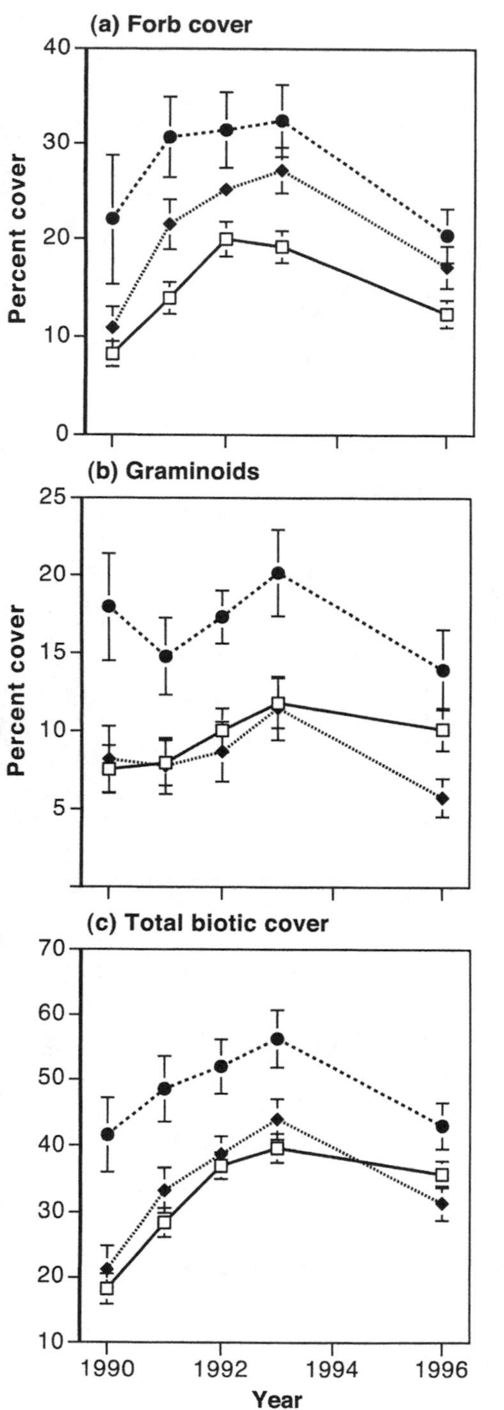

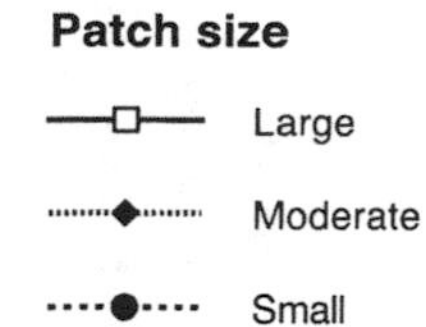

Figure 14.2. Percent cover of forbs, graminoids (grasses and sedges), and all plants combined (total biotic cover) in relation to size of burned patch in Yellowstone National Park, Wyoming.

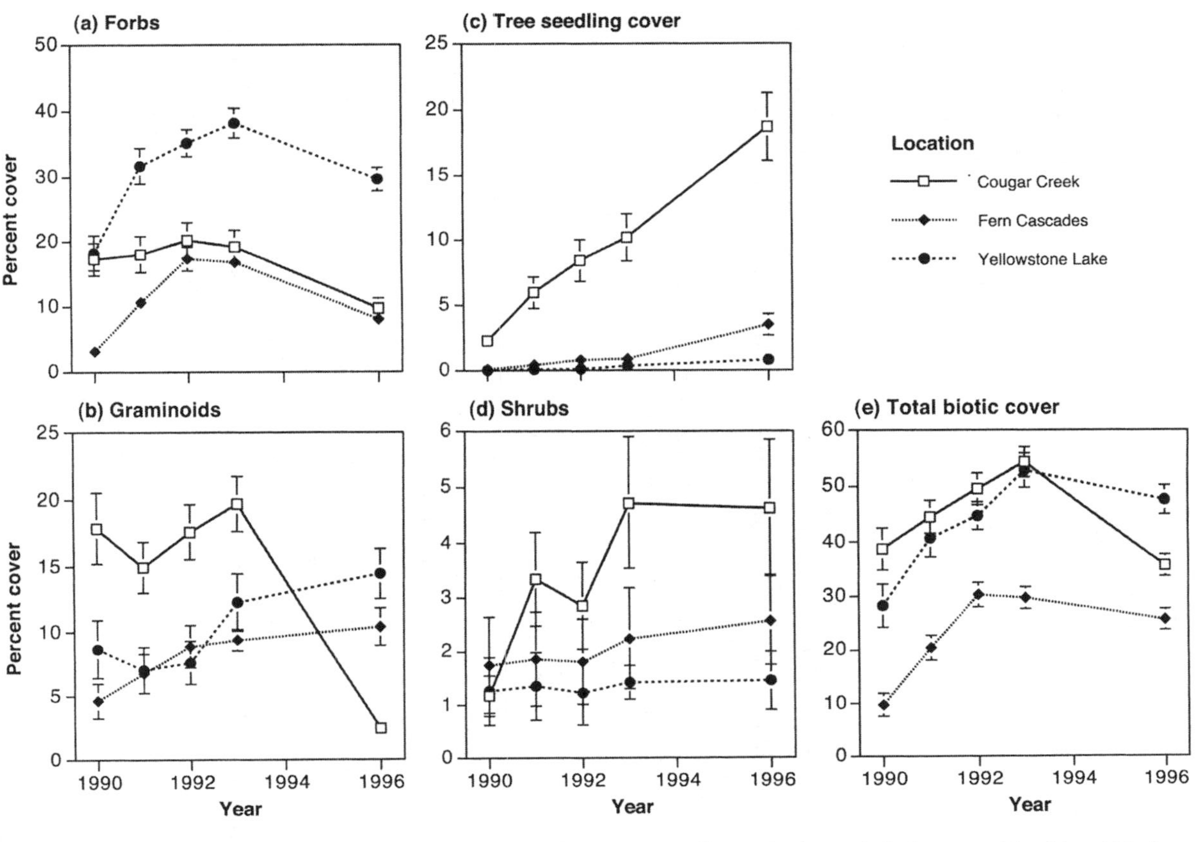

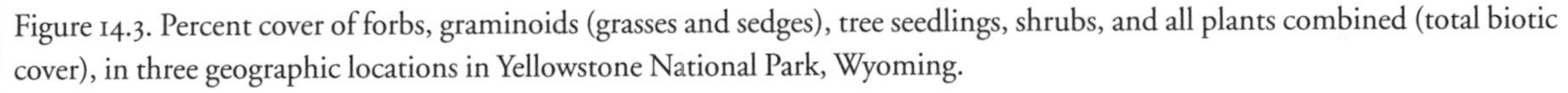
Figure 14.3. Percent cover of forbs, graminoids (grasses and sedges), tree seedlings, shrubs, and all plants combined (total biotic cover), in three geographic locations in Yellowstone National Park, Wyoming.

but in 1996 cover at Cougar Creek was dropping toward the level of Fern Cascades, while cover in the Lake area remained high. It appears that total plant cover is now converging toward similar, relatively low levels in the two areas (Fern and Cougar) underlain by infertile rhyolite substrates (cf. Turner and others 2003b). However, the Lake area, with its fertile andesites and lake bottom sediments, is diverging from the other two areas and developing relatively higher total plant cover (Figure 14.3e). Another striking pattern is the development of very different relative proportions of forbs, graminoids, shrubs, and trees among the three geographic locations (Fig. 14.3a–d). The Yellowstone Lake area has been dominated by forbs and graminoids, while the Cougar Creek site initially had high cover of graminoids but by 1996 was dominated by tree seedlings. Shrubs have constituted a relatively small proportion of cover in all locations in all years.

The density of *opportunistic species* (defined as species that require open habitats and are rare in unburned forests) generally was highest in large patches (Fig. 14.4) and in areas of more severe fire (Fig. 14.5). The density of most opportunistic species peaked three to five years after the fires and then decreased substantially. However, some opportunistic species increased again on some sites in 1996 (for example, *Gayophytum diffusum* and *Cirsium arvense,* Figs. 14.4, 14.5). There were no striking differences among geographic locations in the density of opportunistic species (data not shown).

Species richness (the number of vascular plant species in a 10-m^2 plot) was initially similar in all three locations but had diverged greatly by 1996, with means of 17 species/plot in the Lake area but only 9 and 11 in Fern Cascades and Cougar Creek, respectively (Fig. 14.6a). The greater richness of the Lake area probably reflects the higher precipitation and more fertile substrates in this part of Yellowstone Park. Differences in richness related to burn severity were initially slight, and remained slight through 1996 (Fig. 14.6b). With respect to patch size, richness was highest in small patches and lowest in moderate patches, and these patterns persisted through 1996 (Fig. 14.6c).

Perhaps the most striking of all the patterns we observed were those related to *lodgepole pine seedling density.* Most seedlings became established in 1989 and 1990, and these initial spatial patterns in seedling density then persisted throughout the period of measurement. We refer to these plants as "seedlings" even though by 1996 most were eight to nine years old and up to a meter tall. The highest pine seedling densities were seen in severe surface burns, with lower densities in crown fire and light surface fire areas (Fig. 14.7b). This probably was because crown fires killed much of the seed supply in the canopy (Johnson and

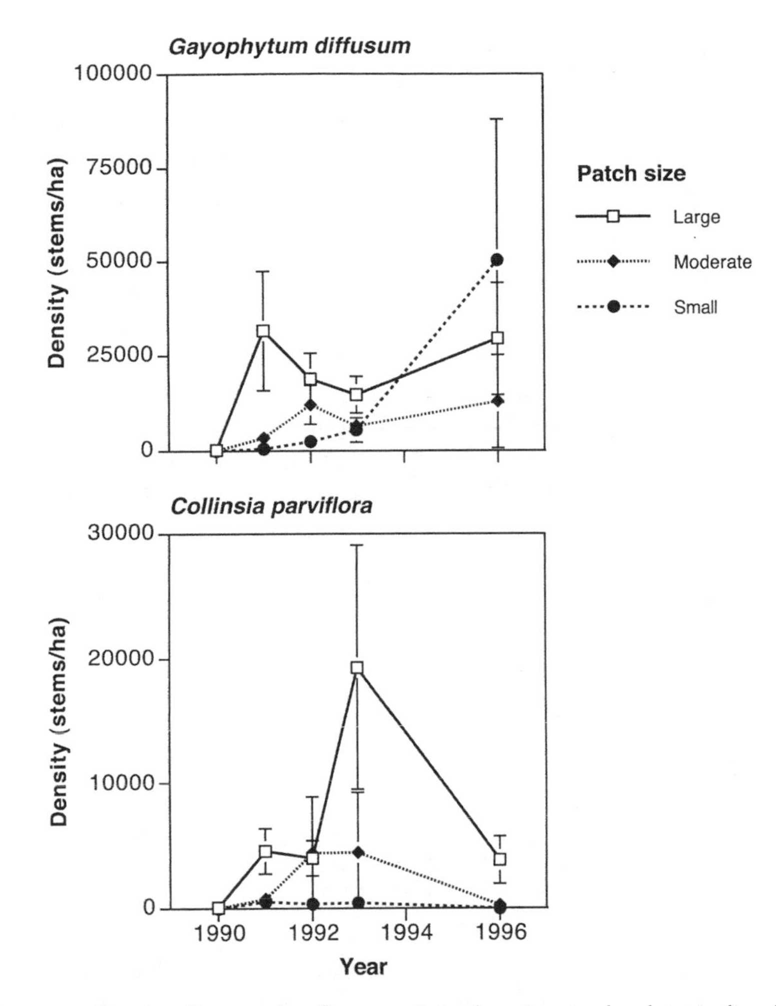

Figure 14.4. Density of two species of opportunistic plants (species that thrive in disturbed areas and are rare in unburned forests) in relation to size of burned patch in Yellowstone National Park, Wyoming. Both species are native annuals.

Gutsell 1993), while light surface fires produced little bare mineral soil for seedling establishment. In contrast, severe surface fires consumed most of the litter on the forest floor and killed the adult trees but did not consume cones and needles in the canopy. Seedling density in crown fire patches also decreased significantly with distance from the nearest patch of severe surface fire (data not shown). Patch size also influenced pine seedling density, with highest densities in large patches and lowest densities in small patches (Fig. 14.7c) (see also Chapter 4).

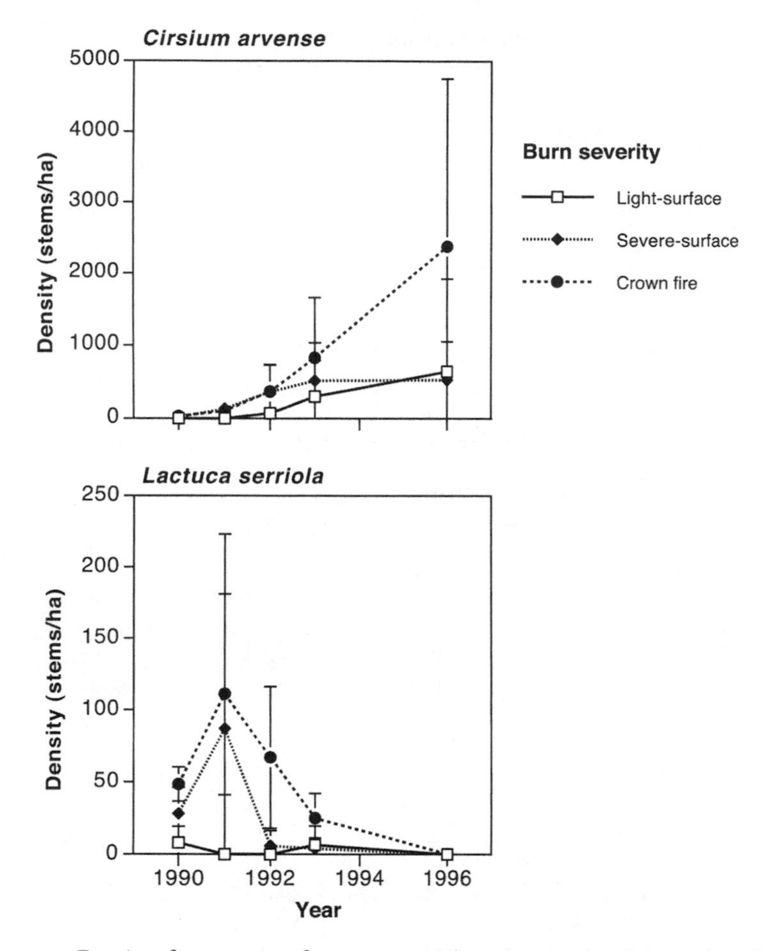

Figure 14.5. Density of two species of opportunistic plants (species that thrive in disturbed areas and are rare in unburned forests) in relation to burn severity in Yellowstone National Park, Wyoming. Both species are nonnative invaders from Eurasia.

By far the greatest differences in lodgepole pine seedling density were seen among geographic locations. Density in the Cougar Creek area was two orders of magnitude greater than in the Fern Cascades or Lake area (Fig. 14.7a). The extremely high densities in the Cougar area were the result of very high levels of cone serotiny in the forests that burned in 1988. Indeed, percent cone serotiny in the prefire forest was the best predictor of postfire pine seedling density. At Cougar, 65 percent of the trees were serotinous, compared with only 5 percent in the Fern Cascades area and 2 percent in the Lake area (Tinker and others 1994, Chapter 4). Correspondingly, mean pine seedling density in 1996 was 43,000

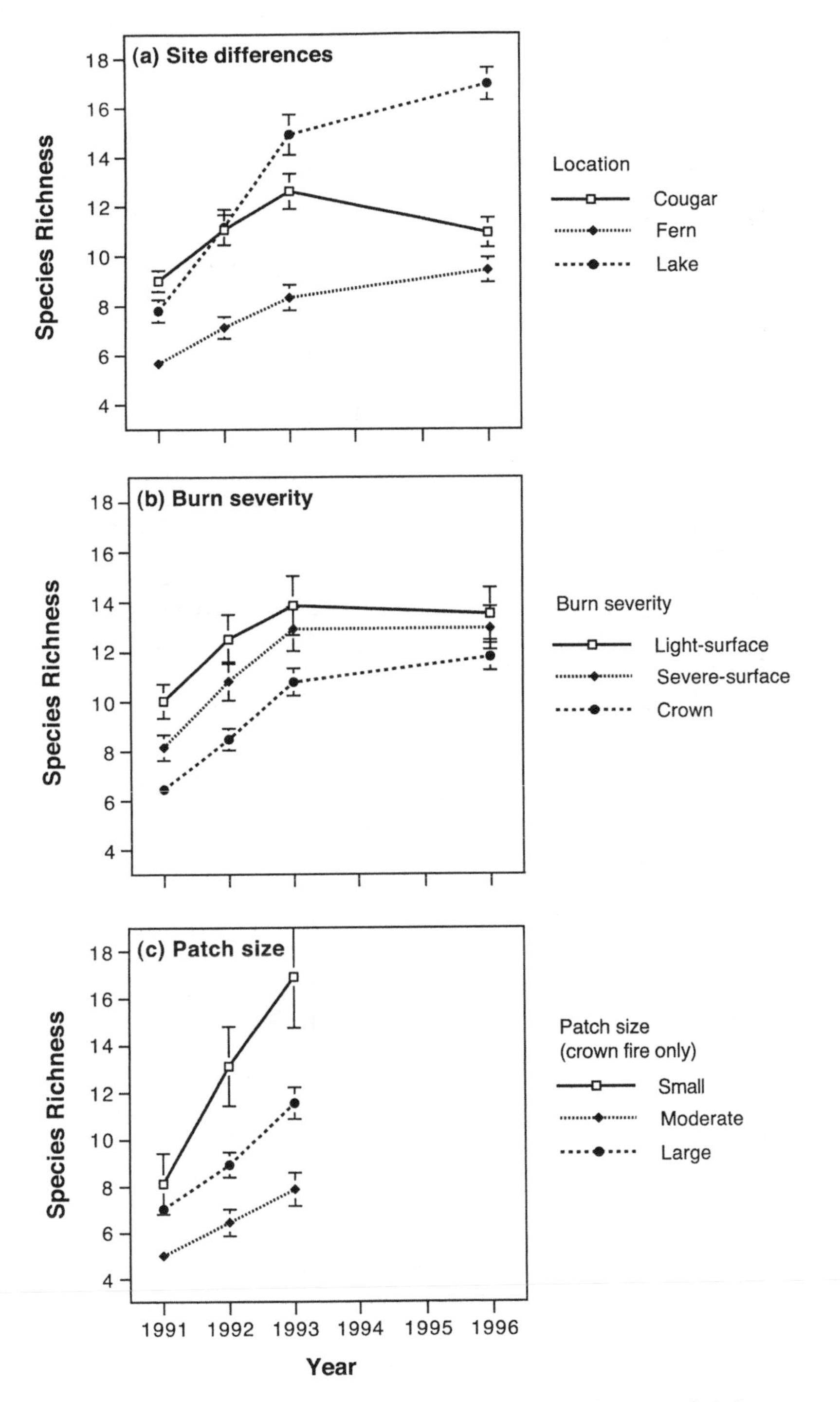

Figure 14.6. Species richness (number of vascular plant species within a 10-m² plot) in relation to geographic location (site differences), burn severity, and size of burned patch, in Yellowstone National Park, Wyoming.

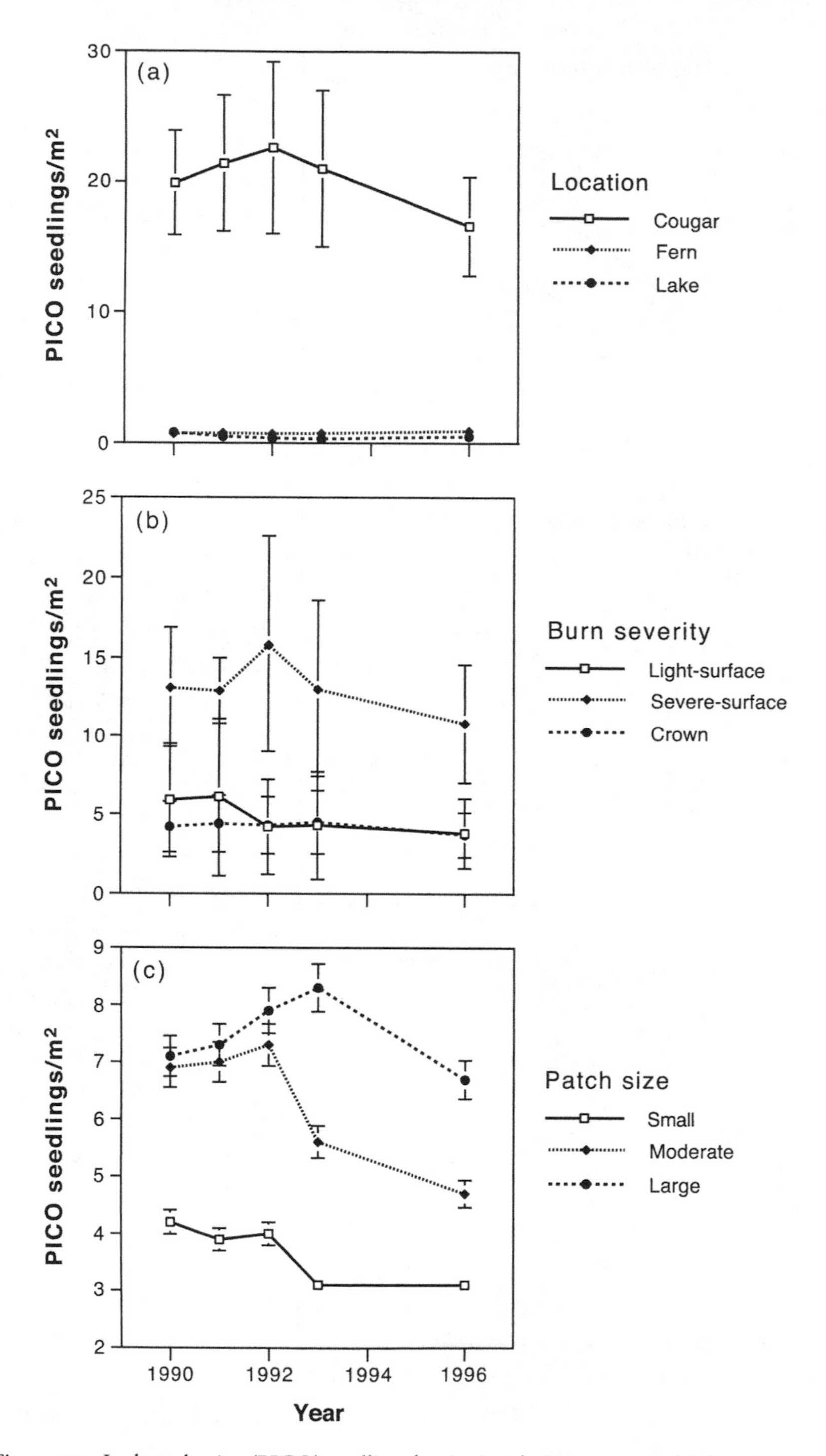

Figure 14.7. Lodgepole pine (PICO) seedling density in relation to geographic location, burn severity, and size of burned patch, in Yellowstone National Park, Wyoming.

stems/ha in the large Cougar patch, but only 4,700 and 14 stems/ha in the large patches at Fern Cascades and Lake, respectively.

Discussion

Differences among the three geographic locations became increasingly prominent during the first eight years of postfire succession, reflecting the powerful influence of abiotic gradients in soils and climate on plant community structure and composition. In 1996, total biotic cover and plant species richness were highest on the more fertile substrates of the high-elevation Yellowstone Lake region, and lowest on the infertile substrates of the lower-elevation Fern and Cougar regions. Plant cover was largely herbaceous in the Yellowstone Lake region but consisted mostly of trees and shrubs in the Cougar Creek region. These results suggest a general pattern of initial similarity in community structure and composition, followed within a decade by divergence along abiotic gradients of microclimate and soils conditions.

In contrast, among burn severity classes we observed a general pattern of large initial differences in total plant cover and richness followed by convergence within the first decade. Total plant cover and richness were still significantly higher in 1996 in less severely burned areas, and densities of some opportunistic species were significantly higher in more severely burned areas—but the magnitude of differences had become relatively small. The important exception to this general pattern of convergence was seen in lodgepole pine seedling densities. Differences among burn severity classes, patch sizes, and geographic locations became established early on, and have persisted with little change. Where the burned forest contained a high proportion of serotinous trees, postfire pine seedling density is orders of magnitude greater than in places where few or no trees were serotinous (Chapter 4). For any given level of prefire serotiny, areas of severe surface fire support approximately 2.5 times as many seedlings per hectare as adjacent areas of crown fire or light surface fire, and pine seedling density within crown fire areas decreases with distance from an area of severe surface fire.

The patterns in lodgepole pine seedling density that became established within the first two years after the fires—patterns related to prefire serotiny, fire severity, and patch size—apparently will determine the course of stand development for many decades to come. Although we continue to detect a few new pine seedlings each year, total seedling densities in any given location were not significantly different among the years 1991, 1992, 1993, and 1996. Moreover, we have observed very few dead or dying pine seedlings—even in extremely dense stands (> 100,000 seedlings/ha) (Chapter 4). Self-thinning in dense stands will

Table 14.5. Comparison of three areas (large patches described in Table 14.4) illustrating alternative initial pathways of succession following the 1988 Yellowstone fires

Geographic Location	Initial Successional Pathway	Pine Seedling Density (Mean Stems ha^{-1})	Mean Cover of Trees + Shrubs	Mean Cover of Forbs + Graminoids	Mean # Vascular Plant Species in 10-m^2 Plots
Cougar Creek	High-density lodgepole pine	43,000	32%	7%	10
Fern Cascades	Low-density lodgepole pine	4,700	6%	19%	10
Yellowstone Lake	Nonforest	14	2%	40%	16

Note: All three areas were forested at the time of the fires. Data are from 1996 in areas of crown fire (areas of light or severe surface fire excluded).

surely occur eventually, but no significant thinning had yet begun as of 2003 (Turner and others 2003b). Thus at least three major successional trajectories now can be identified in the forested areas burned by the 1988 fire (Table 14.5). These alternative trajectories, distinguished mainly by huge differences in lodgepole pine seedling density, create a tremendous heterogeneity of stand structure throughout the areas burned in 1988. The functional implications of these alternative trajectories (for example, for productivity and nutrient cycling) are a major focus of our ongoing research, and our findings to date are presented later in this chapter (see Chapter 11, as well).

Returning to the central question of this chapter, whether restoration is needed after the 1988 fires, we find that postfire community development varies enormously across the Yellowstone Plateau—but nowhere do we find evidence of abnormal or dysfunctional response to the 1988 fires. On the contrary, the documented patterns in plant cover, richness, and pine density generally can be explained by variation in climate, soils, fire size, fire severity, and prefire serotiny. Observations by Romme and Despain (1989 and unpublished data), as well as more recent studies (Kashian 2002) suggest that earlier large fires were followed by similar variability in development of community structure.

An especially significant aspect of postfire community development on Yellowstone's subalpine plateaus is the relative scarcity of alien weeds. Invasive alien plant species have dramatically altered postfire succession in many other areas, and are considered to be one of the major threats to the ecological integrity of natural areas (for example, Mooney and Drake 1986, Drake and others 1989, Cole and Landres 1996, D'Antonio 2000). We have detected several alien species in our study areas (for example, *Cirsium arvense, Filago arvense, Lactuca serriola*), but generally these species are patchily distributed (especially along maintained trails) and one finds only native species throughout extensive areas of the Yellowstone Plateau.

HAVE RECENT DISTURBANCES BEEN ACCOMPANIED BY NORMAL RETURN OF ECOSYSTEM FUNCTION?

Major disturbances may disrupt processes of energy flow and nutrient cycling (Chapter 5, 8). For example, numerous studies have reported increased nutrient loss and runoff following fire or clear-cutting (Vitousek and others 1979, Vitousek and Melillo 1979, Vitousek and Matson 1985, Binkley and others 1992b, Parsons and others 1994, Jewett and others 1995, Likens and Bormann 1995). To evaluate the extent of "damage" to the Yellowstone ecosystem caused by the 1988 fires, we report on studies initiated in 1996 to determine the effects of the fires and subsequent heterogeneity in postfire succession, on three key parameters of ecosystem function. These parameters include aboveground net primary productivity (ANPP), leaf area index (LAI), and rate of nitrogen mineralization. ANPP is important because it reflects total energy flow through the ecosystem and forms the base of the food web. LAI is highly correlated with rates of photosynthesis and transpiration, which in turn are related to rates of nutrient uptake (Knight and others 1985). Nitrogen is a limiting nutrient in many boreal ecosystems (Fahey and others 1985, Fahey and Knight 1986, Binkley and others 1995, Chapter 5). Our research on ecosystem function after fire is ongoing. We report here our results through 1998, and relate these preliminary findings to the question of whether ecosystem processes were abnormally altered by the 1988 fires.

ANPP and LAI

In 1997, nine years after the 1988 fires, we measured ANPP and LAI of lodgepole pine seedlings and ANPP of herbaceous plants. Measurements were made at four sites representing a wide range of lodgepole pine seedling densities (100–62,800 stems/ha) and two different soil fertility classes (relatively fertile and infertile). Standard techniques of dimension analysis were used to measure ANPP and LAI; see Reed and others (1999) for details.

Table 14.6. Aboveground net primary production and leaf-area index in four postfire stands, nine years after the 1988 Yellowstone fires

	Study Site			
	Infertile Nonforest	Fertile Nonforest	Low-Density Pine Forest	High-Density Pine Forest
Lodgepole pine density (stems/ha)	100	1,000	20,100	62,800
Elevation (m)	2750	2560	2560	2070
Substrate	infertile rhyolite	relatively fertile andesite	infertile rhyolite	infertile rhyolite
Mean seedling height in 1997 (m)	0.62	0.44	0.42	0.89
Tree ANPP (mg/ha/yr)	0.0021	0.0164	0.1431	3.8134
Herbaceous ANPP (mg/ha/yr)	0.2434	0.7374	0.7121	0.1998
Total ANPP (mg/ha/yr)	0.2455	0.7537	0.8552	4.0133
Tree LAI (m2/m2)	0.002	0.012	0.138	1.822

Source: Reed and others 1999.
Note: "Nonforest" sites were forested at the time of the 1988 fires, but tree density after the fires is extremely low.

Tree ANPP and LAI generally increased as pine density increased, as would be expected (Table 14.6). Interestingly, total ANPP (trees plus herbs) was comparable in the low-density pine stand and the nonforest stand on a fertile substrate, because of the high herbaceous ANPP in the latter stand (Table 14.6). The highest ANPP and LAI were seen in the high-density pine stand, in part because of the dense trees, but possibly also because of a longer growing season at this lower elevation site.

What do these findings tell us about damage and recovery from the 1988 Yellowstone fires? We have detected considerable variability in ANPP and LAI across the landscape a decade after the fires. However, this variability can be explained primarily by differences in abiotic conditions (local substrate and climate) and in lodgepole pine seedling density (which is a function of fire size, severity, and prefire serotiny, as described in the previous section). We find no evidence of inadequate recovery from the fires. Rather, recovery of ecosystem function appears remarkably rapid. After only nine years, ANPP and LAI in the high-density pine stand was close to or within the range of values measured in mature coniferous forests elsewhere in North America (Reed and others 1999, Chapter 4).

NITROGEN DYNAMICS

Fire has a profound effect on nitrogen (N) cycling in forested ecosystems. Nitrogen is lost directly to the atmosphere through combustion of organic matter, and substantial loss of N may follow because of the disruption of biotic controls over N cycling. Nitrogen is a useful indicator of ecosystem function for several reasons: N limits primary productivity (Reich and others 1997); the presence of nitrate-N in soil water and streamwater can be used as an indicator of disturbances that lead to N leaching (Bormann and Likens 1979, Vitousek and Melillo 1979, Parsons and others 1994, Chapter 8); and relative rates of nitrogen mineralization are fairly easily estimated. Nitrate in western coniferous forests is present in very low concentrations except after major disturbances (Brown and others 1973, Helvey and others 1976, Tiedemann and others 1978, Binkley 1984a, Gosz and White 1986, Covington and others 1991, Covington and Sackett 1992, Parsons and others 1994, Jurgensen and others 1997). Disturbances lead to short-term declines in N uptake by vascular plants and increased nitrification by soil microbes, thereby enhancing nitrate production. Therefore, nitrate levels can be used as indicators of whether or not ecosystem development since the 1988 fires has produced a sufficient amount of biomass (especially sufficient root biomass to fill root gaps; Parsons and others 1994) to immobilize the ammonium-N made available through mineralization after disturbance.

The consequences for nitrogen availability resulting from the spatial variation in burn severity and successional pathways are not known. Within different successional pathways (Table 14.5), how long will it be before rates of nitrate production no longer differ from intact forest? Within a stand, N availability could simply be related to leaf area development and ANPP, regardless of composition. However, N availability might be strongly influenced by the relative amount of herbaceous and coniferous leaf area, in which case the variability in successional pathways across the landscape may have important effects. Nitrogen mineralization rates were higher in Canadian boreal forest stands with deciduous species than in conifer stands (Pare and others 1993), and large differences in net N mineralization rates have been associated with successional changes from poplar to white spruce in an Alaskan chronosequence (Van Cleve and others 1993). We measured N mineralization in 1997–1998 to determine (1) whether nitrogen availability ten years after the 1988 fires was similar to that observed in similar mature stands, and (2) whether nitrogen availability in the post-1988 stands varied with plant community structure, particularly the density of lodgepole pine seedlings. In addition, we sampled streamwater at the end of snowmelt during the summer of 1997 to determine whether there was any evidence of elevated nitrate concentrations in areas burned during the 1988 fires.

Methods

Nitrogen availability was determined for two incubation periods. First, we conducted three sequential monthlong incubations during the 1997 growing season at two sites burned in 1988 and differing in postfire pine seedling density. Second, we conducted yearlong incubations at six sites, including three sites burned in 1988 but having different postfire pine seedling densities, one site that burned in 1996 (the Pelican fire), and two mature lodgepole pine stands, one located near the Pelican burn with similar topographic and soils conditions, and the other located near the areas burned in 1988 (Table 14.7).

In situ incubations of ion exchange resin (IER) (Binkley and Matson 1983, Binkley 1984b, Hart and Binkley 1985, Binkley and others 1986, Binkley and Hart 1989, Binkley and others 1992a, Giblin and others 1994) were used to determine rates of net nitrification and net nitrogen mineralization in our study areas. IER is useful for examining spatially extensive patterns of nitrogen availability, particularly in remote locations that cannot be visited frequently (Binkley and Hart 1989, Hart and Firestone 1989, Giblin and others 1994). The resin bag method does not provide absolute estimates of N availability, but rather it provides an index of relative availability (Hart and Firestone 1989, Binkley and Hart 1989, Binkley and others 1992a,b, Giblin and others 1994).

Incubations used unsieved soil (20 cm depth) in open 5-cm diameter PVC tubes with resin bags at the bottom (resin cores, after DiStefano and Gholz 1986, Strader and others 1989, Binkley and others 1992a,b) at each site (n = 10 resin cores per site). The resin cores were open to water flow, thereby allowing the products of mineralization to leach from the soil column into the resin bags. Resin bags contained 20 g of mixed bed exchange resin (J. T. Baker #4631) in commercial nylon stocking material. Resin bags were saturated with 2 M KCl for 24 hr before use to reduce interference with colorimetric analyses. Initial soil samples were obtained adjacent to each resin core. At the end of each incubation, soil cores were removed and 20-g subsamples of soil were extracted in 100 ml 2 M KCL; the resin bags were refrigerated, washed with deionized water prior to extraction to remove soil particles, and then extracted with 100 ml of 2 M KCl. Extracts were analyzed for nitrate and ammonium using standard colori-

Table 14.7. Study sites where soil nitrogen availability was measured in 1997

Site	Fire History	Substrate	Vegetation
Pelican burn	Burned in 1996	Relatively fertile lake sediments	Burned forest was an old stand of spruce and fir (>300 yr old). . . . In 1997 the soil was covered by charred litter, with 5–10% cover of resprouting herbs.
Mallard Lake Trail	Burned in 1988	Infertile hot springs deposits	Burned forest was an old stand of spruce, fir, and pine (300 yr old). . . . In 1997 the vegetation was nonforest (ca. 100 pine seedlings/ha) dominated by relatively sparse herbs; sparse litter.
Pitchstone Plateau nonforest	Burned in 1988	Infertile rhyolite	Burned forest was an old stand of spruce, fir, and pine (300 yr old). . . . In 1997 the vegetation was nonforest (ca. 100 pine seedlings/ha) dominated by relatively sparse herbs; sparse litter.
Pitchstone Plateau low-density pine	Burned in 1988	Infertile rhyolite	Burned forest was an old pine stand (300 yr old). . . . In 1997 the vegetation was low-density pine forest (20,100 stems/ha) with sparse herbaceous cover; sparse litter.
Mount Haynes	Burned in 1988	Infertile rhyolite/ tuff	Burned forest was a mature pine stand (130 yr old). . . . In 1997 the vegetation was high-density pine forest (62,800 stems/ha) with sparse herbaceous cover; moderate litter layer.
Divide Trail	Unburned	Infertile rhyolite	Old forest of pine, spruce, and fir (>300 yr old) with well-developed herbaceous stratum; deep litter layer.
Pelican unburned	Unburned	Fertile lake sediments	Mid-successional pine forest (ca. 150 yr old) with ca. 1,000 stems/ha, ca. 25% herbaceous cover, and 1–5 cm deep litter layer with almost no bare ground.

metric autoanalyzer methods (Alpchem Corp. 1990, 1991) at the Wisconsin Soil and Plant Analysis Laboratory in Madison. Soil moisture content was determined for the pre- and postincubation soils. Net N mineralization was calculated as the postincubation quantity of ammonium and nitrate in both the soil and resin bags minus the quantities in the preincubation soil.

Streamwater was sampled July 14–15, 1997, near the end of the snowmelt period, in first-order ephemeral streams with little or no riparian vegetation. The snowpack was high during the previous winter, and thus snowmelt was late that year. Replicate samples of streamwater were collected from nine streams at six sites. Two of the streams were from watersheds that burned in 1988, and these were compared to two reference streams in unburned forest (Table 14.8). Two other streams were from the area burned in 1996, and these were compared with one nearby reference stream in an unburned area.

Results

The summer 1997 monthlong soil incubations indicated that net nitrogen mineralization in the nonforest stand was nearly double that at the high-density pine stand ($P < 0.05$) (Table 14.9). Nitrogen mineralization rate was highest during the middle interval when precipitation was unusually high. Nitrogen mineralization rates during the first interval were relatively low, possibly because of greater immobilization of N early in the season and/or the unusually cold temperatures that occurred during June. While these studies need to be repeated at more sites to determine their generality, the results suggest that nitrogen mineralization may decline with increasing ANPP and/or tree LAI, and that mineralization rates are influenced by ambient moisture and temperature conditions.

The yearlong incubation data revealed that 1997–1998 nitrate production was low at all sites except for the 1996 Pelican burn (Fig. 14.8). Among the three sites burned in 1988, mean annual nitrate production ranged only from 0.066 to 0.624 mg NO_3-N/kg soil, constituting from 1.6 to 19 percent of the total nitrogen available. The site with the lowest pine seedling density had the highest mean NO_3-N production, but the amount was still low. There was no significant difference in nitrate production between the three 1988 burn sites and the two mature lodgepole pine sites. Among all sites, nitrate production was highest in the 1996 Pelican burn (4 mg NO_3-N/kg soil), and nitrate constituted 94 percent of the total nitrogen availability.

Ammonification was highly variable across sites. The old-growth lodgepole pine stand had significantly higher mean annual NH_4-N (12.4 mg NH_4-N/kg

Table 14.8. Study sites where streamwater nitrate was measured in 1997

Site	Fire History	Substrate and Elevation	Sampling Site(s)	Vegetation
Pelican burn	Burned in 1996	Fertile lake sediments and infertile rhyolite	Two tiny streams plus the stream below their confluence	Burned forest consisted of mature lodgepole pine stands (ca. 200 yr old) plus old spruce-fir stands (>300 yr old) on a steep slope. . . . In 1997 there was a sparse cover of resprouting herbs in most burned areas, plus a narrow band (1m) of dense grasses and sedges along the stream.
Pitchstone Plateau Trail	Burned in 1988	Infertile rhyolite	Two ephemeral streams	Burned forest was an old stand of spruce, fir, and pine (300 yr old) on gentle topography. . . . In 1997 the vegetation was nonforest (ca. 100 pine seedlings/ha) dominated by relatively sparse herbs; sparse litter.
Lewis River Gorge	Burned in 1988	Infertile rhyolite	One ephemeral stream	Burned forest was an old stand of spruce, fir, and pine (300 yr old) on a moderate slope. . . . In 1997 the vegetation was low-density pine forest (ca. 500 seedlings/ha) dominated by relatively sparse resprouting herbs; sparse litter.
Elephant-back Mountain	Unburned	Infertile rhyolite	Two tiny streams	Mature lodgepole pine forest (ca. 200 yr old) on a steep slope. . . . Moderately dense herbaceous cover and litter; dense growth of grasses and sedges along the streams.
Spring Creek Picnic Area	Unburned	Infertile rhyolite	Three points along a tiny stream	Old lodgepole pine forest (>300 yr) on a steep slope. . . . Moderately dense herbaceous cover and litter.
Divide Roadside	Unburned	Infertile rhyolite	Ephemeral stream	Old lodgepole pine forest (>300 yr) on gently rolling topography. . . . Moderately sparse herbaceous cover except for dense grasses along stream channel.

Table 14.9. Mean net nitrogen mineralization (mg N/kg soil/day) during three intervals of the 1997 growing season at two sites burned in 1988 but with different postfire densities of lodgepole pine saplings

	Early (May 18–July 9)			Mid (July 9–30)			Late (July 30–Sept 20)		
Location	NO_3-N	NH_4-N	Total N	NO_3-N	NH_4-N	Total N	NO_3-N	NH_4-N	Total N
High-density pine	0.001 (0.00006)	0.0078 (0.002)	0.009 (0.002)	0.0018 (0.0004)	0.0265 (0.01)	0.0259 (0.009)	−0.001 (0.0001)	0.0047 (0.0017)	0.0039 (0.002)
Non-forest	0.0031 (0.0012)	0.0148 (0.005)	0.019 (0.006)	0.0044 (0.001)	0.0479 (0.011)	0.0479 (0.010)	0.0018 (0.0007)	0.0479 (0.011)	0.0107 (0.003)

Note: The high-density pine site is Mount Haynes and the nonforest site is the Mallard Lake Trail (Table 14.7). Standard error is indicated in parentheses ($n = 15$).

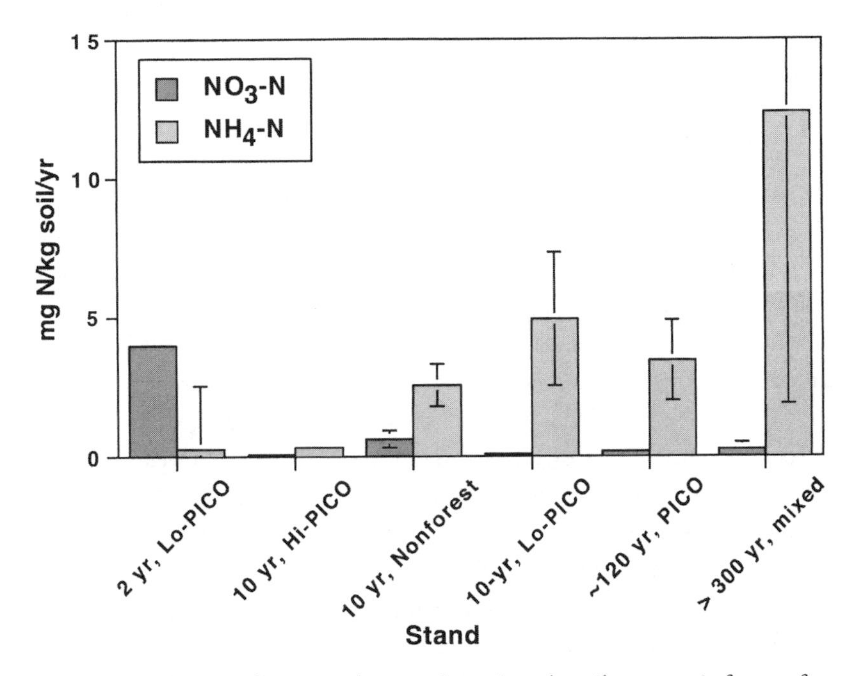

Figure 14.8. Production of nitrate and ammonium nitrate by soil processes in forests of Yellowstone National Park, Wyoming. The 2- and 10-year-old stands were developing after crown fires in 1996 and 1988, respectively. The 120 and > 300 stands had not burned in more than a century, and their ages were estimated from tree size, density, and species composition. The 120-year-old stand appears to represent a forest that is still accumulating biomass, whereas the > 350-year-old stand appears to be in a steady-state condition. The density of lodgepole pine seedlings in the burned stands (Lo-PICO, Hi-PICO, nonforest) is from the classification in Table 14.5.

soil), and the lowest rates of ammonium production were observed in the 1996 Pelican burn and the post-1988 high-density pine seedling sites (0.26 and 0.32 mg NH_4-N/kg soil, respectively). (See Chapter 5 for comparisons of nitrogen in burned and unburned grasslands.)

Streamwater nitrate concentrations, 1997—Samples taken at the end of snowmelt revealed very low streamwater nitrate concentrations (0.04–0.12 mg/L) across all nine first-order ephemeral streams, with no consistent differences between burned and unburned watersheds or among burned watersheds differing in vegetation structure and composition (data not shown). These consistently low values indicate that nitrate concentrations were not elevated even in streams flowing out of the 1996 Pelican burn, as was found by Minshall and others (Chapter 8).

ASSESSMENT: IS ECOSYSTEM FUNCTION RECOVERING NORMALLY?

Our studies to date indicate that ecosystem processes related to ANPP, LAI, and nitrogen dynamics are recovering very rapidly throughout the areas burned in 1988. ANPP and LAI are highly correlated with tree seedling density, and where seedlings are sparse, ANPP and LAI are still relatively low. However, in stands having moderate to high seedling density, by nine years after the fires ANPP and LAI were approaching or already within the range of values observed in mature coniferous forests (Reed and others 1999). Although nitrification was likely elevated immediately following the 1988 fires (as indicated by the data from the 1996 Pelican burn), our data indicate that nitrification rates ten years following the fires were indistinguishable from rates in mature lodgepole pine forests.

We found no evidence of extensive nitrogen loss in streamwater, even in the Pelican burn one year after fire, probably because of rapid uptake and incorporation into plant biomass of this limiting nutrient. Other studies in Yellowstone and in boreal forests (McColl and Grigal 1975, Wright 1976, Albin 1979, Minshall and others 1989) have found similar low nutrient concentrations in streams flowing out of burned watersheds. Franke (2000:94–95) has summarized the studies on lake, stream, and groundwater nutrient concentrations that were conducted during the first few years after the 1988 fires. Results were mixed and inconsistent. Several studies found little or no changes in water chemistry in response to the fires, even in heavily burned watersheds. Other studies detected changes in water chemistry after 1988 but suggested that the changes were due to other factors (for example, drought) rather than direct effects of fire. Two studies reported that nitrate concentrations were 3–33 times as high in streams flowing out of burned watersheds as in unburned watersheds. We conclude from these mixed results that nutrient dynamics related to fire probably vary in relation to watershed characteristics, fire size and severity, postfire weather conditions such as localized storms (Chapter 3, 8), and postfire vegetation responses. More research is needed to clarify these relationships; but the evidence to date suggests that nutrient loss associated with the 1988 fires was localized and of relatively small magnitude overall.

In sum, the 1988 Yellowstone fires appear to have had relatively small and transient effects on ecosystem processes related to energy flow, leaf area, and nutrient dynamics. We see no evidence that the fires "damaged" the system in any important way, nor do we see any need for restoration of ecosystem processes. Interestingly, however, our studies have documented substantial spatial variation in ANPP, LAI, and N mineralization, in relation to the very wide range in postfire density of lodgepole pine seedlings across the burned landscape. Understanding the implications of landscape heterogeneity and disturbance dynamics for ecosystem processes is an important challenge for ecologists (Schimel and others 1997, Turner and Carpenter 1999). A major focus of our ongoing research is to characterize this variability and its consequences for ecosystem function at both stand and landscape scales.

HAVE RECENT DISTURBANCES BEEN ASSOCIATED WITH ANY NOVEL OR UNEXPECTED EFFECTS?

Most of the areas that burned in 1988 appear to be on a successional trajectory returning to vegetation structure and composition very similar to what existed before the fires—for example, where dense mature pine stands burned, we generally see dense stands of pine seedlings today; where open stands burned, we see sparse stands of seedlings; and where sagebrush or meadows burned, we see reestablishment of sagebrush or herbaceous species. However, there was one big surprise after 1988. Quaking aspen *(Populus tremuloides),* which in this region usually reproduces exclusively by vegetative root sprouting, established seedlings throughout much of the burned subalpine forest area (Kay 1993, Romme and others 1997). Turner and others (1998) predicted that novel successional pathways are most probable following disturbances of large spatial extent and high intensity, because such disturbances leave fewer biological legacies and are more slowly filled by propagules migrating from undisturbed areas. The rare episode of sexual reproduction in aspen following the 1988 Yellowstone fires appears to support this prediction.

Extensive surveys across the subalpine plateaus of Yellowstone Park in 1993 and 1996 revealed that aspen seedlings were found only in burned areas, and that they were patchily distributed with densities of 1 to more than 300 stems/ha (Romme and others 1997, Turner and others 2003a). Seedlings were most abundant in the west-central portions of Yellowstone Park, and least abundant in the north-central part of the park. More severely burned sites generally supported more aspen seedlings than less severely burned sites, and small or moderate-sized burned patches had higher densities than large patches (Romme and others 1997, Turner and others 2003a,b). Seedlings were found in many places that had no aspen before 1988, suggesting a possible range expansion. The seedling populations on the Yellowstone Plateau exhibited greater genetic diversity than adult aspen clones in Yellowstone's northern winter ungulate range (Tuskan and others 1996), but there were no strong genetic patterns among seedling populations that could be explained by differences in elevation, substrate, or geographic location (Stevens and others 1999). The genetic studies indicated that initial seedling establishment patterns were largely random with respect to genetic makeup, and that strong selection pressures have not yet shaped these new aspen populations. However, such selection is likely to occur in the near future.

Continuing surveys indicate that the aspen seedlings are persisting throughout most of the areas where they originally occurred, but densities are decreasing in many places (Romme and Turner, in preparation). Mortality appears largely due to intense ungulate herbivory, but many seedlings are failing to thrive even in exclosures—suggesting that local climate or soils may be unsuitable for aspen in many places (Romme and Turner, in preparation). Most of the seedlings will likely perish over the next few decades, leaving no lasting ecological legacy. However, at least a few new aspen genets are likely to survive and produce new aspen clones. These individuals will inject new genetic diversity into the regional aspen population, and will change the local ecology of sites where aspen was absent before 1988. Especially at lower elevations and in places where seedlings are protected from elk browsing by piles of fallen fire-killed tree boles, many aspen seedlings now exhibit heights greater than 1 m and a robust growth form that bodes well for their long-term survival. Here we see another surprise—the abundant coarse woody debris created by the fires (Tinker and Knight 2000, Chapter 12) may enhance long-term aspen persistence and genetic diversity (Ripple and Larsen 2000). Had the National Park Service "salvaged" the great numbers of dead trees in the aftermath of the fires (fortunately the service had neither the ability nor the will to accomplish this), the establishment of new aspen genets might have been greatly reduced.

Tuskan and others (1996) suggest that episodes of new aspen genet establishment may typically exhibit just such a pattern of initially high seedling densities, followed by extensive mortality but survival of a few individuals that are best adapted to local conditions and which go on to produce clonal structures (Eriksson 1993). Additional seedling establishment events of this kind in the future may be crucial to long-term survival of aspen in the Yellowstone region if climate change in the next century shifts the elevational zones of tolerance for plant species (Romme and Turner 1991, Bartlein and others 1997) and if continued browsing pressure combined with drought and other stresses causes local extirpation of some old aspen clones (Romme and others 2001).

The widespread establishment of aspen seedlings after the 1988 fires was unexpected, but consistent with current ecological theory (for example, Turner and others 1998). Although postfire succession will be different in some local areas because of this unexpected consequence of fire size and severity, we do not regard aspen seedling establishment to be any kind of threat to the integrity of the Yellowstone ecosystem. On the contrary, the addition of new, genetically novel individuals to the regional aspen population may enhance the resilience of this keystone plant species to ecological changes expected in the next century.

In summary, we posed six general questions to guide assessments of the need for active intervention to restore damaged ecosystem elements or processes (Table 14.1). Within this framework, we evaluated the effects of the 1988 Yellowstone fires and of pre-1988 fire suppression. Our conclusion is that the Yellowstone subalpine ecosystem is still within or nearly within its range of natural variability with respect to (1) disturbance regime, (2) community composition and structure at both stand and landscape scales, and (3) ecosystem function as measured by aboveground net primary productivity, leaf area index, and nitrogen dynamics. The 1988 fires produced quantifiable and spatially heterogeneous effects on ecosystem composition, structure, and function, but none of these effects appears to be either abnormal or a threat to long-term ecological integrity. The Yellowstone subalpine landscape requires no active intervention or restoration. On the contrary, it provides an outstanding example of a naturally functioning system. Yellowstone's high-elevation forests provide a valuable benchmark for comparison of other subalpine forest systems in western North America that have been altered by such activities as timber harvest and extensive road building (for example, Knight and others 2000).

There are at least three reasons why the large and severe fires of 1988 did not cause serious or long-lasting damages to the organisms or ecological processes of the Yellowstone subalpine landscape. First, this area's disturbance history is characterized by similar infrequent but large, severe fires. This disturbance history leads to the second reason why fire damage was so slight: past disturbances have selected for biological characteristics that enable the organisms to withstand large, severe fires (Chapters 4, 5, 15). Many herbs and shrubs are able to resprout from roots and rhizomes, and then to produce prolific seeds within two or three years after the fire (for example, *Epilobium angustifolium, Carex geyeri, Lupinus argenteus* [see Chapter 5 for more detail on the ecological roles of *Lupinus*], *Calamagrostis canadensis*). Others have long-persistent seed banks in the soil (for example, *Dracocephalum parviflorum, Geranium bicknellii, Ceanothus velutinus*) or in the canopy (for example, *Pinus contorta*). The rapid reestablishment of plant biomass probably also involves rapid uptake of nutrients and therefore minimal nutrient loss from the system. The third reason why the system was so resilient to the 1988 fires is that there were very few sources of alien plant species capable of invading the burned areas and preempting the space from the native species. We urge the managers of Yellowstone Park to continue an aggressive program of controlling alien weeds so that future large fires do not result in an explosion of nonnative species and potentially very different consequences from those that we measured after 1988.

Many systems can withstand a single disturbance, even if it is large and intense, if the organisms are adapted to that kind of disturbance (Paine and others 1998). However, where multiple disturbances occur within the time frame required for recovery, the system may be pushed into a new state (also see Turner and others 1993). The wilderness landscapes of Yellowstone National Park apparently were not significantly altered either by the 1988 fires or by pre-1988 fire suppression. However, serious future changes could occur if fire frequency increases due to global climate change (Chapters 2, 13), or if the system becomes further stressed by drought or invasion of alien species. Indeed, other Rocky Mountain subalpine ecosystems have been substantially altered by twentieth-century changes in disturbance regimes and by invasion of alien plant species. In national forests near Yellowstone Park, for example, intensive clear-cutting and road building during the past fifty years have dramatically altered landscape patterns (for example, Tinker and others 1998, Baker and Knight 2000). Some of these intensively logged areas also burned in 1988. The combination of intensive logging, road building, and severe fire may constitute the kind of compounded, interacting disturbances that can push an ecosystem into a qualitatively different state (Paine and others 1998). Many subalpine landscapes throughout the Rocky Mountains, where intensive timber harvest and road-building have been conducted, may be at risk of serious degradation if large fires occur in the near future. As a baseline reference for natural ecological structure and processes, the value of Yellowstone's largely untrammeled subalpine system will only increase as we proceed through the next century.

Acknowledgments
The research described here would not have been accomplished without contributions from our collaborators and all the members of our field research crews over the past two decades. Participants are too numerous to list here, but we especially thank Bob Gardner, Dan Tinker, Don Despain, and Dennis Knight for their many contributions and discussions. The staff of Yellowstone National Park and the University of Wyoming–National Park Service Research Center

have provided valuable logistical support. Research funding provided by the National Science Foundation (grants in 1990, 1991, and 1997), USDA National Research Initiative Competitive Grants Program, Forest/Range/Crop/Aquatic Ecosystems Program (1996), and the National Geographic Society (1989 and 1996) is gratefully acknowledged. We also thank Greg Aplet for his helpful comments on an early draft of the manuscript.

References

Albin, D. P. 1979. Fire and stream ecology in some Yellowstone Lake tributaries. California Fish and Game 65:216–238.

Allen, C. D., M. Savage, D. A. Falk, K. F. Suckling, T. W. Swetnam, T. Schulke, P. B. Stacey, P. Morgan, M. Hoffman, and J. T. Klingel. 2002. Ecological restoration of southwestern ponderosa pine ecosystems: A broad perspective. Ecological Applications 12:1418–1433.

Alpchem Corp. 1990. Ammonia. RFA Method no. A303-S021-2, rev. A.

———. 1991. Nitrate/Nitrite. RFA Method no. A303-S173.

Baker, W. L., and R. L. Knight. 2000. Roads and forest fragmentation in the Southern Rocky Mountains. Pages 97–122 in R. L. Knight, F. W. Smith, S. W. Buskirk, W. H. Romme, and W. L. Baker, eds., Forest fragmentation in the Southern Rocky Mountains. University Press of Colorado, Boulder.

Balda, R. P. 1975. The relationship of secondary cavity nesters to snag densities in western coniferous forests. Southwest Wildlife Habitat Technical Bulletin no. 1, USDA Forest Service, Albuquerque, N.M.

Balling, R. C., Jr., G. A. Meyer, and S. G. Wells. 1992. Climate change in Yellowstone National Park: Is the drought-related risk of wildfires increasing? Climatic Change 22:35–45.

Barrett, S. W. 1994. Fire regimes on andesitic mountain terrain in northeastern Yellowstone National Park, Wyoming. International Journal of Wildland Fire 4:65–76.

Bartlein, P. J., C. Whitlock, and S. L. Shafer. 1997. Potential future environmental change in the Yellowstone National Park region. Conservation Biology 11:782–792.

Binkley, D. 1984a. Does forest removal increase rates of decomposition and nitrogen release? Forest Ecology and Management 8:229–233.

———. 1984b. Ion exchange resin bags: Factors affecting estimates of nitrogen availability. Soil Science Society of America Journal 48:1181–1184.

Binkley, D., J. Aber, J. Pastor, and K. Nadelhoffer. 1986. Nitrogen availability in some Wisconsin forests: Comparisons of resin bags and on-site incubations. Biol. Fertil. Soils 2:77–82.

Binkley, D., R. Bell, and P. Sollins. 1992a. Comparison of methods for estimating soil nitrogen transformations in adjacent conifer and alder-conifer forests. Canadian Journal of Forest Research 22:858–863.

Binkley, D., and S. C. Hart. 1989. The components of nitrogen availability assessments in forest soils. Advances in Soil Science 10:57–112.

Binkley, D., and P. Matson. 1983. Ion exchange resin bag method for assessing forest soil nitrogen availability. Soil Science Society of America Journal 47:1050–1052.

Binkley, D., D. Richter, M. B. David, and B. Caldwell. 1992b. Soil chemistry in a loblolly/longleaf pine forest with interval burning. Ecological Applications 2:157–164.

Binkley, D., F. W. Smith, and Y. Son. 1995. Nutrient supply and declines in leaf area and production in lodgepole pine. Canadian Journal of Forest Research 25:621–628.

Bormann, F. H., and G. E. Likens. 1979. Biogeochemistry of a forested ecosystem. Springer-Verlag, New York.

Brown, G. W., A. R. Gahler, and R. B. Marston. 1973. Nutrient losses after clear-cut logging and slash burning in the Oregon Coast Range. Water Resources Research 9:1450.

Christensen, N. L., J. K. Agee, P. F. Brussard, J. Hughes, D. H. Knight, G. W. Minshall, J. M. Peek, S. J. Pyne, F. J. Swanson, J. W. Thomas, S. Wells, S. E. Williams, and H. A. Wright. 1989. Interpreting the Yellowstone fires of 1988. BioScience 39:678–685.

Cole, D. N. 2000. Paradox of the primeval: Ecological restoration in wilderness. Ecological Restoration 18:77–86.

Cole, D. N., and P. B. Landres. 1996. Threats to wilderness ecosystems: Impacts and research needs. Ecological Applications 6:168–184.

Covington, W. W., L. F. DeBano, and T. G. Huntsberger. 1991. Soil nitrogen changes associated with slash pile burning in pinyon-juniper woodlands. Forest Science 37:347–355.

Covington, W. W., and M. M. Moore. 1994. Southwestern ponderosa forest structure: Changes since Euro-American settlement. Journal of Forestry 92:39–47.

Covington, W. W., and S. S. Sackett. 1992. Soil mineral nitrogen changes following prescribed burning in ponderosa pine. Forest Ecology and Management 54:175–191.

Dahms, C. W., and B. W. Geils, tech. eds. 1997. An assessment of forest ecosystem health in the Southwest. USDA General Technical Report RM-GTR-295.

D'Antonio, C. M. 2000. Fire, plant invasions, and global change. Pages 65–93 in H. Mooney and R. Hobbs, eds., Invasive species in a changing world. Island Press, Washington, D.C.

Dayton, P. K., M. J. Tegner, P. B. Edwards, and K. L. Riser. 1998. Sliding baselines, ghosts, and reduced expectations in kelp forest communities. Ecological Applications 8:309–322.

Despain, D. G. 1990. Yellowstone vegetation: Consequences of environment and history. Roberts Rinehart, Boulder, Colo.

DiStefano, J., and H. Gholz. 1986. A proposed use of ion exchange resin to measure nitrogen mineralization and nitrification in intact soil cores. Commun. Soil Sci. Plant Anal. 17:989–998.

Drake, J. A., H. A. Mooney, F. diCastri, R. H. Groves, F. J. Kruger, M. Rejmanek, and M. Williamson, eds. 1989. Biological invasions: A global perspective. Scope 37. Wiley, Chichester, England.

Eriksson, O. 1993. Dynamics of genets in clonal plants. TREE 8:313–316.

Fahey, T. J., and D. H. Knight. 1986. Lodgepole pine ecosystems. BioScience 36:610–617.

Fahey, T. J., J. B. Yavitt, J. A. Pearson, and D. H. Knight. 1985. The nitrogen cycle in lodgepole pine ecosystems. Biogeochemistry 1:257–275.

Flannigan, M. D., and C. E. Van Wagner. 1991. Climate change and wildfire in Canada. Canadian Journal of Forest Research 21:66–72.

Floyd, D. A., and J. E. Anderson. 1987. A comparison of three methods for estimating plant cover. Journal of Ecology 75:221–228.

Foster, D. R., D. H. Knight, and J. F. Franklin. 1998. Landscape patterns and legacies resulting from large, infrequent forest disturbances. Ecosystems 1:497–510.

Franke, M. A. 2000. Yellowstone in the afterglow: Lessons from the fires. Yellowstone Center for Resources, Yellowstone National Park, Wyo.

Friederici, P., ed. 2003. Ecological restoration of southwestern ponderosa pine forests. Island Press, Washington, D.C.

Fule, P. Z., W. W. Covington, and M. M. Moore. 1997. Determining reference conditions for ecosystem management of southwestern ponderosa pine forests. Ecological Applications 7:895–908.

Gardner, R. H., W. W. Hargrove, M. G. Turner, and W. H. Romme. 1996. Global change, disturbances, and landscape dynamics. Pages 149–172 in B. Walker and W. Steffen, eds., Global change and terrestrial ecosystems: The first GCTE Science Conference, 23–27 May 1994, Woods Hole, MA. Cambridge University Press, Cambridge.

Giblin, A. E., J. A. Laundre, K. J. Nadelhoffer, and G. R. Shaver. 1994. Measuring nutrient availability in arctic soils using ion exchange resins: A field test. Soil Sci. Soc. Am. J. 58:1154–1162.

Gosz, J. R., and C. S. White. 1986. Seasonal and annual variation in nitrogen mineralization and nitrification along an elevational gradient in New Mexico. Biogeochemistry 2:281–297.

Grissino-Mayer, H. D. 1999. Modeling fire interval data from the American southwest with the Weibull distribution. International Journal of Wildland Fire 9:37–50.

Hardy, C. C., and S. F. Arno, eds. 1996. The use of fire in forest restoration. USDA Forest Service General Technical Report INT-GTR-341.

Harig, A. L., and M. B. Bain. 1998. Defining and restoring biological integrity in wilderness lakes. Ecological Applications 8:71–87.

Hart, S. C. 1999. Nitrogen transformations in fallen tree boles and mineral soil of an old-growth forest. Ecology 80:1385–1394.

Hart, S. C., and D. Binkley. 1985. Correlations among indices of forest soil nutrient availability in fertilized and unfertilized loblolly pine plantations. Plant Soil 85:11–21.

Hart, S. C., and M. K. Firestone. 1989. Evaluation of three in situ nitrogen availability assays. Canadian Journal of Forest Research 19:185–191.

Helvey, J. D., A. R. Tiedemann, and W. B. Fowler. 1976. Some climatic and hydrologic effects of wildfire in Washington State. Proceedings of the Tall Timbers Fire Ecology Conference (1974) 15:201–222.

Hemstrom, M. A., and J. F. Franklin. 1982. Fire and other disturbances of the forests in Mount Rainier National Park. Quaternary Research 18:32–51.

Houston, D. B. 1973. Wildfires in northern Yellowstone National Park. Ecology 54:1111–1117.

Jewett, K., D. Daugharty, H. H. Krause, and P. A. Arp. 1995. Watershed responses to clear-cutting: Effects on soil solutions and stream water discharge in central New Brunswick. Canadian Journal of Soil Science 75:475–490.

Johnson, E. A. 1992. Fire and vegetation dynamics. Cambridge University Press, Cambridge.

Johnson, E. A., and S. L. Gutsell. 1993. Heat budget and fire behavior associated with the opening of serotinous cones in two *Pinus* species. Journal of Vegetation Science 4:745–750.

Johnson, E. A., and D. R. Wowchuk. 1993. Wildfires in the southern Canadian Rocky Mountains and their relationship to mid-tropospheric anomalies. Canadian Journal of Forest Research 23:1213–1222.

Jurgensen, M. F., A. E. Harvey, R. T. Graham, D. S. Page-Dumroese, J. R. Tonn, M. J. Larsen, and T. B. Jain. 1997. Impacts of timber harvesting on soil organic matter, nitrogen, productivity, and health of inland northwest forests. Forest Science 43:234–251.

Kashian, D. M. 2002. Landscape variability and convergence in forest structure and function following large fires in Yellowstone National Park. Ph.D. diss., University of Wisconsin, Madison.

Kay, C. E. 1993. Aspen seedlings in recently burned areas of Grand Teton and Yellowstone National Parks. Northwest Science 67:94–104.

Knight, D. H., T. J. Fahey, and S. W. Running. 1985. Water and nutrient outflow from contrasting lodgepole pine forests in Wyoming. Ecological Monographs 55:29–48.

Knight, D. H., and L. L. Wallace. 1989. The Yellowstone fires: Issues in landscape ecology. BioScience 39:700–706.

Knight, R. L., F. W. Smith, S. W. Buskirk, W. H. Romme, and W. L. Baker, eds. 2000. Forest fragmentation in the Southern Rocky Mountains. University of Colorado Press, Boulder.

Landres, P. B., P. Morgan, and F. J. Swanson.1999. Overview of the use of natural variability concepts in managing ecological systems. Ecological Applications 9:1179–1188.

Likens, G. E., and F. H. Bormann. 1995. Biogeochemistry of a forested ecosystem. 2d. ed. Springer-Verlag, New York.

Lynch, D. L., W. H. Romme, and M. L. Floyd. 2000. Ecology and economics in forest restoration: An operational study in southwest Colorado. Journal of Forestry.

McColl, J. G., and D. F. Grigal. 1975. Forest fire: Effects on phosphorus movement to lakes. Science 185:1109–1111.

Meagher, M., and D. B. Houston. 1998. Yellowstone and the biology of time: Photographs across a century. University of Oklahoma Press, Norman.

Millspaugh, S. H., C. Whitlock, and P. J. Bartlein. 2000. Variations in fire frequency and climate over the past 17000 yr in central Yellowstone National Park. Geology 28:211–214.

Minshall, G. W., J. T. Brock, and J. D. Varley. 1989. Wildfire and Yellowstone's stream ecosystems. BioScience 39:707–715.

Mooney, H. A., and J. A. Drake, eds. 1986. Ecology of biological invasions of North America and Hawaii. Springer-Verlag, Berlin.

Moore, M. M., W. W. Covington, and P. Z. Fule. 1999. Reference conditions and ecological restoration: A southwestern ponderosa pine perspective. Ecological Applications 9:1266–1277.

Overpeck, J. T., D. Rind, and R. Goldberg. 1990. Climate-induced changes in forest disturbance and vegetation. Nature 343:51–53.

Paine, R. T., M. J. Tegner, and E. A. Johnson. 1998. Compounded perturbations yield ecological surprises. Ecosystems 1:535–545.

Pare, D., Y. Bergeron, and C. Camire. 1993. Changes in the forest floor of Canadian southern boreal forest after disturbance. Journal of Vegetation Science 4:811–818.

Parsons, W. F. J., D. H. Knight, and S. L. Miller. 1994. Root gap dynamics in lodgepole pine forest: Nitrogen transformation in gaps of different size. Ecological Applications 4:354–362.

Pickett, S. T. A., and V. T. Parker. 1994. Avoiding the old pitfalls: Opportunities in a new discipline. Restoration Ecology 2:75–79.

Reed, R. A., M. E. Finley, W. H. Romme, and M. G. Turner. 1999. Aboveground net primary production and leaf area index in initial postfire vegetation communities in Yellowstone National Park. Ecosystems 2:88–94.

Reich, P. B., D. F. Grigal, J. D. Aber, and S. T. Gower. 1997. Nitrogen mineralization and productivity in 50 hardwood and conifer stands on diverse soils. Ecology 78:335–347.

Renkin, R. A., and D. G. Despain. 1992. Fuel moisture, forest type, and lightning-caused fire in Yellowstone National Park. Canadian Journal of Forest Research 22:37–45.

Ripple, W. J., and E. J. Larsen. 2000. The role of postfire coarse woody debris in aspen regeneration. Western Journal of Applied Forestry 16(2).

Romme, W. H. 1982. Fire and landscape diversity in subalpine forests of Yellowstone National Park. Ecological Monographs 52:199–221.

Romme, W. H., and D. G. Despain. 1989. Historical perspective on the Yellowstone fires of 1988. BioScience 39:695–699.

Romme, W. H., L. Floyd-Hanna, D. D. Hanna, and E. Bartlett. 2001. Aspen's ecological role in the West. Proceedings of the aspen symposium, Grand Junction, CO, June 2000. USDA Forest Service, Rocky Mountain Experiment Station.

Romme, W. H., and D. H. Knight. 1982. Landscape diversity: The concept applied to Yellowstone Park. BioScience 32:664–670.

Romme, W. H., and M. G. Turner. 1991. Implications of global climate change for biogeographic patterns in the Greater Yellowstone Ecosystem. Conservation Biology 5:373–386.

Romme, W. H., M. G. Turner, R. H. Gardner, W. W. Hargrove, G. A. Tuskan, D.G. Despain, and R. Renkin. 1997. A rare episode of sexual reproduction in aspen (*Populus tremuloides* Michx.) following the 1988 Yellowstone fires. Natural Areas Journal 17:17–25.

Rothermel, R. C. 1991. Predicting behavior of the 1988 Yellowstone fires: Projections versus reality. International Journal of Wildland Fire 1:1–10.

Schimel, D. S., VEMAP Participants, and B. H. Braswell. 1997. Continental scale variability in ecosystem processes: Models, data, and the role of disturbance. Ecological Monographs 67:251–271.

Schullery, P. 1989. The fires and fire policy. BioScience 39:686–694.

———. 1997. Searching for Yellowstone: Ecology and wonder in the last wilderness. Houghton Mifflin, Boston.

Singer, F. J., W. Schreier, J. Oppenheim, and E. O. Garton. 1989. Drought, fires, and large mammals. BioScience 39:716–722.

Stevens, M. T., M. G. Turner, G. A. Tuskan, W. H. Romme, L. E. Gunter, and D. M. Waller. 1999. Genetic variation in postfire aspen seedlings in Yellowstone National Park. Molecular Ecology 8:1769–1780.

Strader, R., D. Binkley, and C. Wells. 1989. Nitrogen mineralization in high elevation forests of the Appalachians, part 1, Regional patterns in southern spruce-fir forests. Biogeochemistry 7:131–145.

Street, R. B. 1989. Climate change and forest fires in Ontario. Pages 177–182 in Proceedings from the tenth conference on fire and forest meteorology, Forestry Canada, Environment Canada.

Strong, W. E. 1968. A trip to the Yellowstone National Park in July, August, and September, 1875. University of Oklahoma Press, Norman.

Swetnam, T. W., C. D. Allen, and J. L. Betancourt. 1999. Applied historical ecology: Using the past to manage for the future. Ecological Applications 9:1189–1206.

Swetnam, T. W., and J. L. Betancourt. 1998. Mesoscale disturbance and ecological response to decadal climatic variability in the American Southwest. Journal of Climate 11:3128–3147.

Tiedemann, A. R., J. D. Helvey, and T. D. Anderson. 1978. Stream chemistry and watershed nutrient economy following wildfire and fertilization in eastern Washington. Journal of Environmental Quality 7:580–588.

Tinker, D. B., and D. H. Knight. 2000. Coarse woody debris following fire and logging in Wyoming lodgepole pine forests. Ecosystems 3:472–483.

Tinker, D. B., C. A. C. Resor, G. P. Beauvais, K. F. Kipfmueller, C. I. Fernandes, and W. L. Baker. 1998. Watershed analysis of forest fragmentation by clear-cuts and roads in a Wyoming forest. Landscape Ecology 13:149–165.

Tinker, D. B., W. H. Romme, W. W. Hargrove, R. H. Gardner, and M. G. Turner. 1994. Landscape-scale heterogeneity in lodgepole pine serotiny. Canadian Journal of Forest Research 24:897–903.

Tomback, D. F., S. F. Arno, and R. E. Keane, eds. 2001. Whitebark pine communities: Ecology and restoration. Island Press, Washington, D.C.

Turner, M. G., W. L. Baker, C. J. Peterson, and R. K. Peet. 1998. Factors influencing succession: Lessons from large, infrequent natural disturbances. Ecosystems 1:511–523.

Turner, M. G., and S. R. Carpenter. 1999. Spatial variability in ecosystem function: Introduction to special feature. Ecosystems 2:383.

Turner, M. G., W. H. Hargrove, R. H. Gardner, and W. H. Romme. 1994. Effects of fire on landscape heterogeneity in Yellowstone National Park, Wyoming. Journal of Vegetation Science 5:731–742.

Turner, M. G., and W. H. Romme. 1994. Landscape dynamics in crown fire ecosystems. Landscape Ecology 9:59–77.

Turner, M. G., W. H. Romme, and R. H. Gardner. 2000. Prefire heterogeneity, fire severity, and plant reestablishment in subalpine forests of Yellowstone National Park, Wyoming. International Journal of Wildland Fire 9:21–36.

Turner, M. G., W. H. Romme, R. H. Gardner, and W. W. Hargrove. 1997. Effects of fire size and pattern on succession in Yellowstone National Park. Ecological Monographs 67:411–433.

Turner, M. G., W. H. Romme, R. H. Gardner, R. V. O'Neill, and T. K. Kratz. 1993. A revised concept of landscape equilibrium: Disturbance and stability on scaled landscapes. Landscape Ecology 8:213–227.

Turner, M. G., W. H. Romme, R. A. Reed, and G. A. Tuskan. 2003a. Post-fire aspen seedling recruitment across the Yellowstone (USA) landscape. Landscape Ecology 18:127–140.

Turner, M. G., W. H. Romme, and D. B. Tinker. 2003b. Surprises and lessons from the 1988 Yellowstone fires. Frontiers in Ecology and the Environment 1:351–358.

Tuskan, G. A., K. E. Francis, S. L. Russ, W. H. Romme, and M. G. Turner. 1996. RAPD markers reveal diversity within and among clonal and seedling stands of aspen in Yellowstone National Park, U.S.A. Canadian Journal of Forest Research 26:2088–2098.

Van Cleve, K. J., J. Yarie, R. Erickson, and C. T. Dyrness. 1993. Nitrogen mineralization and nitrification in successional ecosystems on the Tanana River floodplain, interior Alaska. Canadian Journal of Forest Research 23:970–978.

Vitousek, P. M., J. R. Gosz, C. C. Grier, J. M. Melillo, W. A. Reiners, and R. L. Todd. 1979. Nitrate losses from disturbed ecosystems. Science 204:469–474.

Vitousek, P. M., and P. A. Matson. 1985. Disturbance, nitrogen availability, and nitrogen losses in an intensively managed loblolly pine plantation. Ecology 66:1360–1376.

Vitousek, P. M., and J. M. Melillo. 1979. Nitrate losses from disturbed forests: Patterns and mechanisms. Forest Science 25:605–619.

Wright, R. F. 1976. The impact of forest fire on the nutrient influxes to small lakes in northeastern Minnesota. Ecology 57:649–663.

Chapter 15 Epilogue: After the Fires. What Have We Learned?

Linda L. Wallace and Norman L. Christensen

Yellowstone National Park is in the same position as our nation, which also stands at a crossroad, faced by fearful decisions.
—*Aubrey L. Haines,* The Yellowstone Story

For well over a century, disturbance and the changes deriving from it have been central objects of study for ecologists. The Yellowstone fires have most certainly provided an ideal laboratory for the expansion of such studies. That said, it is also true that ecologists have historically not always communicated results of such studies effectively to natural resource managers (Peters 1991), with the result that management protocols and practices are not always informed by the most up-to-date science. Such communication is a central goal of this volume.

What sorts of questions do resource managers and the general public have in the wake not only of the Yellowstone fires of 1988 but also of several subsequent severe fire years in the western United States? First on most people's lists would be the likelihood that a large, uncontrollable fire would occur (or recur) in their area of interest. Second, how do ecosystems recover from such events and what sorts of management might improve the recovery process? Other questions of interest would include how specific organisms or areas of interest are affected by fire including rare and endangered species or charismatic organisms, and favorite habitats such as hunting, fishing, or scenic areas. Embedded in these latter questions is the concern about the time required to recover prefire conditions. Finally, important questions would include how might we manage our landscapes to either keep future fires like this from recurring or how might we manage them to ensure maximum resiliency when they do occur. These management questions should be posed in the context of a changing environment, not only with respect to climate, but also with respect to increased human population pressures on these landscapes. The preceding chapters provide a wealth of information relevant to each of these questions.

HOW LIKELY IS IT THAT LARGE FIRES WOULD RECUR IN THE YELLOWSTONE ECOSYSTEM?

To calculate the likelihood of recurrence of large fires in the Yellowstone Ecosystem, we need to know what previous fire frequencies were and how those frequencies related to climate and vegetation and, perhaps, previous human management. In Chapter 2, Millspaugh and Whitlock show that the Yellowstone Plateau has experienced a variety of climatic regimes during the Holocene (the past 12,000 years). During wetter periods, large-scale fires were less frequent, and during drier periods such fires were more frequent. It is important to note that the park is very heterogeneous and that climatic conditions in one portion of the park may be quite different from those in another area. There is a natural divide between two major regions, the northern range (Yellowstone-Lamar Province, Chapter 3) and the subalpine plateau (Central Plateau Province, Chapter 3). Using sediment cores that yielded vegetation and charcoal deposition records, Millspaugh and Whitlock found that fire frequency in the northern range varied between 5 events/1000 years and more than 15 events/1000 years, with a gradual increase in fire frequency from about 6000 years B.P. to the present (Chapter 2). In the subalpine plateau fires are less frequent because this area's climate is moister and less variable than at the lower-elevation, northern sites. However long-term, fire frequencies also are closely correlated with climate, varying from 3 to 10 events/1000 years. Fire frequency has been decreasing recently in this portion of the park (Chapter 2).

Meyer (Chapter 3) used the geomorphology of debris flows to date fires and establish prehistoric fire frequencies and occurrence patterns across the park, finding that at low-elevation sites fire frequencies varied between 9 and 17 events/1000 years and that frequencies were lower at higher elevations. Again, fire frequency was correlated quite closely with climatic regimes, with few fires occurring during the "Little Ice Age" (100–700 years B.P.) (Chapters 3, 14).

It is clear that large fires—on the scale of those in 1988—have occurred many times and that large fires will likely recur in Yellowstone National Park in the future. Furthermore, their occurrence will likely depend on climatic variables. It appears that the fire suppression activities of the past contributed little to the occurrence of the 1988 event (Chapters 3, 14). If various models of global warming that predict warmer, drier conditions in the Yellowstone region in the decades ahead are accurate, then fires may be more likely in some parts of the park (Chapter 13).

In addition to climatic factors, fire is dependent upon the availability of fuel (in other words, vegetation). If similar, relatively flammable vegetation (for example, lodgepole pine forest) regrows following a fire, then the likelihood of a fire recurring in that area is greater than if prefire vegetation is replaced by vegetation that is less likely to burn either because of lower amount of fuel production or because of some inherent quality of the vegetation that reduces its flammability. In both Yellowstone Park (Chapter 4) and in Grand Teton National Park (Chapter 11), the type of vegetation found in burned areas after the fire was strongly positively correlated with the type of vegetation found prior to the fires. Interestingly, both Anderson and others (Chapter 4) and Doyle (Chapter 11) found that postfire vegetation development did not follow the "rules" of standard Clementsian succession; in other words, the forests were not growing back as a series of different plant communities, one replacing the other over time. Rather, the dominant species found in the area prior to the fire was in most cases the species that regrew quickly and occupied that same space. This, too, would lead to the conclusion that large fires are likely to recur.

In addition, the legacy of large fires includes lots of downed trees and dead logs (coarse woody debris) (Chapter 12). In addition to being important insect habitats and nutrient "sponges," when these materials dry, they, too, provide potential fuel for future fires.

HOW CAN THE ECOSYSTEM RECOVER FOLLOWING A LARGE FIRE?

Despite a cool, dry, and relatively nutrient-poor environment, recovery to prefire conditions is occurring relatively rapidly in the Greater Yellowstone Ecosystem (Chapters 4, 5, 7, 8, 9, 11, 13). This is surprising for most people given that the visual effects of the fire are still quite dramatic, even ten to fifteen years following the fire. However, many aspects of species composition (Chapters 4, 5, 6, 8, 10, 11, 14) and ecosystem function (Chapter 14) returned to their prefire conditions within the first decade following the fire.

Leaf area (usually expressed as Leaf Area Index, LAI, the amount of leaf area per unit area of ground) is an especially important feature affecting ecosystem functions. In particular, it regulates the amount of water used by the forest as well as the amount of nutrients taken up by the roots. When leaf area is lost via fire, excess water (Chapter 3) and nutrients (Chapter 14) may then move into aquatic systems (Chapters 7, 8). Romme and Turner (Chapter 14) showed that within nine years, the leaf area of lodgepole pine forests had recovered to prefire levels. Further, they showed that even one year after fire, nutrient losses from terrestrial systems were generally insignificant. There was variation in the amounts of nutrients reaching streams, but this variation was due to local topography, the local levels of fire severity, and localized storms (Chapter 3). Nutrient retention was probably improved by the abundant coarse woody debris left following a fire (Chapter 12), as well as by the rapid regrowth of herbaceous species. In nonforested areas, Tracy (Chapter 5) found that fire stimulated invasion and growth of lupine with a subsequent increase in soil moisture and nutrients in lupine patches.

Local variations in hydrology also affect the amount of nutrient and sediment loss from terrestrial to aquatic. Farnes and others (Chapter 10) show that water yield is not solely a function of fire. Rather much of the variation in water flow is a function of annual climate (Chapter 8). By examining gaging stations on the Yellowstone River north of the park, Farnes and others were able to determine that the amount of burned area had little influence on stream flow at the scale of the entire park. Fire effects were much more pronounced at smaller, individual stream scales, but even these responses were ameliorated over time.

The specific character of ecosystem responses and recovery following fire is clearly scale dependent. The short and small-scale responses to fire are quite dramatic whereas the long and large-scale responses seem to be less dynamic. For example, Minshall and others (Chapter 8) note that the initial impact on the ecology of smaller streams was much more pronounced than that experienced by larger-order streams. Changes in nutrient levels were greatest in the small streams, for instance. Intermediate-term effects occurred in both small and mid-sized streams. For example, the accumulation of coarse woody debris occurred in all orders of streams. Small headwater streams experienced a narrowing and deepening of their channels during this intermediate time scale, as well. These sorts of changes have important implications for aquatic organisms.

Mihuc (Chapter 9) found that entire food web structure was immediately altered in small-order streams, changing from a web whose energy is based on coarse particulate organic matter to one whose energy is based on periphyton (algae and diatoms). This will change the relative abundances of different insect taxa that feed on these different food sources. In turn, changes in the insect fauna can cause changes in food availabilities for fish. However, as streamside vegetation regrows and again shades these small watercourses, the food web returns to its prefire structure.

This return means that the effects of the fires on fish populations were transitory in streams; at larger spatial scales, such as in Yellowstone Lake, they were not discernable at all. Gresswell found that growth increments in cutthroat trout in Yellowstone Lake were governed more by angler harvest, hatchery operations, and the introduction of nonnative lake trout than by fire (Chapter 7). The influences of the alien fish species are actually quite significant. In terrestrial systems, the park has been relatively more successful in restricting the movement of alien species into fire-affected areas. This has likely contributed to the rapid recovery by these areas following the fire (Chapter 14).

DID INDIVIDUAL SPECIES SUFFER IN RESPONSE TO THE FIRES?

Although many individual organisms died, no species population appears to have been jeopardized as a result of the fires (Chapters 4, 5, 6, 13). That is because the plant and animal species that exist in this environment have unique adaptations to fire and other ecological disturbances that occur in this system.

Many of the herbaceous plants are perennials and can regrow from belowground structures such as roots, rhizomes, or corms (Chapters 5, 14). In addition, many species are prolific seed producers, thus revegetating fire-affected areas using either sexual or asexual means of propagation (Chapters 5, 14). Therefore the year following the fires saw a tremendous proliferation of herbaceous plants in burned grasslands and meadows (Chapters 4, 13).

Woody species take longer to respond. One of the intermediate-term effects of the fire was the reduction in sagebrush cover in the northern range of the park (Chapter 5). Herbaceous species (for example, lupine, Chapter 5) rapidly filled the void left by sagebrush loss. However, this does not mean that sagebrush is lost from the system, as high populations of sagebrush seedlings were found as well (Chapters 4, 5). Immediately following the fires, overall grass and meadow production was stimulated and remained higher than prefire levels for at least five years (Chapters 6, 13).

The effect of the fires on elk and bison populations attracted enormous public attention. Only a few animals died as a direct result of fire and smoke (Chapter 6), but there was substantial mortality the winter following the fire due to winter severity (Chapters 6, 13). Both species possess important adaptations that allowed them to survive in the face of this great disturbance. Both are quite mobile; neither is restricted by territorial boundaries. Thus if one area was burned and forage levels were reduced, they simply moved to an area with adequate levels of forage. Both can tolerate a wide range of forage quality (Chapter 6), and both can survive off of fat stores for some time during winter (Chapter 13). As a consequence, although population levels were reduced by a relatively severe winter in 1988–1989, both populations recovered rapidly to prefire levels.

One trophic level that was not discussed in depth was that of the large vertebrate predators. This includes such groups as grizzly bears *(Ursus arctos)* and wolves *(Canis lupus)*. Wolf reintroduction to the park did not occur until after the fires. However, there is a great deal of information concerning grizzly bear populations prior to, during, and following the fires. Just as for elk and bison, grizzly bears are highly mobile species and were able to avoid the fires. Indeed, following the fires, grizzly bear populations have gotten larger, due to a variety of factors. Whitebark pine *(Pinus albicaulis)* is a crucial food source for grizzly bears in the fall, when they gain weight prior to hibernation (Mattson and others 1992). There were four mast years for this species immediately following the fires, yielding one positive influence on grizzly bear populations (Pease and Mattson 1999). Grizzly bears also get a great deal of energy from scavenging carcasses (Green and others 1997), which were abundant following the severe winter of 1988–1989 (Chapter 6). Blanchard and Knight (1990) noted no net change in movement patterns (Mattson 1997) or food habits due to the fires. In terms of conservation of this important and endangered species, such issues as the introduction of exotic species (Reinhart and others 2001), and human activities (Mace and others 1999, Carroll and others 2001), including poaching (Mattson and Merrill 2002), are of far greater importance than the natural disturbance, fire. Loss of grizzly bears from this ecosystem could have far-ranging and devastating impacts (Terborgh and others 2001) including alterations in the abundance of herbivores, especially elk (Chapter 6) and initiation of trophic cascade effects (Schmitz and others 2000, Snyder and Wise 2001).

WHAT ARE THE MANAGEMENT LESSONS TO BE LEARNED FROM THE FIRES?

Management of natural preserves ultimately comes down to addressing three questions.

1. What should we preserve? By this we mean what are the specific elements, species, and communities that are to be preserved?
2. What form should preserves have? Here we refer to issues of size and locations of boundaries.
3. How should preserves be maintained? What management protocols should be followed and actions taken to ensure conservation of key preserve elements?

The studies presented here provide important information relevant to each of these questions.

What should we preserve? The 1916 National Park Service Organic Act provides what seems unambiguous guidance on this question, namely that the National Park Service should "conserve the natural and historic objects." That this is to be done for "the enjoyment of the people" certainly does introduce ambiguity and conflict that was exemplified in the media and public reactions to the fires and the changes that proceeded from them.

Historically, much of preserve management has been "object" oriented, as if such management were the outdoor equivalent of museum curation (Christensen 1995). In the words of the 1963 Leopold Report, the primary goal of our national parks should be to preserve "vignettes of primitive America." Decades before the 1988 fires, however, park managers became aware of the fact that the landscapes they were charged to conserve were undergoing constant change—that management needed to shift from an "object" to a "process" focus. The implementation of the so-called natural prescribed fire program in Yellowstone in the early 1970s was an exemplar of that shift, although the 1988 fires caused some to question its wisdom.

Although the fires of 1988 were quite large, it is clear from the results presented here that they were not outside the historic range of variation for such events. Furthermore, some of the strongest ecological effects of them were felt at only small scales and existed only for short times. Not only were no elements ("natural objects") lost, but the data presented here indicate that the changes proceeding from the fires have had positive effects with regard to the regeneration of the diversity of this landscape. All of these studies support the assertion that sustainable management of preserves such as Yellowstone depends on our ability to accommodate natural disturbances and the processes of change that derive from it.

What form should preserves have? This question is all about preserve design, and it would be fair to say that the boundaries of Yellowstone National Park were set with virtually no reference to the ecological processes that challenge managers most today. The 1988 fires drive home two important lessons in this regard. First, although Yellowstone is one of our largest parks, it is small relative to the spatial scale of important ecological processes (such as fire, animal migration patterns, and so on). There is no question that fires on that scale of those in 1988 occurred in prehistoric times (Chapters 2, 3, 13), but it is also true that they burned in the context of a much more extensive and contiguous wilderness landscape. Second, the fires and management challenges that derive from them make clear that straight-line borders that separate park from nonpark are arbitrary relative to key management concerns, such as animal migrations and proliferation of exotic species. With regard to this latter issue, it is reassuring to note that recently burned sites appear to be comparatively resistant to invasion (Chapter 14).

How should preserves be maintained? In the best of all worlds, "wilderness management" would be an oxymoron. However, given the boundary issues described above, and the ubiquity of human influences, management intervention seems necessary and inevitable. Immediately after the fires, concerns were expressed in many quarters that the ecological impacts on species populations, sediment movement, and water quality would be enormous. The data presented here make clear that these responses have been small in spatial scale and short in duration. The management lesson is obvious. Wholesale, large-scale responses may end up being out of proportion to the responses of the ecosystem. Reseeding of plants, fishes, and ungulate feeding programs may end up exacerbating other problems (in other words, alien species invasions, diseases) and may act to slow ecosystem recovery following fire (Chapters 4, 6, 14). This is particularly important if global circulation models calling for warming and drying of regions of the park are to be believed. This could increase fire frequency, yielding a shorter recovery time between large-scale fires (Chapters 2, 13).

The "enjoyment of the people" will remain an important part of the National Park Service mission. Given the importance of parks to public understanding of ecological processes, this is wholly desirable, even if it presents important challenges and conflicts. In the years ahead, human resident populations immediately adjacent to park boundaries are expected to increase, as are visitation rates. This will affect several features that currently allow for rapid and complete recovery following fire. First, more alien species may be introduced to the system (Chapter 14), thus slowing recovery or completely altering community change trajectories. Second, more humans could translate to mean more ignition sources, thus increasing fire frequency substantially. Third, human demands for water and fishing could alter management regimes of aquatic resources in the ecosystem. Finally, as humans come to occupy important habitats outside of the park, those critical "overflow" areas become less available to mobile species such as ungulates and large predators. This habitat loss could become critical to the response of these species to fire. As these species appear to be keystone elements to the Greater Yellowstone Ecosystem, any alteration in their ability to respond to fire could have strong influences on whole system recovery.

Acknowledging these challenges, the Yellowstone fires present a learning moment at a time when the inexorable role of fire on many landscapes (rural and suburban) has become a matter of national significance. The Yellowstone experience makes clear that sustainable management of landscapes is all about maintaining the capacity to change and to accommodate inevitable surprise.

References

Blanchard, B. M., and R. R. Knight. 1990. Reaction of grizzly bears, *Ursus arctos horribilis*, to wildlife in Yellowstone National Park, Wyoming. Canadian Field Naturalist 104:592–594.

Carroll, C., R. F. Noss, and P. C. Paquet. 2001. Carnivores as focal species for conservation planning in the Rocky Mountain region. Ecological Applications 11:961–980.

Christensen, N. L. 1995. Plants in dynamic environments: Is "wilderness management" an oxymoron? Pages 63–72 in P. Schullery and J. Varley, eds., Plants and the environment. U.S. Department of the Interior, National Park Service, Washington, D.C.

Green, G. I., D. J. Mattson, and J. M. Peek. 1997. Spring feeding on ungulate carcasses by grizzly bears in Yellowstone National Park. Journal of Wildlife Management 61:1040–1055.

Haines, A. L. 1977. The Yellowstone story. Yellowstone Library and Museum Association, Yellowstone National Park, Wyo.

Mace, R. D., J. S. Waller, T. L. Manley, K. Ake, and W. T. Wittinger. 1999. Landscape evaluation of grizzly bear habitat in western Montana. Conservation Biology 13:367–377.

Mattson, D. J. 1997. Use of lodgepole pine cover types by Yellowstone grizzly bears. Journal of Wildlife Management 61:480–496.

Mattson, D. J., B. M. Blanchard, and R. R. Knight. 1992. Yellowstone grizzly bear mortality, human habituation, and whitebark pine seed crops. Journal of Wildlife Management 56:432–442.

Mattson, D. J., and T. Merrill. 2002. Extirpations of grizzly bears in the contiguous United States. Conservation Biology 16:1123–1136.

Pease, C. M., and D. J. Mattson. 1999. Demography of the Yellowstone grizzly bears. Ecology 80:957–975.

Peters, R. H. 1991. A critique for ecology. Cambridge University Press, Cambridge.
Reinhart, D. P., M. A. Haroldson, D. J. Mattson, and K. A. Gunther. 2001. Effects of exotic species on Yellowstone's grizzly bears. Western North American Naturalist 61:277–288.
Schmitz, O. J., P. A. Hamback, and A. P. Beckerman. 2000. Trophic cascades in terrestrial systems: A review of the effects of carnivore removals on plants. American Naturalist 155:141–153.
Snyder, W. E., and D. H. Wise. 2001. Contrasting trophic cascades generated by a community of generalist predators. Ecology 82:1571–1583.
Terborgh, J., L. Lopez, P. Nunez, M. Rao, G. Shahabuddin, G. Orihuela, M. Riveros, R. Ascanio, G. H. Alder, T. D. Lambert, and L. Balbas. 2001. Ecological meltdown in predator-free forest fragments. Science 294:1923–1926.

Contributors

Jay E. Anderson, deceased, Center for Ecological Research and Education, Department of Biological Sciences, Idaho State University, Pocatello, ID 83209

Patrick J. Bartlein, Department of Geography, University of Oregon, Eugene, OR 97403

Norman L. Christensen, School of the Environment, Duke University, Box 90328, Durham, NC 27708

Michael B. Coughenour, Natural Resources Ecology Lab, Colorado State University, Fort Collins, CO 80523

Kathleen M. Doyle, Department of Environmental Studies, Ackley 37, Green Mountain College, Poultney, VT 05764

Marshall Ellis, North Carolina Division of Parks and Recreation, P.O. Box 27687, Raleigh, NC 27611

Phillip E. Farnes, Department of Earth Sciences, Montana State University, Bozeman, MT 59717

Robert E. Gresswell, USGS Forest and Rangeland Ecosystem Science Center, 3200 SW Jefferson Way, Corvallis, OR 90731

Katherine J. Hansen, Department of Earth Sciences, Montana State University, Bozeman, MT 59717

Dennis H. Knight, Department of Botany, University of Wyoming, Laramie, WY 82070

Ward W. McCaughey, USDA Forest Service, Rocky Mountain Research Station, Forestry Sciences Laboratory, Montana State University, Bozeman, MT 59717

Grant A. Meyer, Department of Earth and Planetary Sciences, University of New Mexico, Albuquerque, NM 87131

Timothy B. Mihuc, Lake Champlain Research Institute and Center for Earth, and Environmental Science, Plattsburgh State University of New York, 102 Hudson Hall, 102 Broad Street, Plattsburgh, NY 12901

Sarah H. Millspaugh, Department of Geography, University of Oregon, Eugene, OR 97403

G. Wayne Minshall, Department of Biology, Idaho State University, Pocatello, ID 83209

Jack E. Norland, Department of Animal and Range Science, North Dakota State University, Box 5727, Fargo, ND 58105

Christopher T. Robinson, Department of Limnology, Swiss Federal Institute for Environmental Science and Technology, CH-8600 Duebendorf, Switzerland

William H. Romme, College of Natural Resources, Colorado State University, Fort Collins, CO 80523

Todd V. Royer, Department of Natural Resources and Environmental Sciences, W-503 Turner Hall, 1102 S. Goodwin Ave., University of Illinois, Urbana, IL 61801

Paul Schullery, Center for Resources, Yellowstone National Park, Yellowstone National Park, WY 82170

Francis J. Singer, Midcontinent Ecological Science Center, USGS, Natural Resources Ecology Laboratory, Colorado State University, Fort Collins, CO 80523

Daniel B. Tinker, Department of Botany, University of Wyoming, Laramie, WY 82070

Benjamin F. Tracy, Department of Crop Sciences, W-201B Turner Hall, University of Illinois, Urbana, IL 61801

Monica G. Turner, Department of Zoology, University of Wisconsin, Birge Hall, Madison, WI 53706

Carol D. von Dohlen, Department of Biology, Utah State University, Logan, UT 84322

Linda L. Wallace, Department of Botany and Microbiology, University of Oklahoma, Norman, OK 73019

Cathy Whitlock, Department of Geography, University of Oregon, Eugene, OR 97403

Index

Note: Italic page numbers refer to illustrations.